PREFACE

This book is intended as a text in introductory courses at the junior, senior, and graduate levels in probability and mathematical statistics. It assumes no previous contact with either probability or statistics, but does assume that the student possesses a good background in calculus and is comfortable with the basic concepts and operations of linear algebra. Given this background, the book is intended for all students just beginning the study of statistics, whether they be in the social, biological, or physical sciences, or budding professional statisticians. It is especially with this broad audience in mind that the study guide accompanying the text, prepared by Robert Kushler, provides a wide-ranging list of illustrative examples and problems from economics, psychology, sociology, physics, engineering, biology, and medicine, as well as mathematical statistics per se.

In content, the subject matter included forms the background that I (ideally) would like my own graduate students in economics to bring into the beginning graduate course in econometrics. The material goes beyond what will normally be covered in one semester and with some supplementation, is adequate for most courses lasting a full year. In presentation, I have aimed at an exposition that is informal yet rigorous, and with emphasis on foundations as opposed to applications. With very few exceptions, all theorems and propositions are proved, and derivations are presented in detail. Results from pure mathematics are stated and proven wherever they are central to the argument and are felt to be ones with which students are unlikely to be familiar. However, with the exception of Chapter 10, which deals with change in variables, their statements and proofs are relegated to appendixes.

The presentation follows what is by now a fairly traditional format.

Chapter 2 begins with the motion of a random event and then proceeds to the basic laws of probability. Altogether, the discussion of probability occupies Chapters 2–4. The next 9 chapters, Chapters 5–13, focus on probability distributions, both discrete and continuous, and their fundamental properties. Sampling theory is treated in Chapters 14 and 15, while Chapters 16–19 are devoted to statistical inference. Chapter 20 discusses the least squares regression model, and Chapter 21 delves briefly into order statistics. Finally, Chapter 22 discusses some of the many unsettled issues surrounding statistical inference and includes a brief introduction to inference from a Bayesian perspective.

In general, my presentation with regard to both organization and subject matter, has been greatly influenced by the fine textbooks of Freeman (*Introduction to Statistical Inference*, Addison Wesley), Hogg and Craig (*Introduction to Mathematical Statistics*, Macmillan), and Mood and Graybill (*Introduction to the Theory of Statistics*, McGraw-Hill). In addition, my views regarding probability have been shaped in great part by the presentation of B. V. Gnedenko in Chapter 1 of his *The Theory of Probability* (Chelsea). Indeed, the first several paragraphs of Chapter 2 are essentially undistilled Gnedenko.

This book had its genesis in lecture notes that were developed in 1967–1968 while I was with the Harvard Development Advisory Service in Bogotá, Colombia and teaching a course each semester in statistics and econometrics at the Universidad de los Andes. It was completed during the 1972–1973 academic year while I was a visiting professor in the Department of Economics at the University of Arizona. I am grateful to my students at both of these institutions and at the University of Michigan for many comments (critical and otherwise) that have led to a material improvement in the final product. The manuscript has also benefited substantially from careful readings by several reviewers and by Robert Kushler and Eugene May, both of the University of Michigan. In addition, I am grateful to Professor Peter S. Landweber of the Department of Mathematics at Rutgers University for providing the derivation given in Appendix 4 of a limit used in the proof of the classical central limit theorem, to Carol A. Taylor for facilitating the derivations in Chapter 13 of the distributions associated with the normal distribution, to Marilyn Spencer and Mary Chavez, both of the University of Arizona, for general assistance, and to Donna Hoff for helping to prepare the Index. Finally, I am much indebted to Elsa Bermudez, my secretary at the University of Arizona, and Genoveva Torres at the Universidad de los Andes for their superb talents and cheerful dispositions in typing the manuscript.

Ann Arbor, Michigan LESTER D. TAYLOR
April 1974

PROBABILITY AND MATHEMATICAL STATISTICS

CHAPTER 1

INTRODUCTION

Even the experts are unable to agree on a single, all-encompassing definition of statistics. To the lay public, statistics ordinarily includes the collection of large masses of data and their presentation in tables and charts, plus possibly the calculation of totals, averages, percentages, and the like. These routine calculations are indeed a part, and an important part, of statistics, but they are by no means the whole story. The design of experiments and of sample surveys, sampling procedures, inference to a population from a sample, data reduction, data processing, and many other things are also key concerns of statistics.

Statistics forms a fundamental part of the technology of the scientific method. It provides tools for making decisions under conditions of uncertainty, for testing observable implications of scientific theories, and for quantifying models that are constructed to describe certain aspects of reality. These tools are of quite general applicability and are useful to the business executive and the government policymaker, as well as to the research worker in any field of science.

For our purposes, statistics can be divided into two broad categories: (1) descriptive statistics, which is concerned with summarizing, describing, and the reduction of data for analysis, and (2) inferential statistics, which is concerned with the process whereby data are used to make statements or decisions about the universe from which the data were obtained. We will be concerned with inferential statistics in this book, and henceforth when we refer to statistics we will have inferential statistics in mind.

Within inferential statistics, we can make a further division into applied statistics and theoretical statistics. The applied statistician is the one that is "out in the field," so to speak, either as a researcher himself or as a consultant to one needing statistical advice. Faced with a practical problem, the applied statistician constructs a mathematical model that describes the situation as closely as possible, analyzes the model by mathematical methods, and then devises procedures to deal with the problem. In devising these procedures, he may be able to use an already existing model or methodology, or he may have to develop a methodology that is entirely new. However, in doing so he will be guided by the principles of the theory of statistics.

The theoretical statistician, on the other hand, generally does not "dirty" his hands with actual data, but rather is concerned with developing the general theory of statistics. He may be motivated by practical problems, but there is no need of this, for his motive may be nothing more than a pure and simple pursuit of knowledge. In general, however, the broad advances in the application of statistical methods in the past three decades were made possible by the remarkable developments in statistical theory which immediately preceded them.

The theory of statistics is a branch of applied mathematics with its roots in the theory of probability. Indeed, the broad structure of statistical theory can be said to include the theory of probability, although interestingly enough, the importance of probability to the development of statistical theory was not fully recognized until early in this century. Until then, probability theory, which was first systematically developed in the seventeenth century in the works of Fermat (1602–1665), Pascal (1623–1662), Huygens (1629–1695), and especially J. Bernoulli (1654–1705), had largely been inspired by the study of games of chance.

Because statistics is essentially a mathematical discipline, this is the view that we shall adopt. The purpose of the book is to acquaint the reader who has a good background in the calculus and linear algebra with a detailed and wide-ranging introduction to the laws of probability, their application to mathematical statistics, to the basic results of the latter, and to statistical inference. With very few exceptions, all results stated in the book are derived and proven. In some places, purely mathematical theorems are proven if it is felt that they are ones with which the typical reader is unlikely to be familiar. However, with exception of Chapter 10, these theorems are for the most part presented in appendixes.

The 21 chapters that follow approach the material of probability and mathematical statistics in fairly traditional fashion. We begin in the next chapter with the notion of a random event and then proceed to derive the basic laws of probability. Altogether, the discussion of probability occupies Chapters 2–4. The next 9 chapters, Chapters 5–13, concentrate on probability distributions, both continuous and discrete, and their fundamental properties. Sampling is treated in Chapters 14 and 15. Finally, Chapters 16–22 are devoted to statistical inference and some of the issues surrounding it.

CHAPTER 2

RANDOM EVENTS AND THE BASIC CONCEPTS OF PROBABILITY

2.1 THE CONCEPT OF RANDOMNESS

On the basis of observation and experiment science attempts to formulate the natural laws of the phenomena that it studies. The simplest and most widely used such scheme of laws is the following:

$$\text{Whenever a certain set of conditions } K \text{ is realized, the event } A \text{ occurs.} \tag{2.1}$$

Thus, for example, if water under 760-mm atmosphere pressure is heated to over 212°F (the set of conditions K), it is transformed into steam (the event A). The law of gravity and the law of the conservation of mass are other examples of such laws, and the reader can undoubtedly supply still other examples in chemistry, physics, biology, and the other sciences.

An event whose occurrence is inevitable whenever the set of conditions K is realized is said to be *certain* (or *sure*). On the other hand, an event that can never occur whenever the set of conditions K is realized is said to be *impossible*. And an event that may or may not occur whenever the set of conditions K occurs is said to be *random*.

From these definitions, it is clear that when we speak of the certainty, impossibility, or randomness of an event, it is with respect to a particular set of conditions. In particular, an event may be certain with respect to one set of conditions, impossible with respect to another, and random with respect to still a third set.

Moreover, it will be noted that with respect to a particular set of conditions the certain occurrence of event A is equivalent to the impossibility of the event that is complementary to A. As a consequence, this means that an impossible event can be reduced to a statement of type (2.1).

In contrast to certain and impossible events, the information conveyed by the assertion that an event is random is only of limited value. Because the event is neither certain nor impossible, we only know that the set of conditions K do not reflect all of the necessary and sufficient conditions for the event to occur.

A wide range of phenomena exists, however, for which if the set of conditions K are realized *repeatedly*, the proportion of occurrences of event A seldom deviates significantly from some average value, and this number can therefore serve as a characteristic index of the mass phenomenon (the repeated realization of the set of conditions K) with respect to event A. For such phenomena it is possible not only simply to state that event A is random but also to estimate in quantitative terms the chance of its occurrence. This estimate can be expressed by propositions of the form

The probability that the event A will occur
whenever the set of conditions K is realized (2.2)
is equal to p, where $0 \leq p \leq 1$.

Laws of this kind are called *probabilistic*, or *stochastic*, laws. Probabilistic laws play an important part in nearly every field of science. In physics, for example, no methods exist for predicting whether or not a given atom of radium will disintegrate within a given period of time, but on the basis of experimental data it is possible to determine the probability of this disintegration: An atom of radium disintegrates in a period of t years with the probability

$$p = 1 - e^{-0.000436t}. \tag{2.3}$$

The set of conditions K in this example consists in the atom of radium not being subjected to any unusual external conditions, such as bombardment by fast particles, during the time under consideration. Event A is the disintegration of the radium atom during the given period of t years.

For a second, and more complicated, example, we can note what is one of the best established empirical regularities in all of economics. It has been observed on numerous occasions in numerous countries that as the level of income per capita increases, the proportion of income spent on food decreases. This is referred to as Engel's law, after Ernst Engel, a German economist and statistician, who first enunciated the finding in a paper published in 1857. This "law" does not say that everyone spends a smaller proportion of his income on food as his income rises or even that every time there is an increase in an economy's per capita income the proportion of income spent on food falls. Rather, it simply states that there are forces in operation that cause consumers on the average to spend a decreasing proportion of a growing income on food. The set of conditions K is an increasing income with other circumstances held reasonably constant (e.g., absence of drought, abrupt changes in social structure, etc.), while the event A is the proportion of the increase in income that is spent on food.

And as a third example, we can take the classic case of the rolling of a numerical, homogeneous die. When such a die has been rolled repeatedly, it has been observed on a large number of occasions that each of the six faces appears about $\frac{1}{6}$ of the time. The set of conditions K here is the roll of the die, while the event A is the appearance of any particular face.

It is important to emphasize that the p in statement (2.2) is a characteristic of the mass phenomenon in question, and is not a reflection of the state of mind of the observer. In other words, the probability is an objective quantity in a scientific sense, for it is independent of the observer. This means that questions of the type:

1. What is the probability that every natural number greater than two can be expressed as the sum of two primes? or
2. What is the probability that on June 25, 1977 Joe Smith will marry Judy Jones?

do not come under the purview of statement (2.2), for there is no objective basis for the probabilities involved. About the former, little is known at present, except that many mathematicians believe that it is very likely to be true; about the latter one can only assume that a definitive answer will be given on June 25, 1977.

On the other hand, it is also important to note that there is a school of probability, whose members are willing, if their analyses require it, to assign numbers to the probabilities in the two statements above. Indeed, some members of this, the so-called subjective school of probability, maintain that *all* probabilities must reflect judgments of the observer. Except for Chapter 22 where subjective probability and a variety of its implications will be discussed, we will for the most part dispense with such a view, instead considering probabilities to be objective characteristics of the mass phenomenon in question.

With the concepts of a random event and probability established, let us now inquire into how the probabilities are assigned or discovered. Historically, there have been two approaches, the *classical* (or *a priori*) method and the relative *frequency* (or *statistical*) method, and these were joined in this century by the axiomatic approach. The axiomatic approach, because it is consistent with both the classical and relative frequency definitions and does not suffer their defects, has essentially carried the day, and will be the approach that we will follow. Nevertheless, it will be useful to the sequel if we briefly mention the other two methods. But before doing this we need to appropriate some concepts from the algebra of sets.

2.2 TERMS AND DEFINITIONS FROM THE ALGEBRA OF SETS

The mathematics of probability is essentially the mathematics of the algebra of the subsets of a set, and it will accordingly be useful at this time to develop the terms and definitions from the algebra of sets that are used in the study of probability. It is assumed in the text that the reader has had exposure to the basic results of the algebra of sets. Readers for which this is not the case should turn to Appendix 1 before continuing.

We begin with the notion of mutually exclusive. Two sets A and B are *mutually exclusive* if their intersection is the empty set; that is if

$$A \cap B = \varnothing. \tag{2.4}$$

A set and its complement are, of course, mutually exclusive by definition. Note, however, that sets that are mutually exclusive are not necessarily complements.

If we have n sets A_i, $i = 1, n$, that are pairwise mutually exclusive; that is, if

$$A_i \cap A_j = \varnothing \qquad i, j = 1, n \qquad i \neq j, \tag{2.5}$$

and if

$$\bigcup_{i=1}^{n} A_i = S, \tag{2.6}$$

we say that the sets A_i form an *exhaustive partition* of S. This means that each element in S is in one, and only one, A_i. Obviously, a set and its complement form an exhaustive partition. Another example for $n = 3$ is given in the Venn diagram (Figure 2.1).

Suppose now that we have n sets E_i, $i = 1, n$, that form an exhaustive partition of S. We will call the E_i's *elementary events* and S the *space* containing the events E_i. (Later on we will refer to S as a *sample space*.)

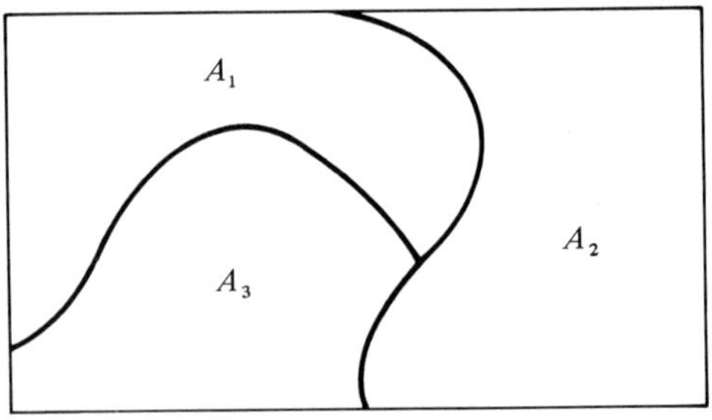

FIGURE 2.1 An exhaustive partition.

Assume that S contains three elementary events E_1, E_2, and E_3. From these three elementary events we can construct the events: $\varnothing$, E_1, E_2, E_3, $E_1 \cup E_2$, $E_1 \cup E_3$, $E_2 \cup E_3$, and $E_1 \cup E_2 \cup E_3$. It is seen that each of these events can be decomposed into elementary events and that there are eight such events in all. The empty set is obtained by the intersection of any two of the elementary events, while $E_1 \cup E_2 \cup E_3$, it will be noted, is equal to S. Therefore, these eight events, or subsets, exhaust the non-redundant subsets.

What the foregoing has illustrated is the following fact: If a set S can be decomposed into n sets that are mutually exclusive, S contains 2^n subsets in all.

Two Games of Chance Interpreted in the Algebra of Sets

For future reference and to illustrate the application of the algebra of sets to the study of games of chance, it will be useful to characterize the flipping of a coin and the rolling of a die in the language of set algebra.

Flipping a coin. There are two possible outcomes to the flipping of a coin: If it does not come up heads, it will come up tails. These are the only possibilities, and in this case the sample space S is comprised of two elementary events: E_1 = heads and E_2 = tails. Using H and T for short, we can enumerate the four possible events when a coin is flipped as follows:

1. H (heads show)
2. T (tails show)
3. H $\cup$ T (heads or tails show)
4. $\varnothing$ (neither heads nor tails show).

The empty set is equivalent to H $\cap$ T, or the simultaneous (i.e., joint) occurrence of heads and tails. This is, of course, impossible, and so we associate $\varnothing$ with the impossible event. On the other hand, either heads or tails must occur, and so it is natural to associate $H \cup T$ (or S) with the certain event.

Rolling a die. A die has six faces, so the sample space S associated with the rolling of a die contains six elementary events, namely:

1. E_1: ⚀ shows
2. E_2: ⚁ shows
3. E_3: ⚂ shows
4. E_4: ⚃ shows
5. E_5: ⚄ shows
6. E_6: ⚅ shows.

As we already know, from these six elementary events, 2^6 or 64 events can be constructed in all, including the impossible event (the simultaneous showing of two or more faces) and the certain event (the showing of one of the six faces). Included among the other 62 events are the following:

$$\begin{array}{ll} E_2 \cup E_4 & \text{(the showing of 2 or 4)} \\ E_1 \cup E_5 \cup E_6 & \text{(the showing of 1 or 5 or 6)} \\ E_3 & \text{(the showing of 3).} \end{array} \tag{2.7}$$

If we recall our definition of a random event, it is apparent that the events in these two examples are random. The conditions K are, respectively, the flipping of a coin and the rolling of a die, while the random events are, respectively, the showing of heads or tails and the showing of one of the six faces of the die. Moreover, it is well established for these two phenomena that if the conditions K are realized repeatedly—that is, if the coin is flipped or the die rolled repeatedly—each event occurs a proportion of the time that seldom deviates very much from some average value. In other words, a probability can be associated with each event. We will now begin to study how these probabilities are determined.

2.3 CLASSICAL DEFINITION OF PROBABILITY

The classical definition of probability reduces to the concept of equal likelihood, which is taken as a primitive notion and therefore not subject to formal definition. For example, if a die is a perfect cube and made of a homogeneous substance, there is no objective basis for assuming that any one face is more probable than any other. Similarly, if a coin is balanced, each face is considered to be equally probable.

In the general case, suppose that we have n pairwise mutually exclusive elementary events E_i, $i = 1, n$, that are equally likely. Let us now form the family F of events consisting of the impossible event $\varnothing$, the certain event S, the elementary events E_i, and all events A which can be decomposed into a union of elementary events. We have already shown that S contains 2^n such events in all.[1]

The classical definition of probability is given in terms of the events of the family F and may be formulated as follows:

> If an event A is decomposable into the sum of m events belonging to a complete group of n pairwise mutually exclusive and equally likely events, then the probability $P(A)$ of the event A is equal to

$$P(A) = \frac{m}{n}. \tag{2.8}$$

For example, the six faces of a homogeneous, perfectly cubical die are equally likely, and so the probability of any one face is $\frac{1}{6}$. Therefore, the probability of the event

$$C = E_1 \cup E_3 \cup E_5, \tag{2.9}$$

that is, the probability of obtaining an odd face is $\frac{1}{6} + \frac{1}{6} + \frac{1}{6} = \frac{1}{2}$. Similarly, the probability of obtaining a 1 or 3 is $\frac{1}{3}$, and so on.

As another example, suppose that we draw a card from a well-shuffled bridge deck. The pairwise mutually exclusive elementary events in this case are the 52 cards, and if the deck is well shuffled there is no objective basis for asserting that the probability of drawing any one card is any higher than drawing any other card. Consequently, all 52 cards are equally likely.

As still a third example, let us consider the rolling of a pair of dice. If the dice are true, each of the 36 possible combinations of faces of the two dice should be considered equally likely, and therefore given the probability $\frac{1}{36}$. The sum of the two faces can take on the values 2 through 12, and it is evident from the principle of equal likelihood that the probability of the sum being 4, say, is $\frac{3}{36}$. Similarly, the reader can easily verify that the other probabilities are given as follows:

Sum of Faces:	2	3	4	5	6	7	8	9	10	11	12
Probability:	$\frac{1}{36}$	$\frac{2}{36}$	$\frac{3}{36}$	$\frac{4}{36}$	$\frac{5}{36}$	$\frac{6}{36}$	$\frac{5}{36}$	$\frac{4}{36}$	$\frac{3}{36}$	$\frac{2}{36}$	$\frac{1}{36}$

[1] It can be shown that if A and B are included in F, then so are AB, $A \cup B$, and $A - B$. Such a family F of events is called a *field*.

In the study of probability, the following terminology is standard and we shall frequently use it. Let us imagine that in order to clear up the question as to whether or not a particular event will occur it is necessary to make a *trial* (i.e., to realize the set of conditions K) which would yield the answer to our question. (In the immediately preceding example a pair of dice have to be thrown.) The complete group of pairwise mutually exclusive and equally likely events which may occur when this experiment is performed is called the complete group of *possible outcomes* of the trial. The possible outcomes into which the event A can be decomposed are called the outcomes (or cases) favorable to A. Using this terminology, we can say that *the probability $P(A)$ of the event A is equal to the number of possible outcomes favorable to A divided by the total number of possible outcomes of the trial.*

According to the definition given, every event belonging to the family of events F, as just defined, has a well-defined probability

$$P(A) = \frac{m}{n}$$

assigned to it, where m is the number of mutually exclusive and equally likely events E_i of F into the union of which the event A can be decomposed. Thus, the probability $P(A)$ can be regarded as *a real-valued function of the event A defined over the field of events F.*

This function possesses the following properties:

1. For every event A of the field F,

$$P(A) \geq 0.$$

2. For the certain event S,

$$P(S) = 1.$$

3. If the event A can be decomposed into the mutually exclusive events B and C and all three of the events A, B, and C belong to the field F, then

$$P(A) = P(B) + P(C).$$

This property is called the *theorem on the addition of probabilities.*

Property 1 is obvious, since the ratio m/n can never be negative, and property 2 is equally obvious, since the n possible outcomes are all favorable to the certain event S and hence,

$$P(S) = \frac{n}{n} = 1.$$

As for property 3, let us suppose that k is the number of outcomes favorable to the event B and that r is the number of outcomes favorable to event C. Because the events B and C are, by definition, mutually exclusive, the outcomes that are favorable to the event $A = B \cup C$ are equal to $k + r$. Hence,

$$P(A) = \frac{k + r}{n} = \frac{k}{n} + \frac{r}{n} = P(B) + P(C).$$

A few further properties of probability are:

4. The probability of the event A' complementary to the event A is equal to

$$P(A') = 1 - P(A).$$

PROOF. Because $A \cup A' = S$, it follows from property 2 that

$$P(A \cup A') = 1,$$

and since A and A' are mutually exclusive, property 3 implies

$$P(A \cup A') = P(A) + P(A').$$

Consequently,

$$P(A') = 1 - P(A).$$

5. The probability of the impossible event $\varnothing$ is 0.

PROOF. Because the events S and $\varnothing$ are mutually exclusive,

$$P(S \cup \varnothing) = P(S) + P(\varnothing) = P(S),$$

whence

$$P(\varnothing) = 0.$$

6. If the event A implies the event B (i.e., if $A \subset B$), then

$$P(A) \leq P(B).$$

PROOF. Because $A \subset B$, B can be represented by the union of two mutually exclusive events A and $A'B$. Therefore,

$$P(B) = P(A \cup A'B) = P(A) + P(A'B),$$

from which it follows that

$$P(A) \leq P(B).$$

Properties 2, 5, and 6, in turn, imply:

7. The probability of any event lies between 0 and 1.

8. For any events A and B included in F,

$$P(A \cup B) = P(A) + P(B) - P(AB).$$

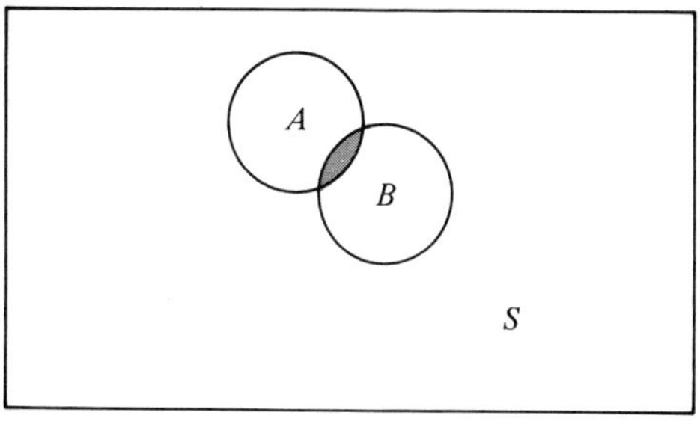

FIGURE 2.2

PROOF. From Figure 2.2 it is evident that in adding the probability of A to the probability of B we are counting the shaded area twice; $P(AB)$ must therefore be subtracted.

Property 8 can be extended to the union of any number of events, so long as each event is included in F. For three sets A, B, and C, the formula is

$$P(A \cup B \cup C) = P(A) + P(B) + P(C) \\ - P(AB) - P(AC) - P(BC) + P(ABC),$$

and so on.

If for no reason other than its elegant simplicity, the classical definition of probability has a great deal of appeal, and so long as the study of probability was the study of the outcomes of games of chance in finite sample spaces, there was little apparent need for any other definition. Nevertheless, there are some rather troublesome defects in the classical definition. It is obvious, for example, that the classical definition has to be modified somehow when the number of possible random events is infinite. For instance, we might seek the probability that a positive integer drawn at random is even. The intuitive answer to this is that the probability is $\frac{1}{2}$, and if pressed to justify this answer on the basis of the classical definition, we might reason as follows: Of the first 10 positive integers, 5 are even, hence the probability is $\frac{1}{2}$; of the first 100, 50 are even, again the probability is $\frac{1}{2}$, and so on. In general, of the first $2N$ positive integers, N are even, so that if we form the ratio $N/2N$ and let $N \to \infty$ so as to encompass all positive integers, the ratio, and therefore the probability of an even integer, remains at $\frac{1}{2}$.

Although such an argument is plausible, it is not an easy matter to make it stand up. For one thing, it depends on the natural ordering of the positive integers, and a different ordering would produce a different result. We could, for example, take the following ordering: 1, 3, 2, 5, 7, 4, 9, 11, 6, . . . , in which case the probability would appear to be $\frac{1}{3}$. Indeed, we could, just by choosing the right ordering, obtain any probability that we desired. In fact, the integers can even be so ordered that the ratio will oscillate and never approach a definite value as N increases.

However, there is an even more fundamental difficulty with the classical definition of probability than that encountered with an infinity of possible outcomes. Consider the flipping of a coin that is known to be biased in favor of heads, or the rolling of a die that is known to be made out of an inhomogeneous material. The two possible outcomes of flipping and the six possible outcomes of rolling the die are no longer equally likely. How do we calculate the probabilities (for it is still reasonable to suppose that they exist)? The classical definition is completely helpless in these situations.

Still another difficulty with the classical definition of probability is when we try to answer the following questions: What is the probability that a man living in Sweden will live to age 60? What is the probability that a light bulb will burn 100 hours or less? What is the probability that two cars going through the same intersection will collide? What is the probability that a man's income known to be above \$100,000 per year is above \$150,000?

All of these are legitimate questions that we would like to bring into the realm of probability, but which are impossible to deal with using the classical definition because questions of "symmetry" and "equally likely" cannot be answered as in games of chance. Hence, it is obvious that the classical definition must be altered or extended if these and similar questions are to be brought into the framework of the theory. The relative frequency (or statistical) definition of probability allows for this to be done.

2.4 RELATIVE FREQUENCY DEFINITION OF PROBABILITY

Lengthy observations as to the occurrence or nonoccurrence of an event A in a large number of repeated trials under the same set of conditions K show that for a wide class of phenomena the number of occurrences or nonoccurrences is subject to a stable law. Namely, if we denote by μ the number of times that A occurs in a sequence of n independent trials, then it turns out that for sufficiently large n the ratio μ/n in most of such series of observations assumes an almost constant value, with large deviations being less frequent the larger the number of trials.

This kind of stability of the relative frequency (i.e., of the ratio μ/n) was first observed in demographic phenomena. In ancient times it was already noticed that the ratio of male to the total number of births in entire countries and in large cities varied little from one year to the next, and later on, particularly in the seventeenth or eighteenth centuries, a number of fundamental works devoted to the study of population statistics appeared. Laplace, for example, in his extensive analyses of continental birth statistics discovered that during one decade the ratio of male to total births for London, St. Petersburg, Berlin, and all of France all fluctuated around one and the same number, which was approximately $\frac{22}{43}$. Apart from the stability of the ratio of male to female births, the researchers of that period also discovered that demographic laws of another character existed: the percentage of deaths at a particular age in certain groups of the population (of a given economic and social background), the distribution of people (of a definite sex, age, and nationality) according to height, weight, breadth of chest, length of footstep, and so on.

Moreover, it turns out that in all those cases in which the classical definition of probability is applicable, the fluctuation of the relative frequency is in the neighborhood of the probability p of the event. There is a vast amount of experimental evidence in verification of this fact resulting from experiments involving coin tossing, dice rolling, and so on. The results from three such coin tossing experiments are recorded in Table 2.1.

TABLE 2.1

Name of Experimenter	Number of Tosses	Number of Tails	Relative Frequency
Buffon	4,040	2,048	0.5080
Karl Pearson	12,000	6,019	0.5016
Karl Pearson	24,000	12,012	0.5005

The fact that in a number of instances the relative frequency of random events in a large number of trials is almost a constant compels us to presume the existence of certain laws, independent of the experimenter, that govern the course of these phenomena. And the fact that the relative frequency of an event to which the classical definition is applicable is, as a rule, close to its probability whenever the number of experiments is large is strong evidence that, in the general case as well, there exists some constant about which the relative frequency fluctuates. Because this constant is an objective numerical characteristic of the phenomena, it is natural to call it the *probability* of the random event A under investigation.

Therefore we shall say that a random event A has a probability if it satisfies the following conditions:

1. It is possible, at least in principle, to make an unlimited number of independent trials under the same set of conditions K, in which the event A may or may not occur.

2. As a result of a sufficiently large number of trials, the relative frequency in nearly every one of a large group of sequences of trials is observed to deviate only negligibly from a certain (but generally unknown) constant.

We may take as an approximation to this constant the relative frequency of the event in a large number of trials, or else some value reasonably close to the relative frequency. Hence, the probability should have the following properties:

1. The probability of the certain event is one.
2. The probability of the impossible event is zero.
3. If a random event C is the sum of a finite number of mutually exclusive events A_i, $i = 1, n$, each with a probability, then the probability of C exists and is equal to the sum of the probabilities of the events A_i:

$$P(C) = P(A_1) + P(A_2) + \cdots + P(A_n).$$

An important attribute of the relative frequency definition of probability is that it retains a character that is independent of the experimenter. The fact that the probabilities can be inferred only after observation in no way reduces the value of deductions nor lessens the objectivity of the probabilities.

Nevertheless, despite the fact that the relative frequency definition overcomes some of the problems of the classical definition, it has some drawbacks of its own. In the first place, it is more in the way of a description than of a formal mathematical character. Moreover, it does not reveal to any extent, although this is less serious, the nature of the phenomena being investigated and for which the relative frequency is stable. Finally, the relative frequency definition is totally inadequate in situations where probabilities are viewed as representing reasonable degrees of belief.

Thus, from a formal point of view the mathematical foundation of the theory of probability around the turn of this century was unsatisfactory. This had not kept natural scientists with their rather naive concepts of probability from many important successes, but clearly what was needed was a method for systematically studying the fundamental concepts of probability and for clarifying the conditions under which the results of the theory could be applied. The theory needed to be based on a set of premises which generalized the centuries of experience with probability and from which the development of the theory could proceed by logical deduction without recourse to intuitive notions of "common sense." It is to this that we now turn.

REFERENCES

Feller, W., *An Introduction to Probability Theory and Its Applications*, vol. I, 3rd ed., Wiley, 1968, pp. 1–25.

Gnedenko, B. V., *Theory of Probability*, Chelsea Publishing Co., 1962, pp. 9–51.

CHAPTER 3

AXIOMATIC FORMULATION OF THE THEORY OF PROBABILITY

An axiomatic approach to probability theory was first expressed in 1917 in a paper by the Russian mathematician, S. N. Bernstein. However, the approach that we shall follow here is the one of another Russian mathematician, A. N. Kolmogorov, and closely relates probability theory to the theory of sets and the modern measure-theoretic aspects of the theory of functions of a real variable. The strength of Kolmogorov's formation is that it embodies all of the fundamental properties of probability observed in the classical and relative frequency definitions, yet at the same time has met all of the demands placed on the theory by modern science, not only the natural sciences, but the social sciences as well.

3.1 KOLMOGOROV'S AXIOMS

We begin with a set S consisting of *elementary events* (what the elements of this set are is immaterial to the logical development of the theory) and consider a certain family F of subsets of this set. The elements of the family F shall be called *random events.* The following three conditions are imposed on the structure of the family F:

1. F contains the set S as one of its subsets.
2. If the subsets A and B are included in F, then so also are $A \cup B$, A', and B'.

Because S is included in F, the second requirement implies that F also includes S'; that is, *the empty set is included in F.* Moreover, it is easy to see that the second requirement implies that the union and complements of a finite number of random events of F are also included in F. In other words, F is closed under the operations of union and complementation. Such a set, as was noted in footnote 1 of Chapter 2, is called a *field of events*.

3. If the infinite sequence of subsets $A_1, A_2, \ldots, A_n, \ldots$ of S are included in F, then the union $A_1 \cup A_2 \cup \cdots \cup A_n \cup \cdots$ and the intersection $A_1 A_2 \cdots A_n \cdots$ of these subsets are also included in F.

The above method of defining a random event is in complete conformity with the idea we arrived at by concrete examples. It will be useful if we reconsider two of these from the present point of view.

Tossing a coin. A coin is tossed. The set $S = \{E_1, E_2\}$ of the *elementary events* consists of two elements, E_1 and E_2, where E_1 denotes heads and E_2 denotes tails. The set F of *random events* consists of the $2^2 = 4$ elements: $S, \varnothing, E_1, E_2$.

Rolling a die. A die is rolled. The $S = \{E_1, E_2, \ldots, E_6\}$ of *elementary events* consists of the six elements $E_1, E_2, \ldots, E_6$, where E_i denotes the showing of face i. The set F of *random events* consists of the $2^6 = 64$ elements:

$$\varnothing, E_1, E_2, \ldots, E_6, E_1 \cup E_2, \ldots, E_1 \cup E_6, E_2 \cup E_3, \ldots, E_5 \cup E_6,$$
$$E_1 \cup E_2 \cup E_3, \ldots, E_4 \cup E_5 \cup E_6, E_1 \cup E_2 \cup E_3 \cup E_4, \ldots,$$
$$E_3 \cup E_4 \cup E_5 \cup E_6, E_1 \cup E_2 \cup \cdots \cup E_5, E_2 \cup E_3 \cup \cdots \cup E_6,$$
$$E_1 \cup E_2 \cup \cdots \cup E_6.$$

If two random events A and B are such that the elements of S that comprise A are distinct from those that comprise B; that is, if

$$A \cap B = \varnothing,$$

we say that the events A and B are *mutually exclusive*. The random events S and $\varnothing$ are called the *certain* (or *sure*) event and the *impossible* event, respectively.

We now formulate the axioms that define probability.

AXIOM 1

With each random event A in a field of events F, there is associated a nonnegative number $P(A)$, called its probability.

AXIOM 2

$$P(S) = 1.$$

AXIOM 3 (Addition Axiom)
If the events $A_1, A_2, \ldots, A_n$ are pairwise mutually exclusive, then

$$P(A_1 \cup A_2 \cup \cdots \cup A_n) = P(A_1) + P(A_2) + \cdots + P(A_n).$$

With the classical definition of probability, it was not necessary to postulate the second and third axioms, for they can be derived as theorems. The first axiom, however, is contained in the classical definition.

AXIOM 4 (Extended Axiom of Addition)
If the event A is equivalent to the occurrence of at least one of the pairwise mutually exclusive events $A_1, A_2, \ldots, A_n, \ldots$, then

$$P(A) = P(A_1) + P(A_2) + \cdots + P(A_n) + \cdots.$$

The Extended Axiom of Addition is needed for those cases in which the number of elementary events is infinite.

We can now deduce from these axioms several important elementary consequences. First of all, from the fact that

$$S = S \cup \varnothing$$

and Axiom 3, we deduce that

$$P(S) = P(S) + P(\varnothing).$$

Therefore:

1. The probability of the impossible event is zero.

In an analogous way, it is easy to show the following two properties:

2. For any random event A,

$$P(A') = 1 - P(A).$$

3. For any random event A,

$$0 \leq P(A) \leq 1.$$

Next:

4. If the event A implies the event B, then

$$P(A) \leq P(B).$$

PROOF. The event B can be written as the union of the two mutually exclusive events A and $A'B$. Therefore,

$$P(B) = P(A \cup A'B) = P(A) + P(A'B) \geq P(A).$$

5. Let A and B be two arbitrary events. Then

$$P(A \cup B) = P(A) + P(B) - P(AB).$$

PROOF. Consider the following identities:

$$A \cup B = A \cup (B - AB),$$
$$B = AB \cup (B - AB).$$

Because the sets being united in each identity are mutually exclusive, it follows from Axiom 3 that

$$P(A \cup B) = P(A) + P(B - AB),$$
$$P(B) = P(AB) + P(B - AB).$$

Subtracting the second from the first then yields

$$P(A \cup B) = P(A) + P(B) - P(AB).$$

Because $P(AB) \geq 0$, it follows from property 5 that

6. $$P(A \cup B) \leq P(A) + P(B).$$

Finally, by mathematical induction it can be proved that

7. $$P(A_1 \cup A_2 \cup \cdots \cup A_n) \leq P(A_1) + P(A_2) + \cdots + P(A_n).$$

This system of axioms is consistent, since we can find real objects that satisfy the axioms. For example, let S be an arbitrary set containing a finite number of elements, that is, $S = \{a_1, \ldots, a_n\}$ and let $F = \{B_1, \ldots, B_k\}$ be the set of all subsets of S. We can then satisfy all of the axioms by setting

$$P(a_i) = p_i, \qquad i = 1, n,$$

where $p_1, p_2, \ldots, p_n$ are a set of n arbitrary nonnegative numbers for which

$$p_1 + p_2 + \cdots + p_n = 1$$

and

$$P(B_i) = \sum_{j \in i_s} p_j,$$

where i_s is an index set designating the elements a_j that are included in B_i.

It is to be noted, however, that the system of axioms is incomplete, since for a given set S we can select the probabilities in the set F in different ways. For example, in the rolling of a die we can set

$$P(E_1) = P(E_2) = \cdots = P(E_6) = \tfrac{1}{6}, \qquad (3.1)$$

or alternatively,

$$P(E_1) = P(E_2) = \tfrac{1}{6}; \qquad P(E_3) = P(E_4) = \tfrac{1}{4}; \qquad P(E_5) = P(E_6) = \tfrac{1}{12}, \tag{3.2}$$

and so on.

The incompleteness of the set of axioms does not mean that the choice of axioms is an unfortunate one or that not enough thought has gone into their development, but rather is in the nature of things. Often there are situations where it is necessary to study the same random events, but with different probabilities. For example, the system of equations (3.1) would be the probabilities associated with a die that is true. System (3.2), on the other hand, might be the probabilities associated with a "loaded" die. Therefore, the incompleteness of the set of axioms is necessary for the practical usefulness of the theory of probability.

3.2 CONDITIONAL PROBABILITY

Thus far it has been taken for granted that in speaking of the probability of an event A it has been with reference to the entire sample space S. In many instances, however, we will encounter problems of the following type:

> What is the probability of event A given that event B has already occurred?

Some example are:

1. Suppose that a pair of dice are thrown. What is the probability that one of the faces is a 3 given that the sum of the two faces is 9?
2. Suppose that a card is drawn from a well-shuffled bridge deck. What is the probability that the card is an ace given that it is a spade?
3. Suppose that 10 people are drawn at random from a city in Canada. What is the probability that they are of Italian descent given that the city is Toronto?

In each of these examples, it is evident that the probability asked for is with reference to a *restricted* sample space. In the first example, instead of the original sample space of 36 elementary events, the relevant sample space consists only of those 4 events which give a sum of 9. Similarly, the relevant sample space in the second example consists of the 13 spades,

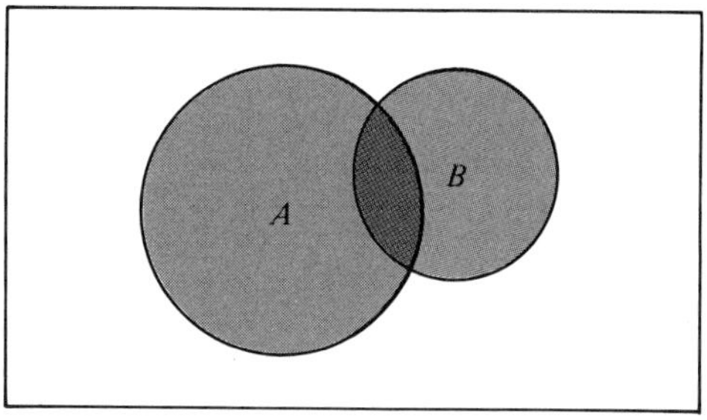

FIGURE 3.1 Conditional probability $P(A \mid B)$.

while in the third we need only consider the distribution of nationalities within Toronto. In each case, the restricted sample space is equal to the intersection of event B with the original sample space S, that is, that part of S that is also B.

Probabilities with reference to a restricted sample space are called conditional probabilities, and with reference to Figure 3.1 it is evident that the probability of A given that B has already occurred, written as $P(A \mid B)$, is equal to

$$P(A \mid B) = \frac{P(AB)}{P(B)}. \tag{3.3}$$

In words, the probability of A given B is equal to the ratio of the probability of the simultaneous (or joint) occurrence of A and B to the probability of B. It is assumed, of course, that $P(B)$ is not 0. It should be noted that we can interpret all probability as conditional probability, since, with reference to the unrestricted sample space S, we have

$$P(A \mid S) = \frac{P(A \cap S)}{P(S)} = P(A). \tag{3.4}$$

We shall consider some examples based on the experiment of throwing a pair of dice, for which the sample space follows (Table 3.1). Assume that all 36 elementary events are equally likely, and let the sum of the two faces be denoted by R.

TABLE 3.1 Sample Space for the Throwing of Two Dice

	Die 2 (F_2)					
	1, 1	1, 2	1, 3	1, 4	1, 5	1, 6
	2, 1	2, 2	2, 3	2, 4	2, 5	2, 6
Die 1	3, 1	3, 2	3, 3	3, 4	3, 5	3, 6
(F_1)	4, 1	4, 2	4, 3	4, 4	4, 5	4, 6
	5, 1	5, 2	5, 3	5, 4	5, 5	5, 6
	6, 1	6, 2	6, 3	6, 4	6, 5	6, 6

1. What is $P(F_1 = 3 \mid R = 9)$?
Because $P(R = 9) = 4/36$ and $P\{(F_1 = 3) \cap (R = 9)\} = 1/36$,

$$P(F_1 = 3 \mid R = 9) = \frac{1}{36} \Big/ \frac{4}{36} = \frac{1}{4}.$$

2. What is $P(F_1 \leq 4 \mid R = 7)$?
Because $P(R = 7) = 6/36$ and $P\{(F_1 \leq 4) \cap (R = 7)\} = 4/6$,

$$P(F_1 \leq 4 \mid R = 7) = \frac{4}{36} \Big/ \frac{6}{36} = \frac{2}{3}.$$

3. What is $P(F_1 \leq 4 \mid R \leq 7)$?
Because $P(R \leq 7) = 21/36$ and $P\{(F_1 \leq 4) \cap (R \leq 7)\} = 18/36$,

$$P(F_1 \leq 4 \mid R \leq 7) = \frac{18}{36} \Big/ \frac{21}{36} = \frac{6}{7}.$$

In a similar fashion to $P(A \mid B)$, we have [assuming $P(A) \neq 0$]

$$P(B \mid A) = \frac{P(BA)}{P(A)}, \tag{3.5}$$

and since $P(BA) = P(AB)$, it follows from Equations (3.3) and (3.5) that

$$P(A \mid B)P(B) = P(B \mid A)P(A). \tag{3.6}$$

As is evident from the examples given above, A and B in Equations (3.3) and (3.5) can be *any* random events with reference to a given sample space S. In particular, we can have $A = CD$ and $B = E$. Therefore,

$$P(E \mid CD) = \frac{P(CDE)}{P(CD)}, \tag{3.7}$$

or

$$P(CDE) = P(E \mid CD) \cdot P(CD). \tag{3.8}$$

But since

$$P(CD) = P(D \mid C) \cdot P(C),$$

it follows that

$$P(CDE) = P(E \mid CD) \cdot P(D \mid C) \cdot P(C). \tag{3.9}$$

Equation (3.9) can be generalized to the intersection of n events (as long as, of course, n does not exceed the number of random events defined with respect to the sample space).

3.3 INDEPENDENT EVENTS

We say that an event A is *independent* of an event B if

$$P(A|B) = P(A), \tag{3.10}$$

that is, if the conditional probability of A given B is equal to the probability of A given S. For $P(B) \neq 0$, Equation (3.10) implies that

$$P(AB) = P(A) \cdot P(B), \tag{3.11}$$

and since $P(AB) = P(BA)$, this in turn implies that

$$P(B|A) = P(B). \tag{3.12}$$

In other words, if A is independent of B, then B is also independent of A, and, moreover, so also are A and B' and A' and B. The proof that A and B' are independent if A and B are is as follows:

By definition of AB', we have

$$P(AB') = P(A - B).$$

But

$$\begin{aligned} P(A - B) &= P(A) - P(AB) \\ &= P(A) - P(A)P(B) \qquad \text{(since } A \text{ and } B \text{ are independent)} \\ &= P(A)\{1 - P(B)\} \\ &= P(A)P(B'). \end{aligned}$$

Therefore,

$$P(AB') = P(A)P(B').$$

The proof that A' and B are also independent is similar.

As an illustration of two events that are independent, we can take the second example of Section 3.2, namely: Given that a card drawn at random is a spade, what is the probability that it is an ace? Taking each of 52 cards to be equally likely, we have $P(\text{spade}) = 13/52$ and $P(\text{ace and spade}) = 1/52$. Therefore,

$$P(\text{ace}|\text{spade}) = \frac{1}{52}\bigg/\frac{13}{52} = \frac{1}{13}.$$

But

$$P(\text{ace}) = \frac{4}{52} = \frac{1}{13}$$

also. Hence, the two events are independent.

In the rolling of a pair of true dice, the nature of the experiment suggests that the face showing on the second die should in no way be influenced by the face on the first, so that the probability of rolling two 6's, say, is equivalent to the products of the probabilities of a 6 on each die. Generally, however, it is impossible to tell whether two events are independent simply by looking at them or considering their nature. Thus, to check independence we usually have to calculate the probabilities, as in the immediately preceding example.

It is important not to confuse independence with mutually exclusive. Two events that are mutually exclusive are *not* independent, as the example of the event A and its complement A' will readily show. A possible source of confusion is the identifying of mutually exclusive events with the common expression "having nothing to do with each other." This is a useful description of independence when applied to everyday events, but is mistaken when applied to sets for it suggests nonoverlapping. For two sets to be independent it is necessary that they have an overlap; that is, $A \cap B$ must not be equal to the empty set.

It is natural to wish to extend the notion of independence to three or more events by requiring that the probability of their intersection be equal to the product of their probabilities, namely,

$$P(A_1 \cap A_2 \cap \cdots \cap A_n) = P(A_1) \cdot P(A_2) \cdots P(A_n). \tag{3.13}$$

However, such a requirement is not sufficient to guarantee the truth of the equations that we would get by replacing some of the events by their complements such as was the case with two events. What we need is *complete independence*.

We say that n events are *completely independent* if and only if every combination of these events, taken any number at a time, is independent. For example, complete independence of the events A_1, A_2, and A_3 means that the following equations are satisfied.

$$P(A_1 A_2 A_3) = P(A_1) \cdot P(A_2) \cdot P(A_3), \tag{3.14}$$

$$\left.\begin{aligned} P(A_1 A_2) &= P(A_1) \cdot P(A_2) \\ P(A_1 A_3) &= P(A_1) \cdot P(A_3) \\ P(A_2 A_3) &= P(A_2) \cdot P(A_3). \end{aligned}\right\} \tag{3.15}$$

If these equations are satisfied, then so are the ones obtained by replacing any of the events by their complements. For example,

$$P(A_1 A_2' A_3') = P(A_1) \cdot P(A_2') \cdot (A_3').$$

3.4 BAYES THEOREM

Suppose that we pose the following situation: A ball is drawn from one of two urns depending on the outcome of a roll of a true die. If the die shows a 1 or 2, the ball is drawn from Urn I which contains 7 red balls and 3 white balls. If the die shows a 3, 4, 5, or 6, the ball is drawn from Urn II which contains 4 red balls and 6 white balls. Given that a red ball is drawn, what is the probability that it came from Urn I? Urn II?

To compute these probabilities, we proceed as follows: First, denote the urns by I and II and the balls by R and W. We therefore have

$$P(\mathrm{I}) = \frac{1}{3}, \qquad P(\mathrm{II}) = \frac{2}{3}, \qquad P(\mathrm{R} \mid \mathrm{I}) = \frac{7}{10},$$

$$P(\mathrm{W} \mid \mathrm{I}) = \frac{3}{10}, \qquad P(\mathrm{R} \mid \mathrm{II}) = \frac{4}{10}, \qquad P(\mathrm{W} \mid \mathrm{II}) = \frac{6}{10},$$

and we want $P(\mathrm{I} \mid \mathrm{R})$ and $P(\mathrm{II} \mid \mathrm{R})$. By definition of conditional probability

$$P(\mathrm{I} \mid \mathrm{R}) = \frac{P(\mathrm{I} \cap \mathrm{R})}{P(\mathrm{R})}. \tag{3.16}$$

But since I and II are mutually exclusive and exhaustive events,

$$P(\mathrm{R}) = P(\mathrm{R} \cap \mathrm{I}) + P(\mathrm{R} \cap \mathrm{II}), \tag{3.17}$$

and also

$$\begin{aligned} P(\mathrm{I} \cap \mathrm{R}) &= P(\mathrm{R} \mid \mathrm{I}) P(\mathrm{I}), \\ P(\mathrm{II} \cap \mathrm{R}) &= P(\mathrm{R} \mid \mathrm{II}) P(\mathrm{II}). \end{aligned} \tag{3.18}$$

Therefore,

$$\begin{aligned} P(\mathrm{I} \mid \mathrm{R}) &= \frac{P(\mathrm{R} \mid \mathrm{I}) P(\mathrm{I})}{P(\mathrm{R} \mid \mathrm{I}) \cdot P(\mathrm{I}) + P(\mathrm{R} \mid \mathrm{II}) \cdot P(\mathrm{II})} \\ &= \frac{\frac{7}{10} \cdot \frac{1}{3}}{\frac{7}{10} \cdot \frac{1}{3} + \frac{4}{10} \cdot \frac{2}{3}} = \frac{7}{15}. \end{aligned} \tag{3.19}$$

The probabilities obtained for the urns after the result of the drawing are often referred to as *posterior* (or after the fact) probabilities in contrast to the unconditional probabilities which are called *prior* (or before the fact) probabilities. Table 3.2 summarizes the above example. In this example, the fact that the ball drawn was red substantially increases the probability that it came from Urn I.

TABLE 3.2

Urn	Prior Probability	Posterior Probability Given That the Ball Was: Red	White
I	$\frac{1}{3}$	$\frac{7}{15}$	$\frac{1}{5}$
II	$\frac{2}{3}$	$\frac{8}{15}$	$\frac{4}{5}$

The generalization of formulas (3.18) and (3.19) to n events is called *Bayes theorem* on inverse probabilities and is stated as follows:

THEOREM 3.1 (Bayes Theorem)
If there are n events $A_i (i = 1, n)$ which are mutually exclusive and exhaustive; that is, if

$$A_i \cap A_j = \varnothing, \qquad i \neq j, \qquad \text{and} \qquad \bigcup_{i=1}^{n} A_i = S,$$

and if B is an event in S, then

$$P(A_i | B) = \frac{P(B|A_i)\, P(A_i)}{\sum_{j=1}^{n} P(B|A_j)\, P(A_j)}. \tag{3.20}$$

PROOF. The key to this theorem is that since the A_i's form an exhaustive partition of the sample space, the event B can occur with one and only one of the A_i. Therefore, we can write $P(B)$ as

$$P(B) = \sum_{j=1}^{n} P(B \cap A_j)$$

$$= \sum_{j=1}^{n} P(B|A_j)P(A_j)$$

since $P(B \cap A_j) = P(B|A_j)P(A_j)$.

Formula (3.20) then follows from the definition of conditional probability.

The use of Bayes theorem forms one of the most controversial areas in all of statistics. As Equation (3.20) stands it is true because it is a mathematical theorem. However, trouble arises when it comes to determining the set of events A_i which form an exhaustive partition of the sample space and assigning them probabilities. In most real life instances where we would like to apply Bayes theorem it is impossible to establish for certain whether the A_i in fact form an exhaustive partition. For example, we may have a situation where it is possible to associate via probabilities certain medical symptoms with certain diseases. We would like to be able to turn this around and calculate via Bayes theorem the probability of a particular disease given some particular symptom. However, to do so would require an exhaustive list of the causes of the symptom and their appropriate probabilities. Such knowledge is usually lacking.

Problems such as the one just noted, however, have not stood in the way of Bayes theorem achieving widespread use. Indeed, there is one entire school of statistics, the so-called Bayesian school, for which Bayes theorem provides the point of departure. Almost by necessity, Bayesian statistics requires the use of subjective probability. However, our discussion of Bayesian concepts is postponed until Chapter 22.

We have now discussed the concept of probability from three points of view and have proved the key theorems, and it is time to begin devising ways of systematically assigning probabilities. In most cases, this is done by means of a *probability function*, and after the next chapter, our study of probability will reduce to the study of particular probability functions and their properties.

REFERENCES

Freeman, H., *Introduction to Statistical Inference*, Addison-Wesley, 1963, chap. 1.
Gnedenko, B. V., *Theory of Probability*, Chelsea Publishing Co., 1962, pp. 51–76.

CHAPTER 4

PROBABILITIES IN A FINITE SAMPLE SPACE: ELEMENTS OF COMBINATORIAL ALGEBRA

In this chapter, we develop the basic notions of combinatorial algebra, since being able to "count" is indispensable to calculating probabilities in finite sample spaces where the classical definition of probability is applicable. Those readers with previous exposure to this topic can skip this chapter without loss of continuity.

In order to compute the probability of an event in a situation where the classical definition of probability applies—that is, when there are n equally likely elementary events—we need to be able to compute the total number of these that are favorable to the event in question. To facilitate these calculations, we shall begin by developing several combinatorial formulas which are based on the two following basic principles:

1. If an event A can occur in a total of m ways and an event B can occur in a total of n ways, then the event $A \cup B$ can occur in $m + n$ ways, provided that A and B cannot occur simultaneously.
2. If an event A can occur in a total of m ways and a different event B in a total of n ways, then the event $A \cap B$ can occur in mn ways.

We can illustrate these two principles by letting the event A correspond to the drawing of a spade from a deck of 52 cards and the event B to the drawing of a heart. Each of these events can be done in 13 ways. Therefore, the event of drawing either a spade or a heart can be done in $13 + 13 = 26$

ways. On the other hand, two cards can be drawn from the deck such that one is a spade and the other a heart in $13 \cdot 13 = 169$ ways, since with each spade 13 hearts can be drawn and there are 13 spades in all.

It is evident that these two principles can be generalized to more than two events. Thus, if three mutually exclusive events A, B, and C can occur in k, m, and n ways, respectively, then the event $A \cup B \cup C$ can occur in $k + m + n$ ways and $A \cap B \cap C$ in kmn ways.

4.1 PERMUTATIONS

We shall use the second of these principles to enumerate the number of arrangements that can be made with a set of n objects. Consider the three letters a, b, and c, and let us ask how many arrangements of these letters can be made. To begin with, any of the three letters can occupy the first position, so that this can be done in 3 ways. Then once the first position is filled, the second position can be filled in 2 ways, and finally there remains only 1 way for the third position to be filled. Hence, the three letters can be arranged in $3 \cdot 2 \cdot 1 = 6$ ways in all. Written out, these are

$$abc \quad acb \quad bac \quad bca \quad cab \quad cba.$$

It is easy to see, therefore, that by generalizing the above reasoning, n objects can be arranged in

$$n(n-1)(n-2)\cdots 2 \cdot 1 \tag{4.1}$$

different ways. For example, 6 objects can be arranged in

$$6 \cdot 5 \cdot 4 \cdot 3 \cdot 2 \cdot 1 = 720$$

different ways. Each arrangement is commonly called a *permutation*.

It is useful to have a symbol to denote the expression in (4.1), and we do this with the symbol $n!$ (read *n factorial*); that is,

$$n! = n(n-1)(n-2)\cdots 2 \cdot 1. \tag{4.2}$$

Since

$$(n-1)! = \frac{n!}{n}, \tag{4.3}$$

it is usual to define $0!$ as 1, so that the relation is consistent when $n = 1$.

Let us consider some examples.

1. Suppose that a man wishes to go round trip from Chicago to New York to London. He wants to go by air between Chicago

and New York, and by sea between New York and London. If there are 6 airlines to choose from between Chicago and New York and 3 shiplines between New York and London how many different ways can he make the round trip without using the same airline or shipline twice on the same trip?

SOLUTION. There are 6 different ways that the man can get to New York and once there, there are 3 ways of getting to London. On the return trip, there remain 2 ways of getting back to New York, and finally there are 5 unused airlines to choose from in returning to Chicago. Therefore, there are $6 \cdot 3 \cdot 2 \cdot 5 = 180$ different ways that he can make the round trip from Chicago to New York to London.

2. Given the digits 1, 2, 3, 4, and 5, how many four-digit numbers can be formed (a) If there is no repetition? (b) If there can be repetition? (c) If the number must be even and no repetition? (d) If the digits 2 and 3 must appear in the number and in that order, and with no repetition?

SOLUTION.

(a) *No repetition.* The first digit can be chosen in 5 ways, the second in 4 ways, and so on, so that the number of four-digit numbers that can be formed is $5 \cdot 4 \cdot 3 \cdot 2 = 120$.

(b) *With repetition.* The first digit can be chosen in 5 ways, as can also the second, and so on. Therefore, $5 \cdot 5 \cdot 5 \cdot 5 = 625$ numbers can be formed in all.

(c) *Even, but no repetition.* The last digit must be either 2 or 4, this can occur in 2 ways. After the last place is filled, the remaining places can be filled in 4, 3, 2 ways, respectively, since no digit can be used more than once. Hence, the number is $4 \cdot 3 \cdot 2 \cdot 2 = 48$.

(d) *2 and 3 must appear in that order, and with no repetition.* Because 2 and 3 must appear together, then two digits become a single digit. Therefore, the problem is reduced to the number of permutations of four digits, or $4!$.

Problem (c) illustrates the fact that when a special operation is called for, this operation should be performed first. Thus, since the last digit must be even, this digit is selected first.

Let us now enumerate the number of permutations that can be made from n objects if only r objects are to be used in any permutation. Reasoning as before, the first place can be filled in n ways, the second in $n - 1$, and so on. When we come to the rth place, there remain $n - (r - 1)$ objects from which to choose, hence the number of permutations of n objects taken r at a time is $n(n-1)\cdots(n-r+1)$. The symbol P_r^n is usually used to denote this number, namely,

$$P_r^n = n(n-1)\cdots(n-r+1) = \frac{n!}{(n-r)!}. \tag{4.4}$$

Problem (a) of Example 2 above is an illustration. On putting $r = n$, it is seen that (4.4) is consistent with Equation (4.2).

4.2 COMBINATIONS

With permutations, order counts in that *ab* is different from *ba* even though the same letters are involved. In many situations, however, our only concern will be whether or not the objects are different, and order will be of no consequence. Therefore, we need to solve the following problem:

> In how many different ways can r objects be selected from n objects?

For the present, let us denote this number by x. Now from formula (4.4), the number of permutations that we can have of n objects taken r at a time is P_r^n. But permutations take order into account, and this we do not wish to do. Therefore, each selection of r different objects will have $r!$ permutations. Hence,

$$P_r^n = x \cdot r! \tag{4.5}$$

and solving for x, we have

$$x = \frac{P_r^n}{r!} = \frac{n!}{(n-r)!\,r!}. \tag{4.6}$$

Usually this number is denoted by $\binom{n}{r}$, namely,

$$\binom{n}{r} = \frac{n!}{(n-r)!\,r!}, \tag{4.7}$$

and is called the number of *combinations* of n objects taken r at a time.

As an example, let us calculate the number of different ways that a committee of 5 can be selected from 9 people. We have

$$\binom{9}{5} = \frac{9!}{5!\,4!} = 126$$

different ways.

To illustrate the use of formula (4.7) in assigning probabilities we will consider several examples from games of chance where the classical definition of probability applies.

1. In drawing 2 cards from a well-shuffled bridge deck, what is the probability that both will be hearts?

SOLUTION. Two cards can be drawn from 52 cards $\binom{52}{2}$ different ways, so that the number of elementary events with equal probability will be $\binom{52}{2}$. The number of events favorable to both cards being hearts will be $\binom{13}{2}$, since this is the number of ways that 2 hearts can be drawn from a total of 13. Hence, the required probability is

$$P(\text{both cards hearts}) = \binom{13}{2}\Big/\binom{52}{2} = \frac{3}{51}.$$

2. Same facts. What is the probability that one card is a heart and one a spade?

SOLUTION. As above, the number of possible outcomes is $\binom{52}{2}$ and these are assigned equal probability. Since there are 13 hearts, a heart can be drawn in $\binom{13}{1}$ ways, and similarly for spades. Therefore, the number of favorable outcomes is

$$\binom{13}{1}\binom{13}{1}$$

and the required probability is given by

$$P(\text{H and S}) = \binom{13}{1}\binom{13}{1}\Big/\binom{52}{2} = \frac{13}{102}.$$

3. In drawing 5 cards, what is the probability that they are either all hearts or all spades?

SOLUTION. The number of equally probable elementary events will be $\binom{52}{5}$. Now 5 hearts can be drawn in $\binom{13}{5}$ ways, and the same is true for spades. Because 5 hearts or 5 spades are mutually exclusive events, the probabilities add, and so the required probability is

$$P(\text{5H or 5S}) = \left\{\binom{13}{5} + \binom{13}{5}\right\}\Big/\binom{52}{5} = 0.000495.$$

4. If 5 cards are drawn and 4 of them are either hearts or diamonds, what is the probability that 2 of them are hearts?

SOLUTION. This is a conditional probability, and the sample space is reduced to the 26 hearts and diamonds. The number of equally likely elementary events is then $\binom{26}{4}$. The 2 hearts can be drawn in $\binom{13}{2}$ ways and for each of these the remaining diamonds can also be drawn in $\binom{13}{2}$. Hence there are

$$\binom{13}{2}\binom{13}{2}$$

favorable events in all, and the required probability is

$$P(2\text{H}\,|\,4 \text{ are H or D}) = \binom{13}{2}\binom{13}{2}\Big/\binom{26}{4} = 0.407.$$

5. Five cards are drawn, what is the probability that none of them is an ace?

SOLUTION. There are $\binom{52}{5}$ possible outcomes, of which $\binom{48}{5}$ are favorable. Hence the probability is

$$P(\text{no aces}) = \binom{48}{5}\Big/\binom{52}{5} = 0.659.$$

It is evident from formula (4.7) that

$$\binom{n}{r} = \binom{n}{n-r}. \tag{4.8}$$

Moreover, it can easily be checked that for $1 \leq r \leq n$

$$\binom{n+1}{r} = \binom{n}{r-1} + \binom{n}{r}. \tag{4.9}$$

Formula (4.9) is known as *Pascal's rule*, and can be used to build up the coefficients in the Pascal triangle.

The number $\binom{n}{r}$ can be given a different interpretation, namely, that it is the number of ways that n objects can be divided into two groups, one containing r items and the other $n - r$. Following this reasoning, let us therefore consider the number of ways that n objects can be divided into three groups with n_1, n_2, and n_3 objects, respectively, with $n_1 + n_2 + n_3 = n$.

First, we can divide n items into two groups containing n_1 and $n_2 + n_3$ items, respectively, in $\binom{n}{n_1}$ ways. Then the second group can be divided into two groups one with n_2 and the other with n_3 objects in $\binom{n_2 + n_3}{n_2}$ ways. Hence, the three groups can be formed in

$$\binom{n}{n_1}\binom{n_2 + n_3}{n_2} = \frac{n!}{n_1!(n_2 + n_3)!} \cdot \frac{(n_2 + n_3)!}{n_2!n_3!} = \frac{n!}{n_1!n_2!n_3!} \tag{4.10}$$

ways. This argument can easily be extended to find the number of ways that n objects can be divided into k groups containing $n_1, n_2, \ldots, n_k$ objects, respectively. This number is readily found to be

$$\frac{n!}{n_1!n_2!\cdots n_k!}. \tag{4.11}$$

Expression (4.11), moreover, has still another interpretation. It can be viewed as the number of permutations of n objects when n_1 are alike and of one kind, n_2 are alike and of a second kind, and so on. To see that this is so, let us consider the case of two groups with n_1 items in the first group, n_2 in the second, and $n_1 + n_2 = n$. Let the number of permutations be x. Then for each permutation, if we temporarily distinguish among the objects within the first group and within the second group, there would be $n_1!n_2!$ permutations in all. Therefore we will have $x \cdot n_1!n_2! = n!$, since we must have $n!$ permutations in all if all n objects are distinguishable. From this it follows that

$$x = \frac{n!}{n_1!n_2!}.$$

This reasoning is easily generalized to yield expression (4.11).

REFERENCES

Feller, W., *An Introduction to Probability and Its Applications*, vol. I, 3rd ed., Wiley, 1968, chap. II.

Mood, A. M. and Graybill, F. A., *Introduction to the Theory of Statistics*, 2nd ed., McGraw-Hill, 1963, chap. 2.

CHAPTER 5

RANDOM VARIABLES AND PROBABILITY DISTRIBUTIONS

In preceding chapters, we have developed the basic notions of a random event and of probability, and have derived the fundamental laws of the latter. In this chapter, we introduce a new concept, that of a *random variable*, and begin a study of probability distributions. We shall start with random variables and probability distributions in a finite sample space, and then proceed to random variables that can take on a continuum of values.

5.1 DISCRETE RANDOM VARIABLES

In formulating the axioms of probability, we defined probabilities over a field of events F which had been constructed as the set of random events for a particular experiment S. The probability function was then taken to be a real-valued function which assigned a number between 0 and 1 (including the endpoints) to each event in F. Before the experiment is performed, however, the outcome is uncertain, which means that we can view the outcome as a variable. Since it is always possible to put the random events into a one-to-one correspondence with the integers, the random events can be defined as numbers, and consequently, the variable also can be viewed as numerical. Accordingly, let us formulate the following general definitions:

Random variable. A variable whose value is a number determined by outcome of an experiment is called a *random variable*.

Probability function. Let X be a random variable with possible outcomes $x_1, x_2, \ldots, x_n$ and associated probabilities $f(x_1), f(x_2), \ldots, f(x_n)$. Then the set G whose elements are the ordered pairs

$$(x_i, f(x_i)), \qquad i = 1, n$$

is called the *probability function* of X.[1]

We will consider some examples.

Flipping a coin. The possible outcomes are heads or tails. If we denote tails by 0 and heads by 1 so that $x_1 = 0$ and $x_2 = 1$, then (assuming the coin to be fair)

$$P(X = 0) = f(x_1) = \tfrac{1}{2},$$
$$P(X = 1) = f(x_2) = \tfrac{1}{2}.$$

Tossing two dice. Let X be the sum of the two faces. Then there are 11 possible outcomes, and, assuming that the dice are true, we know the probabilities to be

x_i:	2	3	4	5	6	7	8	9	10	11	12
$f(x_i)$:	$\frac{1}{36}$	$\frac{2}{36}$	$\frac{3}{36}$	$\frac{4}{36}$	$\frac{5}{36}$	$\frac{6}{36}$	$\frac{5}{36}$	$\frac{4}{36}$	$\frac{3}{36}$	$\frac{2}{36}$	$\frac{1}{36}$

with a graph which looks as shown in Figure 5.1.

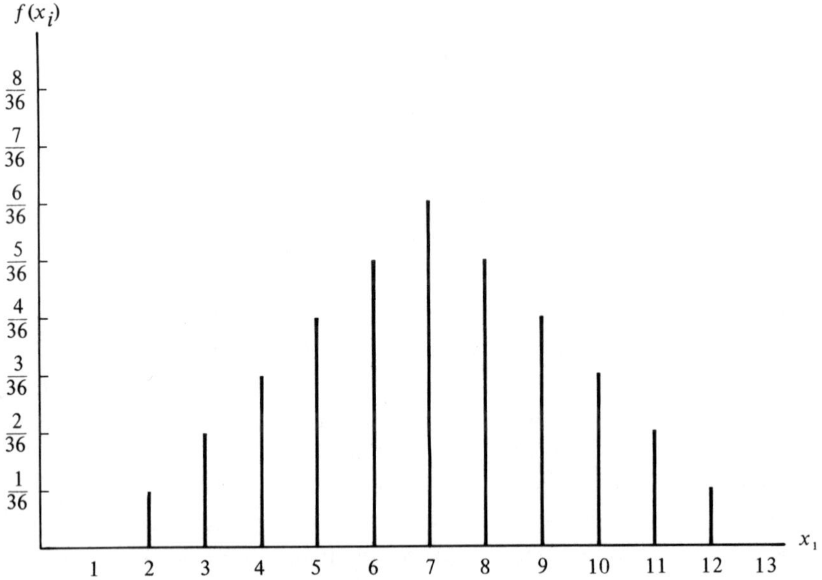

FIGURE 5.1 Discrete density function.

[1] The convention that we shall follow is to represent random variables by upper case letters and particular values of random variables by lower case letters.

Suppose, as above, that X is a random variable that can take the values x_i, $i = 1, n$, with the probabilities $f(x_i)$, $i = 1, n$, and let A be any subset of the points x_i, $i = 1, n$. Then it follows from the axioms of probability that the probability of the event A (the probability that X is in A) is

$$P(A) = \sum_A f(x_i), \tag{5.1}$$

where $\sum_A$ means "sum $f(x_i)$ over all values of x_i that are in A." For example, if X can take on the values: 0, 1, 2, 3, ..., 10, then

$$P(1 \le X \le 4) = f(1) + f(2) + f(3) + f(4).$$

The function $f(x)$ is usually referred to as a *discrete density function* or simply a *density*, and we say alternatively that "X is distributed as $f(x)$," or "$f(x)$ is the probability density of X." It will now be useful to formulate the axioms of probability in terms of the density function $f(x)$. The equivalents of the first two axioms are:

AXIOM 1

$$f(x_i) \ge 0, \qquad i = 1, n,$$

AXIOM 2

$$\sum_i f(x_i) = 1,$$

while for the Addition Axiom we have

AXIOM 3

$$P(a \le X \le c) = \sum_{a \le x_i \le c} f(x_i),$$

which for $a < b < c$

$$= \sum_{a \le x_i \le b} f(x_i) + \sum_{b < x_i \le c} f(x_i)$$

$$= P(a \le X \le b) + P(b < X \le c).$$

Axiom 3 is a formalization of Equation (5.1) above.

Often we will be interested in the event $X \le x_i$, so we will now introduce $F(x_i)$, the *distribution function* (also sometimes called the *cumulative distribution*) of the discrete random variable X, and define it by

$$P(X \le x_i) = F(x_i) = \sum_{x \le x_i} f(x_i). \tag{5.2}$$

Because $f(x_i) \geq 0$, it is evident that $F(x_i)$ is monotonic nondecreasing, and it is easily shown that (allowing x_i to range over the entire real line):

$$F(-\infty) = 0, \qquad F(\infty) = 1, \qquad 0 \leq F(x_i) \leq 1,$$
$$P(a < X \leq b) = \sum_{a < x_i \leq b} f(x_i) = F(b) - F(a).$$

To illustrate the foregoing: For a single die an appropriate correspondence would be

S:

x_i: 1 2 3 4 5 6

and, supposing the die to be true, we would have

$$f(1) = f(2) = \cdots = f(6) = \tfrac{1}{6}.$$

Axioms 1 and 2 are obviously satisfied, and as one example of satisfying Axiom 3, we can take

$$P(1 \leq X \leq 4) = P(1 \leq X \leq 3) + P(3 < X \leq 4) = \tfrac{3}{6} + \tfrac{1}{6} = \tfrac{4}{6}.$$

Moreover, we have

$$F(0) = 0, \qquad F(6) = 1, \qquad 0 \leq F(x) \leq 1,$$

and, for example,

$$P(1 < X \leq 4) = F(4) - F(1) = \tfrac{4}{6} - \tfrac{1}{6} = \tfrac{3}{6}.$$

The preceding example is an illustration of the *uniform distribution* (sometimes also called the *rectangular distribution*) for a discrete random variable. Formally, the uniform distribution is described by

$$f(x_i) = \frac{1}{n}, \qquad i = 1, n, \tag{5.3}$$

where n is the number of possible outcomes. Its graph is a series of n "spikes" each of height $1/n$. The graph of $f(x)$ for the rolling of a true die is given in Figure 5.2.

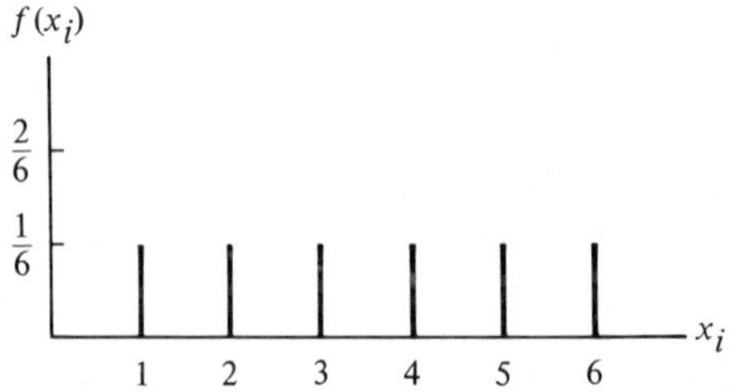

FIGURE 5.2 Graph of $f(x)$ for rolling of true die.

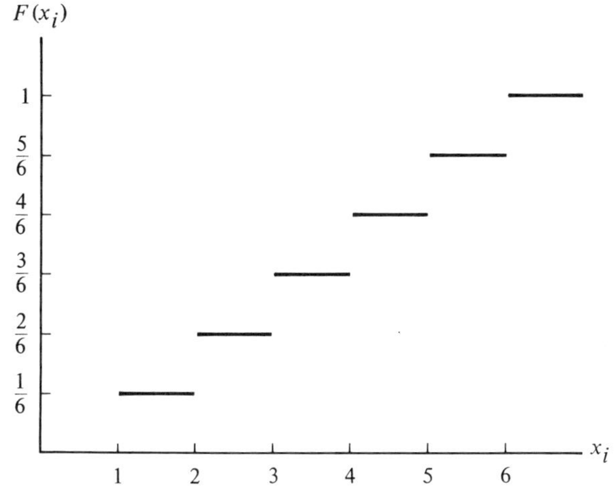

FIGURE 5.3 "Step" function for rolling of true die.

The distribution function for the uniform distribution is a "step" function, with uniform steps at each value of x_i. Figure 5.3 illustrates the distribution function for the rolling of a true die.

Other discrete distributions will be discussed in Chapter 7.

5.2 CONTINUOUS RANDOM VARIABLES

A discrete random variable can assume only a finite, or a countably infinite, number of values—that is, it is restricted to an isolated set of numbers on the real line, in the real plane, and so on. However, in many situations we encounter random variables that can assume any value within an interval or a collection of intervals on the real line, in the real plane, and so on, and we need to extend our definition to cover these situations.

To motivate the definition to be used, it will be useful to consider an example from Mood and Graybill (1963).[2] A rifle is aimed at the center of a target, and after being clamped into position in a vise, is fired several times. The bullets will not all strike the center of the target, because minor variations in the weight of the bullets, in their shape, in the effect of atmospheric conditions on the powder, and other factors will cause the trajectories of the bullets to vary. After a few shots, the pattern of hits on the target might be represented by Figure 5.4. Let X be a variable that measures the horizontal deviation of a hit from the vertical line that is drawn through the

[2] From A. M. Mood and F. A. Graybill, *Introduction to the Theory of Statistics*, 2nd ed., McGraw-Hill Book Company, 1963, pp. 77–79. Copyright © 1963 by A. M. Mood and F. A. Graybill. Used with permission of McGraw-Hill Book Company.

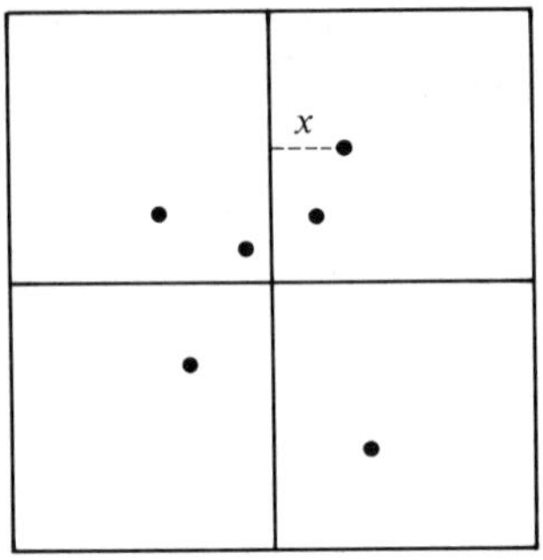

FIGURE 5.4 Pattern of hits on target.

center of the target. It is clear that X is a random variable that can assume a continuum of values.

With discrete random variables, it is possible to have a nonzero probability associated with each admissible event, even in the case where the number of events is countably infinite, and still have the probabilities sum to 1. With a continuous random variable, however, this is not possible. The density will not sum to 1 unless practically all of the admissible events (all but a countable set) are given the probability 0. On the other hand, if we refer to the horizontal deviations in Figure 5.4, it is clear that all values of X within a neighborhood of a particular point will be about equally probable, and so to say that most of these points have probability 0 while a few others have nonzero probabilities does not make much sense. Therefore, it would appear that we have encountered a problem.

It is a problem indeed, but before we establish a way of getting around it, it is well to remark that it is one of a logical nature rather than one of practical importance. From a practical point of view, since we are limited by the accuracy of our measuring devices, a deviation can only be identified within a certain interval. In the above example, we could not, for instance, as a practical matter distinguish between a deviation of 1 in. and 1.00001 in. Thus, if we can measure accurately only to within 0.01 in., a deviation of 1.50 in. would be interpreted to mean that the deviation lies between 1.49 and 1.51 in. and might better be written 1.50 ± 0.01 in. to indicate this fact. It follows, therefore, that as a practical matter only probabilities within an interval have meaning for a continuous random variable.

The logical problem is also met by dealing with intervals instead of points. Suppose that the rifle is fired 100 times at the target in Figure 5.4, and suppose that the target is divided into strips by drawing vertical lines at 1-in. intervals, as in Figure 5.5. Let the random variable X be negative to the left of the central line and positive to the right. Then for a given strip,

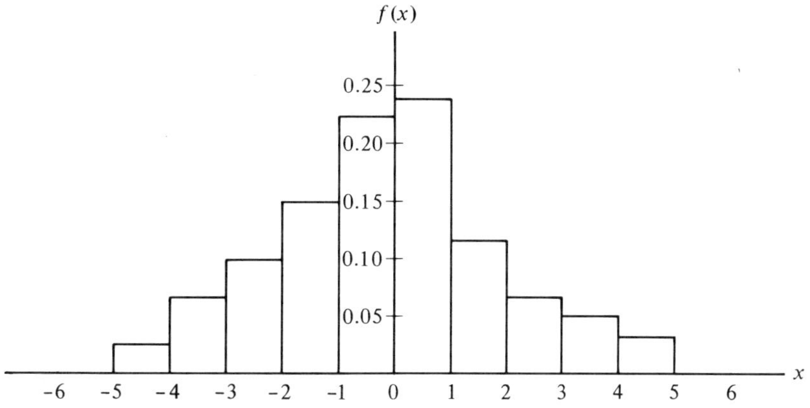

FIGURE 5.5 Histogram for hits on target.

say the one for $0 < X \leq 1$, the number of hits within that interval divided by 100 will be the relative frequency that the deviation is between 0 and 1. We can tabulate a hypothetical distribution of 100 shots and compute the empirical probabilities as in Table 5.1.

TABLE 5.1

Strip	Number of Shots	Relative Frequency
$-5 < X \leq -4$	2	0.02
$-4 < X \leq -3$	6	0.06
$-3 < X \leq -2$	9	0.09
$-2 < X \leq -1$	13	0.13
$-1 < X \leq 0$	21	0.21
$0 < X \leq 1$	23	0.23
$1 < X \leq 2$	11	0.11
$2 < X \leq 3$	7	0.07
$3 < X \leq 4$	5	0.05
$4 < X \leq 5$	3	0.03

This empirical distribution can be plotted, as in Figure 5.5, by a series of rectangles, the height of each rectangle being equal to the relative frequency divided by the width of the interval and width equal to the width of the interval. This is done to indicate that the relative frequency refers to the whole interval and not to particular points within it. Thus, the relative frequency of shots falling within any particular interval is given by the *area* of the rectangle for that interval, whence the sum of the areas for all the rectangles is equal to 1. To obtain the relative frequency for an interval

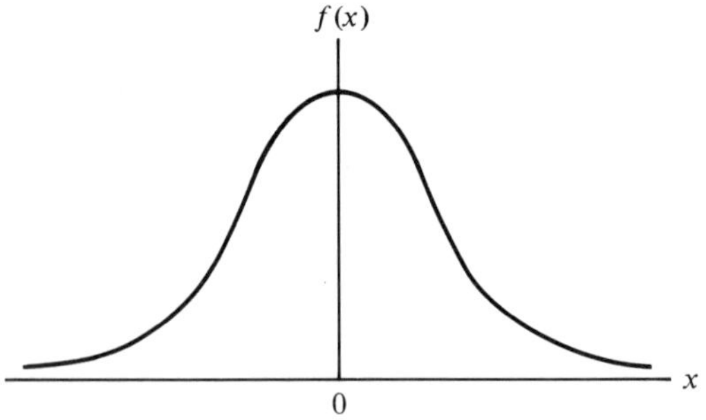

FIGURE 5.6 Graph of the function f.

different from one chosen initially, we add the areas over that interval. For example, for $-1 < X \leq 2$, from Table 5.1 the relative frequency is $0.21 + 0.23 + 0.11 = 0.55$.

If we were to take another 100 shots, we would get a second empirical distribution, which ordinarily would differ from the first, but which nevertheless would have the same general shape, and similarly for a third set of 100 shots, and so on. If we take these observed relative frequencies as estimates of the "true" probabilities, we can then postulate the existence of a function f, such as that plotted in Figure 5.6, which gives the correct probability for any *interval. The probabilities, however, are given by areas under the curve, not by values of the function.*

Therefore, for X, a continuous random variable, we define

$$P(a < X \leq b) = \int_a^b f(x)\,dx \tag{5.4}$$

and this is equal to the area under the curve $f(x)$ and bounded by the points a and b in Figure 5.6. With reference to Figure 5.5,

$$P(0 < X \leq 1) = \int_0^1 f(x)\,dx \tag{5.5}$$

would be the probability estimated by the area of the rectangle over the interval $0 < X \leq 1$.

We must emphasize that with a continuous random variable we cannot think in terms of probability at a point, for from Equation (5.4)

$$P(X = a) = \int_a^a f(x)\,dx = 0. \tag{5.6}$$

In other words, the probability at any point is 0; hence, we can only think in terms of probability being defined over an interval. The shortest interval that we can have is of length dx, and the probability that X is within this interval is defined by

$$P(x < X < x + dx) = f(x)\, dx. \tag{5.7}$$

As with a discrete random variable, we also call $f(x)$ for continuous x the *density* function, and say that "X is distributed as $f(x)$." Reformulating the axioms of probability in terms of the density function for continuous X, we have for the first two axioms:

AXIOM 1″

$$f(x) \geq 0 \qquad \text{all } x,$$

AXIOM 2″

$$\int_{-\infty}^{\infty} f(x)\, dx = 1,$$

and for the addition axiom, we have

AXIOM 3″

$$P(a \leq X \leq c) = \int_a^c f(x)\, dx,$$

which, for $a < b < c$,

$$= \int_a^b f(x)\, dx + \int_b^c f(x)\, dx$$

$$= P(a \leq x \leq b) + P(b \leq X \leq c).$$

It should be noted that for a continuous random variable, any overlap of endpoints in the ranges of the integrals (as at point b) is of no consequence since

$$\int_b^b f(x)\, dx = 0.$$

Hence, $P(a \leq X \leq b)$ is the same as $P(a < X < b)$, and so on.

As with a discrete random variable, we are often interested in the event $X \leq x$, and so we define the *distribution* function for a continuous random variable as

$$F(x) = \int_{-\infty}^{x} f(z)\, dz, \tag{5.8}$$

where $f(x)$ is the density function of X. It is evident that

$$F(-\infty) = 0, \qquad F(\infty) = 1, \qquad 0 \leq F(x) \leq 1. \tag{5.9}$$

Moreover, for those points where $f(x)$ is continuous, we have from elementary calculus that

$$\frac{dF(x)}{dx} = f(x). \tag{5.10}$$

This equation shows the fundamental connection between the distribution and density functions for a continuous random variable.

Next, from the third axiom it follows that

$$P(a < X < b) = \int_a^b f(x)\, dx = F(b) - F(a), \tag{5.11}$$

and finally from the first two axioms it is evident that $F(x)$ is monotonic nondecreasing.

In the chapters which follow we will refer to the density and distribution functions for both discrete and continuous random variables as *probability distributions*. It will always be clear from the context, however, which function we have in mind and whether the random variable is discrete or continuous.

REFERENCES

Freeman, H., *Introduction to Statistical Inference*, Addison-Wesley, 1963, chap. 2.

Mood, A. M. and Graybill, F. A., *Introduction to the Theory of Statistics*, 2nd ed., McGraw-Hill, 1963, chaps. 3 and 4.

CHAPTER 6

EXPECTED VALUES AND MOMENTS

Much of the time the information conveyed in a probability distribution can be effectively summarized by a knowledge of the general shape of the distribution and its location, and for a great many distributions these characteristics can be described by a small number of numerical quantities that are peculiar to the distribution involved. The most common, and much of the time, most useful, of these numerical quantities are the *moments* of the distribution, and we shall begin their study in this chapter. Our point of departure will be the notion of the *expected value* of a random variable.

6.1 EXPECTED VALUES

The expected value of a random variable is closely associated with the every-day concept of the arithmetic average, and indeed, is the mathematical idealization of that concept. The expected value for a continuous random variable is defined as follows:

$$E(X) = \int_{-\infty}^{\infty} xf(x)\,dx, \tag{6.1}$$

where the symbol E denotes "the expected value of" and $f(x)$ is the density function of X. If X is discrete, then its expected value is given by

$$E(X) = \sum_{x} xf(x), \tag{6.2}$$

where $\sum_x$ means that x is summed over all of its possible values and $f(x)$ is the (discrete) density function. It is evident from these two formulas that the expected value of a random variable can be viewed as a weighted average of its possible values, with the weights being the respective probabilities. It should be noted, however, that the expected value is defined only when the integral in (6.1) or the sum in (6.2) exists which is not always the case.

More generally, if $g(x)$ is a function of the random variable X, then we define the expected value of $g(x)$ as

$$E[g(x)] = \int_x g(x) f(x)\, dx \tag{6.3}$$

for continuous X, and

$$E[g(x)] = \sum_x g(x) f(x) \tag{6.4}$$

for the discrete case (again, assuming that the integral or sum exists). For the case of $g(x) = x$, we are back to Equations (6.1) and (6.2).

6.2 SOME PROPERTIES OF EXPECTED VALUES

At this point it will be useful to derive a few simple properties of expected values. We need only do this for the continuous case since the same results hold when the random variable is discrete.

PROPERTY 1

If a is a constant, then

$$E(aX) = aE(X). \tag{6.5}$$

PROOF. By definition

$$E(aX) = \int_x axf(x)\, dx$$

$$= a \int_x xf(x)\, dx = aE(X).$$

PROPERTY 2

If a is a constant, then

$$E(a) = a. \tag{6.6}$$

PROOF. By definition

$$E(a) = \int_x af(x)\,dx$$
$$= a\int_x f(x)\,dx = a.$$

PROPERTY 3

If a and b are constants, then

$$E(a + bX) = a + bE(X). \tag{6.7}$$

PROOF. The proof follows from Properties 1 and 2 and the fact that integration and summation are interchangeable operations.

6.3 MOMENTS

Let $f(x)$ be the density function for a continuous random variable X and suppose that a is an arbitrary constant. Then we define the *rth moment* about the point a for continuous X as

$$E[(X - a)^r] = \int_x (x - a)^r f(x)\,dx \tag{6.8}$$

and for discrete X as

$$E[(X - a)^r] = \sum_x (x - a)^r f(x). \tag{6.9}$$

With $a = 0$ the moments are measured about the origin, and are usually referred to as the *raw* (or *unadjusted*) moments. Henceforth, the raw moments will be denoted by $\mu_r'(r = 0, 1, 2, \ldots)$. The 0th moment about the origin (i.e., $r = 0$) is simply the area under the density function, and this is, of course, equal to 1. The first moment about the origin is also of intrinsic interest, for, as we have just seen, it is simply the expected value of the random variable itself and defines its *mean value* (or *arithmetic mean*, or simply *mean*).

Setting $a = E(X) \equiv \mu$, we next define the moments about the mean, which are usually referred to as the "central" moments and denoted by μ_r $(r = 0, 1, 2, \ldots)$:

$$\mu_r \equiv E[(X - \mu)^r] = \int_x (x - \mu)^r f(x)\,dx \qquad r = 0, 1, 2, \ldots. \tag{6.10}$$

Because $(x - \mu)^0 = 1$, we again have the area beneath the density function for $r = 0$, while for $r = 1$ we have

$$E(X - \mu) = \int_x (x - \mu)f(x)\,dx = \mu - \mu = 0.$$

The second moment about the mean,

$$\begin{aligned} E[(X - \mu)^2] &= \int_x (x - \mu)^2 f(x)\,dx \qquad &\text{(for continuous } X) \\ &= \sum_x (x - \mu)^2 f(x) \qquad &\text{(for discrete } X), \end{aligned} \tag{6.11}$$

defines the *variance* of the distribution. The positive square root of the variance defines, in turn, the *standard deviation* of the distribution. Frequently, the variance is denoted by σ^2 and the standard deviation then by σ. Both the variance and standard deviation are measures of the dispersion of a distribution. However, because of its greater ease of interpretation,[1] the standard deviation is usually preferred in every-day discourse.

Expanding $(X - \mu)^2$, we obtain the relationship between the central and raw second moments:

$$\begin{aligned} E[(X - \mu)^2] &= E(X^2 - 2\mu X + \mu^2) \\ &= E(X^2) - 2\mu E(X) + \mu^2 \\ &= E(X^2) - 2\mu^2 + \mu^2 \\ &= E(X^2) - \mu^2. \end{aligned} \tag{6.12}$$

We thus find the centered second moment to be equal to the raw second moment minus the square of the (raw) first moment.

Generalizing expression (6.12), it is evident from the binomial formula[2] that

$$\begin{aligned} \mu_r \equiv E[(X - \mu)^r] &= E\left[\sum_i \binom{r}{i} X^i \mu^{r-i}\right] \\ &= \sum_i \binom{r}{i} E(X^i)\mu^{r-i} \\ &= \sum_i \binom{r}{i} \mu_i' \mu^{r-i}. \end{aligned} \tag{6.13}$$

Formulas (6.12) and (6.13) hold whether X is continuous or discrete.

It is assumed that when reference is made to the rth moment of a distribution the integral or sum defining it exists for the distribution in question.

[1] The standard deviation, since it is the square root of a weighted sum of squared deviations about the mean, can be interpreted as the distance that the random variable X is "on the average" from its mean.

[2] See Theorem A.2.1 of Appendix 2.

However, there are several distributions in statistics that possess only moments of low order, and a few that do not have them beyond the 0th (which, of course, always exists)—that is, they do not even possess a mean. From (6.13) we see, though, that if the rth moment exists, then all moments of order lower than r also exist.

The final type of moment that we shall discuss is the *absolute moments.* Defined about the point a, these are given by

$$E[|x-a|^r] = \int_x |x-a|^r f(x)\,dx, \qquad r = 0, 1, 2, \ldots. \tag{6.14}$$

A parallel formula also obtains for discrete X. Two facts should be immediately evident about absolute moments. First, since $|x-a|^m \leq |x-a|^r$ for $m < r$, it follows the existence of the rth absolute moment implies the existence of all lower-order absolute moments. Secondly, because $(x-a)^r \leq |x-a|^r$, it follows that the existence of the rth absolute moment implies the existence of the rth ordinary moment.

Moments are of particular interest since the distributions that commonly arise in statistics can be characterized entirely by their moments (assuming that they exist)—that is, complete knowledge of the moments is equivalent to complete knowledge of the distribution. The conditions under which this equivalence holds are discussed in Appendix 3. Moreover, we should expect that if two distributions have a certain number of moments in common, they will bear some resemblance to one another.

For most purposes, however, we need not go beyond the first few in order to obtain a good "feel" for the shape of a distribution. Indeed, in many instances, the mean and the variance suffice. The mean effectively tells where the distribution is located, while the variance is a measure of its dispersion. If, for example, two distributions have the same mean but different variances, then the distribution with the larger variance, generally speaking, has more probability in the "tails" than the distribution with the smaller variance. Figure 6.1 illustrates this fact for two symmetrical

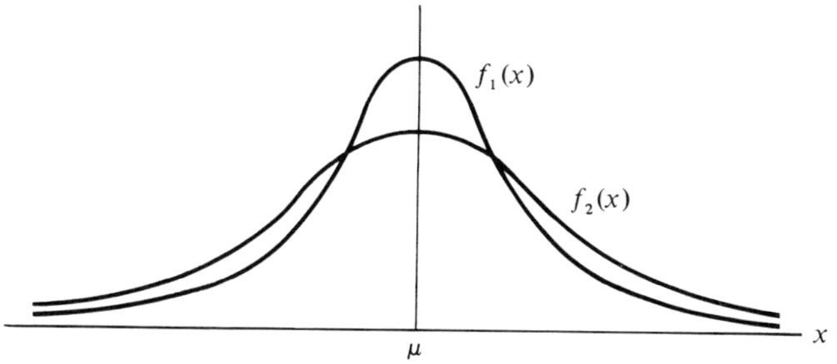

FIGURE 6.1 Two densities with same mean and different variances.

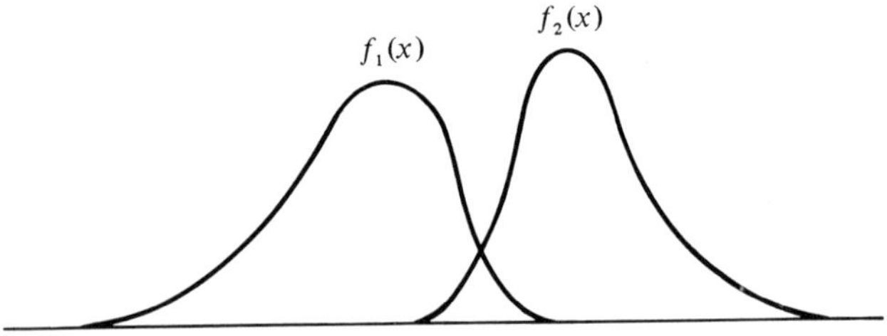

FIGURE 6.2 Skewed distributions.

distributions. [A distribution is *symmetrical* if $f(\mu + x) = f(\mu - x)$ for all x.] $f_2(x)$ obviously has the larger variance.

The third moment about the mean is often used to measure the *skewness* of a distribution, that is, where the bulk of the probability lies in relation to the mean. If the distribution is symmetrical, then μ_3 is zero and there is no skewness. Two skewed distributions are illustrated in Figure 6.2, $f_1(x)$ is skewed to the left, while $f_2(x)$ is skewed to the right. Finally, the fourth moment about the mean has been used historically as a measure of the "peakedness" of a distribution.

6.4 CALCULATION OF MOMENTS: MOMENT-GENERATING FUNCTIONS

Although expression (6.10) gives the rth moment by definition, evaluation of the necessary integral or sum may be difficult, if not impossible. In this section we outline another method which provides an easy alternative for deriving the moments for many distributions. This is through the use of a *moment-generating function.*

Let us return to the expectation in expression (6.3), and in particular consider the expected value of e^{tx}, where t is a real-valued parameter. Then

$$E(e^{tX}) = \int_x e^{tx} f(x)\, dx. \tag{6.15}$$

However, since the Taylor expansion of e^{tx} is

$$e^{tx} = 1 + tx + \frac{t^2x^2}{2!} + \frac{t^3x^3}{3!} + \cdots, \tag{6.16}$$

$$\begin{aligned} E(e^{tX}) &= \int_x \left(1 + tx + \frac{t^2x^2}{2!} + \frac{t^3x^3}{3!} + \cdots\right) f(x)\, dx \\ &= \mu_0' + t\mu_1' + \frac{t^2}{2!}\mu_2' + \frac{t^3}{3!}\mu_3' + \cdots. \end{aligned} \tag{6.17}$$

Thus, we see that the coefficients (apart from the factorials) of the terms in t are simply the raw moments. In view of this, we say that the expression $E(e^{tx})$, assuming it to exist, is called the *moment-generating function* for the random variable X, and will be denoted by $M_X(t)$.

Next, if we differentiate (6.17) with respect to t and then evaluate the resulting expression at $t = 0$,

$$\left.\frac{dM_X(t)}{dt}\right|_{t=0} = \mu_1', \tag{6.18}$$

it is seen that we have obtained the first moment about the origin. Similarly, if we differentiate (6.17) twice with respect to t and then evaluate again at $t = 0$, we have the second moment about the origin:

$$\left.\frac{d^2M_X(t)}{dt^2}\right|_{t=0} = \mu_2'. \tag{6.19}$$

Continuing, it is evident that since all higher-order terms vanish at $t = 0$,

$$\left.\frac{d^rM_X(t)}{dt^r}\right|_{t=0} = \mu_r'\,. \tag{6.20}$$

Consequently, we find that the rth moment about the origin to be obtained simply by differentiating the moment-generating function r times and then evaluating the resulting expression at $t = 0$. Since differentiation is generally an easier operation than integration or summation, formula (6.20), if the moment-generating function exists in a neighborhood of $t = 0$ and can be evaluated, provides a convenient way of deriving the moments. Unfortunately, however, $E(e^{tX})$ does not always exist.

For the moments about an arbitrary point a, we have for the moment-generating function (again, assuming it to exist)

$$\begin{aligned} M_{(X-a)}(t) &= \int_x e^{t(x-a)} f(x)\,dx \\ &= \int_x \left[1 + t(x-a) + \frac{t^2(x-a)^2}{2!} + \frac{t^3(x-a)^3}{3!} + \cdots\right] f(x)\,dx, \end{aligned} \tag{6.21}$$

which for $a = \mu$ reduces to

$$M_{(X-\mu)}(t) = \mu_0 + t\mu_1 + \frac{t^2}{2!}\mu_2 + \frac{t^3}{3!}\mu_3 + \cdots. \tag{6.22}$$

Consequently, the rth moment about the mean is given by

$$\left.\frac{d^rM_{(X-\mu)}(t)}{dt^r}\right|_{t=0} = \mu_r. \tag{6.23}$$

If X is a discrete random variable, then formulas (6.15) and (6.21) are, respectively,

$$M_X(t) = \sum_x e^{tx} f(x), \tag{6.24}$$

$$M_{(X-a)}(t) = \sum_x e^{t(x-a)} f(x). \tag{6.25}$$

It is evident from these two formulas that

$$M_{(X-a)}(t) = e^{-at} M_X(t) \tag{6.26}$$

and in particular for $a = \mu$ that

$$M_{(X-\mu)}(t) = e^{-\mu t} M_X(t). \tag{6.27}$$

Formulas (6.26) and (6.27) hold for continuous as well as discrete X.

As already indicated $f(x)$ must be such that the integrals or sums in formulas (6.15), (6.21), (6.24), and (6.25) converge in a neighborhood of $t = 0$, but sometimes this is not the case. However, if in place of e^{tx} we use e^{itx}, where i denotes $\sqrt{-1}$, then the absolute convergence of $E(e^{itX})$ is assured; the corresponding integral (or sum) is called the *characteristic function* of X, and its use opens up a much broader and deeper methodology for studying distributions and their moments. This technique, though, is beyond our present aspirations.[3]

6.5 OTHER CHARACTERISTICS OF PROBABILITY DISTRIBUTIONS

Moments are not the only useful characteristics of probability distributions and in this section we will discuss two others that are of frequent use, namely, the *median* and the *mode*.

Median

As its name implies, the median is the midpoint of the distribution and is defined as the point at which half of the probability lies to the left and half to the right. Letting this point be denoted by μ_d, we thus have

$$P(X < \mu_d) = 0.5. \tag{6.28}$$

In contrast to the mean, it is evident that the median always exists. For a discrete random variable that can take on n possible outcomes, the median

[3] But see Appendix 3.

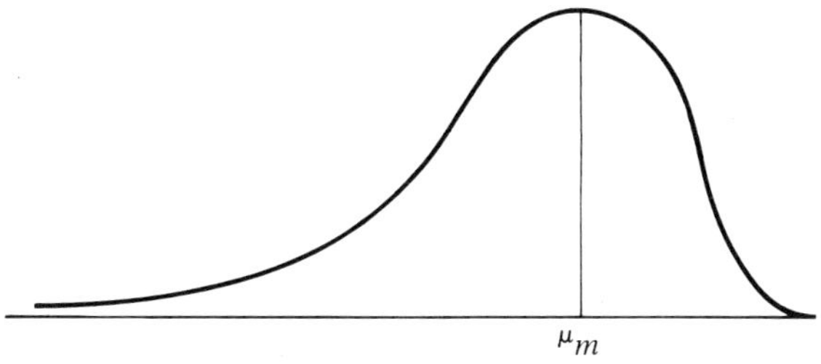

FIGURE 6.3 Mode.

is equal to the $(n + 1)/2$ value (assuming that the values are ordered from smallest to largest) if n is odd, and the midpoint between the $n/2$ and $(n + 2)/2$ values if n is even.

Mode

The mode of a distribution is the value of the random variable which has the highest probability—that is, it is the value of x at the "peak" of the distribution—and we shall denote it by μ_m. The mode is often referred to as the "most likely" or the "most probable" value, and is illustrated for an arbitrary single peaked density function for continuous X in Figure 6.3.

If a distribution has a single peak, then the distribution is said to be *unimodal*; if it has two peaks, it is *bimodal* (even if one peak is higher than the other); and so on. A distribution that is U-shaped, such as in Figure 6.4, is bimodal, with the two modes at the extreme values of the random variable.

Since the maximum value of the density function defines the mode, it can be found by conventional calculus methods in the cases where the density function is continuous and differentiable and the mode is "interior"

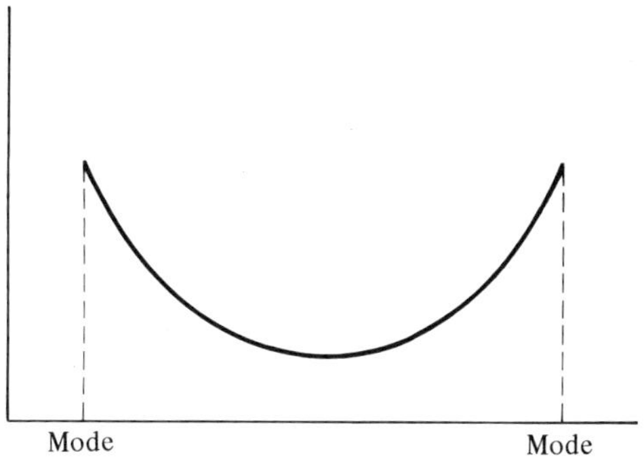

FIGURE 6.4 Bimodal distribution.

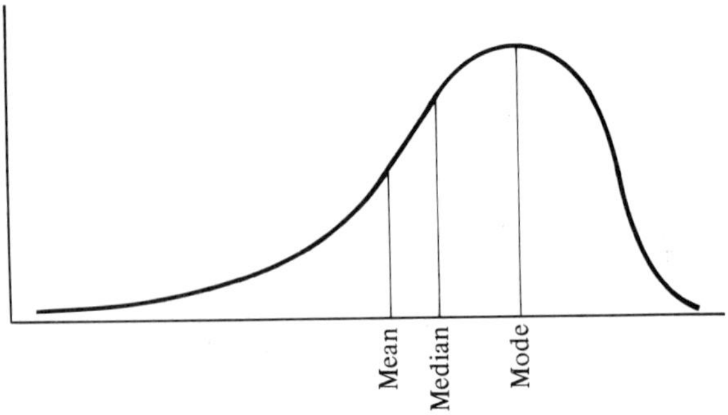

FIGURE 6.5 Mean, median, and mode for left skewed distributions.

to the tails of the distribution. In other words, if $f(x)$ is continuous, then the mode [or the modes if $f(x)$ is multimodal] is given by solving the equation

$$f'(x) = 0. \tag{6.29}$$

The mean, median, and mode are often referred to as *measures of central tendency*, which is to say that, loosely speaking, they give the values about which the random variable fluctuates. For a unimodal distribution (whether continuous or discrete), they bear a definite positional relationship to one another. If the distribution is skewed to the left, that is, where the long tail lies to the left, then

$$\mu < \mu_d < \mu_m, \tag{6.30}$$

while the inequalities are reversed for a right-skewed distribution. (Figure 6.5 illustrates this for a left-skewed distribution.) On the other hand, the mean, median, and mode coincide if the distribution is symmetrical.

An example. In order to illustrate the foregoing, we shall calculate the mean, median, mode, and variance for the following empirical distribution for a random variable X that can take on the following integers between 1 and 10:

$$X: \quad 2, 7, 5, 8, 3, 9, 8.$$

To perform these calculations, it will be useful to construct the following table:

X	Frequency	Relative Frequency
2	1	$\frac{1}{7}$
3	1	$\frac{1}{7}$
5	1	$\frac{1}{7}$
7	1	$\frac{1}{7}$
8	2	$\frac{2}{7}$
9	1	$\frac{1}{7}$

If we take the relative frequencies as estimates of the true probabilities, then for the mean we have

$$\mu = 2 \cdot \frac{1}{7} + 3 \cdot \frac{1}{7} + 5 \cdot \frac{1}{7} + 7 \cdot \frac{1}{7} + 8 \cdot \frac{2}{7} + 9 \cdot \frac{1}{7} = \frac{42}{7} = 6,$$

while the variance σ^2 is

$$\sigma^2 = (2-6)^2 \cdot \frac{1}{7} + (3-6)^2 \cdot \frac{1}{7} + (5-6)^2 \cdot \frac{1}{7} + (7-6)^2 \cdot \frac{1}{7} + (8-6)^2 \cdot \frac{2}{7}$$

$$+ (9-6)^2 \cdot \frac{1}{7} = \frac{(16+9+1+1+8+9)}{7} = \frac{44}{7}.$$

By going down the list until we reach the fourth value, we find the median to be 7, and, with equal ease, the mode is found to be 8.

For many problems a measure of central tendency is all of the information about a distribution that is needed, and the mean is the measure that is most frequently used. If a distribution is very highly skewed, however, the mean can be quite misleading, for it is sensitive to the extreme values in the long tail of the distribution. For example, in summarizing income levels in a country such as Colombia, because of a few individuals with extremely large incomes, the mean gives a false impression of what a typical income level is. In this situation, the median would be more meaningful since it is influenced by only the number of extremely high levels of income not by their values. Therefore, when a distribution is known to be highly skewed the median is a better measure of the "scale" or "location" of the distribution than the mean. And, if the distribution is extremely highly skewed, the mode may be the best measure of all.

6.6 THEOREMS OF THE CHEBYSHEV TYPE

In this section, we shall prove two theorems that relate to the probability bounded by certain specified regions of a probability distribution. The first of these is due to Chebyshev, a famous Russian mathematician of the late nineteenth century, and is one of the most remarkable in all of probability. In particular, it enables questions such as the following to be answered:

1. What proportion of the total probability of a random variable lies in a given interval centered at the mean?
2. How wide an interval about the mean is needed in order to guarantee that 80 percent of the total probability lies within the interval?

THEOREM 6.1 (Chebyshev)
Let X be a random variable with density function $f(x)$ and with mean μ and variance $\sigma^2 < \infty$. Then at least the fraction $1 - 1/h^2$ of the total probability of X lies within h standard deviations of its mean, that is,

$$P(|X - \mu| \leq h\sigma) \geq 1 - \frac{1}{h^2}. \tag{6.31}$$

PROOF. We shall prove this theorem for X continuous; proof for the discrete case is parallel.

By definition,

$$\begin{aligned} \sigma^2 &= \int_{-\infty}^{\infty} (x - \mu)^2 f(x)\, dx \\ &= \int_{-\infty}^{\mu - h\sigma} (x - \mu)^2 f(x)\, dx + \int_{\mu - h\sigma}^{\mu + h\sigma} (x - \mu)^2 f(x)\, dx \qquad (6.32) \\ &\quad + \int_{\mu + h\sigma}^{\infty} (x - \mu)^2 f(x)\, dx, \end{aligned}$$

where h is a positive constant and σ is the standard deviation of X. What we have done in the second part of (6.32) is to define an interval of length $2h\sigma$ centered at μ on the real axis (Figure 6.6). Now, for all values of x to the left of $\mu - h\sigma$ and to the right of $\mu + h\sigma$, the squared term beneath the integral sign will be greater than $h^2\sigma^2$. Therefore, if we replace $(x - \mu)^2$ in the first and third integrals by $h^2\sigma^2$, we will have

$$\begin{aligned} \sigma^2 \geq h^2\sigma^2 &\int_{-\infty}^{\mu - h\sigma} f(x)\, dx \\ &+ \int_{\mu - h\sigma}^{\mu + h\sigma} (x - \mu)^2 f(x)\, dx + h^2\sigma^2 \int_{\mu + h\sigma}^{\infty} f(x)\, dx. \qquad (6.33) \end{aligned}$$

Similarly, 0 can never be larger than $(x - \mu)^2$, so that we can drop out the middle term without upsetting the inequality:

$$\sigma^2 \geq h^2\sigma^2 \int_{-\infty}^{\mu - h\sigma} f(x)\, dx + h^2\sigma^2 \int_{\mu + h\sigma}^{\infty} f(x)\, dx. \tag{6.34}$$

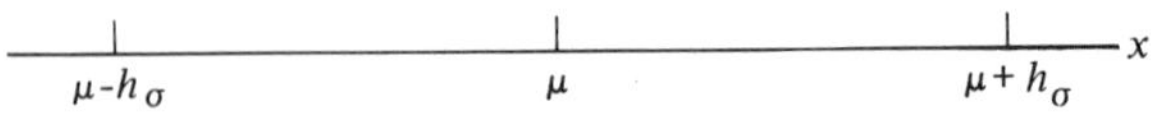

FIGURE 6.6

Next, note that the first integral is simply $P(X \leq \mu - h\sigma)$, and similarly that the second integral is $P(X \geq \mu + h\sigma)$. Consequently, cancelling σ^2, we have

$$1 \leq h^2[P(X \leq \mu - h\sigma) + P(X \geq \mu + h\sigma)], \tag{6.35}$$

or

$$1 \leq h^2 P(|X - \mu| \geq h\sigma). \tag{6.36}$$

Hence,

$$P(|X - \mu| \leq h\sigma) \geq 1 - \frac{1}{h^2}. \tag{6.37}$$

as was to be proved.

To illustrate Chebyshev's theorem with a particular distribution, let X represent the sum of the faces from the roll of two true dice and ask how much of the probability lies within h standard deviations of the mean. The mean in the case is equal to 7, while the variance is equal to 35/6. The standard deviation, therefore, is approximately 2.4. Now let $h = 1.25$, so that $h\sigma = 3$. Then by Chebyshev's theorem,

$$P(|X - 7| \leq 3) \geq 1 - \frac{1}{(1 \cdot 25)^2} = 1 - 0.64 = 0.36.$$

Since we know the distribution, $P(|X - 7| \leq 3)$ can be calculated exactly and this is found to be 30/36, which is well above the Chebyshev limit.

The remarkable feature of the Chebyshev inequality is that it holds for *any* distribution with a finite variance. However, it can be shown, and this we shall now do, that the Chebyshev inequality is simply a special case of a more general inequality which holds for any distribution which possesses an absolute moment of order r.

THEOREM 6.2

Let X be a random variable with density function $f(x)$ and for which $E[|X - a|^r]$ is finite. Then, for any $\lambda > 0$,

$$P(|X - a| < \lambda) \geq 1 - \frac{E[|X - a|^r]}{\lambda^r}. \tag{6.38}$$

PROOF. Again, we shall restrict proof to the case of continuous X. By definition,

$$\begin{aligned} E[|X - a|^r] &= \int_{-\infty}^{\infty} |x - a|^r f(x)\, dx \\ &= \int_{-\infty}^{a-\lambda} |x - a|^r f(x)\, dx + \int_{a-\lambda}^{a+\lambda} |x - a|^r f(x)\, dx \qquad (6.39) \\ &\quad + \int_{a+\lambda}^{\infty} |x - a|^r f(x)\, dx, \end{aligned}$$

so that

$$E[|X-a|^r] \geq \lambda^r \int_{-\infty}^{a-\lambda} f(x)\,dx + \int_{a-\lambda}^{a+\lambda} |x-a|^r f(x)\,dx$$
$$+ \lambda^r \int_{a+\lambda}^{\infty} f(x)\,dx \tag{6.40}$$
$$\geq \lambda^r \int_{-\infty}^{a-\lambda} f(x)\,dx + \lambda^r \int_{a+\lambda}^{\infty} f(x)\,dx$$
$$= \lambda^r P(X \leq a-\lambda) + \lambda^r P(X \geq a+\lambda).$$

Hence,

$$E[|X-a|^r] \geq \lambda^r P(|X-a| \geq \lambda) \tag{6.41}$$

or

$$P(|X-a| \geq \lambda) \leq \frac{E[|X-a|^r]}{\lambda^r}. \tag{6.42}$$

Since $P(|X-a| < \lambda) = 1 - P(|X-a| \geq \lambda)$, we thus have

$$P(|X-a| < \lambda) \geq 1 - \frac{E[|X-a|^r]}{\lambda^r}, \tag{6.43}$$

as was to be shown.

For $r = 2$, $a = \mu$, and $\lambda = h\sigma$, it is seen that the inequality in (6.38) yields Chebyshev's inequality as a special case.

REFERENCES

Freeman, H., *Introduction to Statistical Inference*, Addison-Wesley, 1963, chap. 3.

Kendall, M. G. and Stuart, A., *The Advanced Theory of Statistics*, vol. I, Charles Griffin & Co., 1958, chaps. 2 and 3.

Mood, A. M. and Graybill, F. A., *Introduction to the Theory of Statistics*, 2nd ed., McGraw-Hill, 1963, chap. 5.

CHAPTER 7

DISTRIBUTIONS INVOLVING A DISCRETE RANDOM VARIABLE

In this chapter and the next, we shall turn our attention to the study of particular distributions. Until now, most of our discussion has been with reference to distributions of unspecified functional form. In this chapter, we shall introduce the most important of the discrete distributions—in particular, the binomial, negative binomial, hypergeometric, and Poisson—while in Chapter 8, we shall concentrate on distributions involving a continuous random variable.

7.1 THE BINOMIAL DISTRIBUTION

The binomial distribution is a venerable grandfather of distributions since it arose very early in the study of games of chance. To motivate it, let us consider an experiment in which there are two possible outcomes, such as flipping of a coin. Let us call each realization of the experiment a *trial*, and term the two outcomes *success* and *failure*, respectively. Furthermore, we shall suppose that trials are independent in the sense that the outcome of any trial does not affect the outcome of any other trial. Such trials are called *binomial* (or *Bernoulli*) *trials*.

Suppose now that we have n binomial trials, where the probabilities of success and failure are p and q, respectively ($p + q = 1$), for each trial and we wish to know what the probability is that these n trials give $r (r \leq n)$ successes. For example, suppose that we have a coin for which the probability of heads is p while the probability of tails is q. Then in n flips of the coin, what is the probability of obtaining r heads?

Since the trials are independent, the probability of the event r heads in n flips is equal to $p^r q^{n-r}$, no matter what the order of the r heads and $n - r$ tails. However, r heads and $n - r$ tails can be obtained in $\binom{n}{r}$ ways; hence, the probability of r heads in n flips is equal to

$$P(r \text{ successes}) = \binom{n}{r} p^r q^{n-r}. \tag{7.1}$$

Equation (7.1) defines the density function for the *binomial distribution.* The random variable is the number of successes, which can range from 0 to n, and the distribution is a function of two parameters, p and n. Letting X denote the number of successes, we shall henceforth write the density function for the binomial distribution as

$$P(X = x) = \binom{n}{x} p^x q^{n-x}, \qquad x = 0, 1, 2, \ldots, n. \tag{7.2}$$

The distribution function is given by

$$P(X \leq x) = \sum_{r=0}^{x} \binom{n}{r} p^r q^{n-r}. \tag{7.3}$$

It is easy to check that the probabilities in (7.2) sum to 1, for, by the binomial theorem and since $p + q = 1$,

$$\sum_{x=0}^{n} \binom{n}{x} p^x q^{n-x} = (p + q)^n = 1. \tag{7.4}$$

Consequently, it is clear where the binomial distribution gets its name.

It is evident from (7.2) that for $p = q = 0.5$ the binomial density is symmetrical about the point $n/2$ for n even and about the point $(n + 1)/2$ for n odd. For $p < 0.5$, the distribution is skewed to the right, while for $p > 0.5$, it is skewed to the left. Figure 7.1 illustrates the binomial density for $n = 10$ and $p = 0.4$.

7.2 MEAN AND VARIANCE OF THE BINOMIAL DISTRIBUTION

There are a number of ways to derive the mean and variance of the binomial distribution, and in order to illustrate the different methods we shall derive these two parameters in two alternative ways.

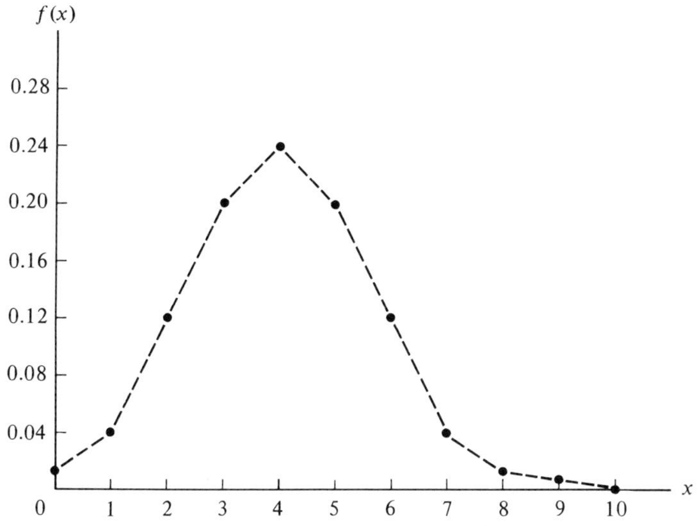

FIGURE 7.1 Binomial distribution for $n = 10$ and $p = 0.4$.

Method 1: From the Definitions

Mean. By definition, we have for the mean number of successes,

$$E(X) = \sum_{x=0}^{n} x \binom{n}{x} p^x q^{n-x}. \tag{7.5}$$

Rewriting the right-hand side and noting that the term for $x = 0$ is also 0, we can write

$$\sum_{x=0}^{n} \frac{xn!}{x!(n-x)!} p^x q^{n-x} = \sum_{x=1}^{n} \frac{n!}{(x-1)!(n-x)!} p^x q^{n-x}.$$

Next, since p and n do not depend on x, we can factor out np:

$$\sum_{x=1}^{n} \frac{n!}{(x-1)!(n-x)!} p^x q^{n-x} = np \sum_{x=1}^{n} \frac{(n-1)!}{(x-1)!(n-x)!} p^{x-1} q^{n-x}.$$

However, if the sum on the right is examined, it is seen to be $(p+q)^{n-1}$, which is equal to 1 because $p + q = 1$. Therefore,

$$E(X) = np. \tag{7.6}$$

In other words, the expected number of successes is equal to the probability of success on any one trial times the number of trials.

Variance. To derive the variance, we shall first find $E(X^2)$ and then use the relationship $E[(X - \mu)^2] = E(X^2) - \mu^2$ derived in Section 6.3.

By definition,

$$E(X^2) = \sum_{x=0}^{n} x^2 \binom{n}{r} p^x q^{n-x}. \tag{7.7}$$

Proceeding as with the mean and noting that the term at $x = 0$ is also 0, we have

$$\sum_{x=0}^{n} \frac{x^2 n!}{x!(n-x)!} p^x q^{n-x} = \sum_{x=1}^{n} \frac{xn!}{(x-1)!(n-x)!} p^x q^{n-x}.$$

Again, factoring out an np:

$$\sum_{x=1}^{n} \frac{xn!}{(x-1)!(n-x)!} p^x q^{n-x} = np \sum_{x=1}^{n} \frac{x(n-1)!}{(x-1)!(n-x)!} p^{x-1} q^{n-x}.$$

Next, subtracting and adding

$$np \sum_{n=1}^{n} \frac{(n-1)!}{(x-1)!\,(n-x)!} p^{x-1} q^{n-x},$$

we obtain

$$np \sum_{x=1}^{n} \frac{x(n-1)!}{(x-1)!(n-x)!} p^{x-1} q^{n-x} = np \sum_{x=1}^{n} \frac{(x-1)(n-1)!}{(x-1)!(n-x)!} p^{x-1} q^{n-x} + np \sum_{x=1}^{n} \frac{(n-1)!}{(x-1)!(n-x)!} p^{x-1} q^{n-x}.$$

But the first sum on the right-hand side is simply the expected number of successes in $n - 1$ trials, which is $(n - 1)p$. [The easiest way to see this is to replace $x - 1$ by y; y is then summed from 0 to $n - 1$ and $n - x = n - 1 - (x - 1) = n - 1 - y$.] Moreover, since the nth term in the sum is zero, the second sum on the right-hand side is $(p + q)^{n-1} = 1$; hence,

$$E(X^2) = np(n-1)p + np = n^2p^2 - np^2 + np. \tag{7.8}$$

Therefore, the variance is equal to

$$\begin{aligned} E[X - \mu)^2] &= n^2p^2 - np^2 + np - n^2p^2 \\ &= np - np^2 \\ &= np(1-p) \\ &= npq. \end{aligned} \tag{7.9}$$

Method 2: Using the Moment-Generating Function

By definition, the moment-generating function for the binomial distribution is equal to

$$E(e^{tX}) = \sum_{x=0}^{n} e^{tx} \binom{n}{x} p^x q^{n-x}. \tag{7.10}$$

But

$$\sum_{x=0}^{n} e^{tx} \binom{n}{x} p^x q^{n-x} = \sum_{x=0}^{n} \binom{n}{x} (pe^t)^x q^{n-x}.$$

Therefore, the moment-generating function is equal to

$$M_X(t) = (pe^t + q)^n. \tag{7.11}$$

Differentiating (7.11) with respect to t,

$$\frac{dM_X(t)}{dt} = n(pe^t + q)^{n-1} \cdot pe^t,$$

and then evaluating the resulting expression at $t = 0$,

$$\left.\frac{dM_X(t)}{dt}\right|_{t=0} = np, \tag{7.12}$$

we see that the mean is np, in agreement with (7.6).

Next, differentiating the moment-generating function twice with respect to t,

$$\frac{d^2M_X(t)}{dt^2} = n(n-1)(pe^t + q)^{n-2} \cdot p^2e^{2t} + n(pe^t + q)^{n-1}\, pe^t,$$

and then evaluating at $t = 0$, we obtain

$$\left.\frac{d^2M_X(t)}{dt^2}\right|_{t=0} = n(n-1)p^2 + np, \tag{7.13}$$

which is $E(X^2)$. To obtain $E[(X - \mu)^2]$, we must subtract n^2p^2 from (7.13), and this yields

$$E[(X - \mu)^2] = -np^2 + np = np(1 - p) = npq \tag{7.14}$$

as before.

7.3 USES OF THE BINOMIAL DISTRIBUTION

As was mentioned earlier, the binomial distribution is one of the oldest in existence and has also been one of the most fundamental in the development of statistics. Since it is applicable to nearly any probabilistic situation in which there are two possible outcomes, its empirical applications have been many. Included among its more unusual applications are those to electrical power, vaccines, the Mendelian theory of heredity, and random walk processes. The binomial distribution has been widely tabulated, and nearly every statistics textbook includes a table for a number of values of p and moderate values of n.

To illustrate the application of the binomial distribution to a practical problem, let us consider an example from Freeman[1]:

Suppose that there is an industrial process which produces a gadget which we shall leave unnamed, and that during a normal shift it will produce upward of 10,000 of these gadgets. Since the production rate is so high, it would be prohibitively expensive for the company involved to inspect each and every gadget produced as to quality, so the company periodically selects a sample and decides on the basis of this sample whether the machine is producing up to snuff. Assume that the process is regarded as satisfactory if the proportion of defective gadgets in the sample is less than or equal to p_1, and unsatisfactory if the proportion of defective output is equal to or greater than p_2. Proportions between p_1 and p_2 are ignored. Let the sampling plan by which decisions are to be made to accept or reject (to be satisfied with or to improve) the process be defined by two quantities: n, the number of gadgets in the sample taken from the process and inspected, and a, the maximum allowable number of defective gadgets in the sample for which the process can still be called satisfactory. Let the probability that the sampling plan and decision rule (n, a) will call unsatisfactory (i.e., reject) the worst of the satisfactory processes (the one with proportion defective p_1) be a small number α; let the probability that the same sampling plan and decision rule will call satisfactory (accept) the best of the unsatisfactory processes (one with proportion defective p_2) be a small number β. It is evident that we should also want the sampling plan and decision rule to be such that if even worse processes were operating (i.e., with proportion of output defective greater than p_2), the probability of calling them *satisfactory* would be even less than β. For the class of binomial sampling plans considered here, these requirements turn out to be satisfied automatically. The problem, then, is as follows: For given p_1, p_2, α, and β, determine the sampling plan and decision rule (n, a) to be employed.

While the statement of the problem is somewhat longwinded, the solution is brief. Assuming that the sample of size n drawn from the process constitutes n Bernoulli trials, we have

$$\alpha = \sum_{x=a+1}^{n} \binom{n}{x} p_1^x(1-p_1)^{n-x}, \qquad \beta = \sum_{x=0}^{a} \binom{n}{x} p_2^x(1-p_2)^{n-x},$$

which, for given p_1, p_2, α, and β, can be approximately (the solution is

[1] From H. Freeman, *Introduction to Statistical Inference*, Addison-Wesley, 1963, p. 99.

exact for only certain α, β combinations) solved for n and a by maneuvering in the binomial tables. For example, for

$$p_1 = 0.02, \qquad p_2 = 0.07, \qquad \alpha = 0.05, \qquad \beta = 0.10,$$

we find $n = 130$, $a = 5$.

The sampling plan and decision rule in this case would thus proceed as follows: A sample of 130 gadgets is drawn from the output; the process is judged satisfactory if and only if 5 or fewer defective gadgets are found. With this sampling plan and decision rule, the risk of rejecting a satisfactory process is 0.05 (the risk of rejecting an even better one is *less* than 0.05), and the risk of approving an unsatisfactory process is 0.10 (the risk of approving an even poorer process is *less* than 0.10).

The logic of this example will play a central role in Chapter 18 when we take up the testing of hypotheses. The probabilities p_1 and p_2 will be described as hypotheses on the parameter p, the true proportion of defective output produced by the process, α will be interpreted as the probability of rejecting the hypothesis $p = p_1$ when it is true and β as the probability of accepting the hypotheses $p = p_1$ when the hypothesis $p = p_2$ is true.

7.4 THE HYPERGEOMETRIC DISTRIBUTION

The second discrete distribution that we shall study is the *hypergeometric*. Suppose that there is a pond known to contain 1000 fish. We remove 400 of these fish, put red tags on a fin, and put them back in the pond. Then, if 100 are removed again, what is the probability that 25 of these 100 will have red tags (assume that no fish have died or have lost their red tags)?

If we assume that each fish stands an equal chance of being selected in the sample of 100, then the classical model of probability is applicable and we can solve the problem as follows. To begin with, the total number of ways that the sample of 100 fish can be selected is $\binom{1000}{100}$. The favorable events are those with 25 fish with red tags and 75 without. The 25 fish with red tags can be selected in $\binom{400}{25}$, while the 75 nontagged fish can be drawn in $\binom{600}{75}$ ways. Therefore, the number of favorable events is $\binom{400}{25} \cdot \binom{600}{75}$ and the required probability is

$$P(25 \text{ fish with red tags}) = \binom{400}{25}\binom{600}{75} \bigg/ \binom{1000}{100}. \tag{7.15}$$

We now generalize this example and let N equal the total number of fish in the lake, R the number with red tags, n the size of the sample, and X the number of fish with red tags in the sample. Then

$$P(X = x) = \binom{R}{x}\binom{N-R}{n-x} \Big/ \binom{N}{n}, \qquad x \leq R, x \leq n, n \leq N, x = 0, 1, \ldots, R. \tag{7.16}$$

Expression (7.16) is the density function for the *hypergeometric distribution*. Its distribution function is given by

$$P(X \leq x) = \sum_{r=0}^{x} \binom{R}{r}\binom{N-R}{n-r} \Big/ \binom{N}{n}. \tag{7.17}$$

The above derivation of the hypergeometric is usually referred to as the "capture-recapture" approach. We shall now give an alternative derivation which shows the relationship between the hypergeometric and binomial distributions. To begin with, we need to change the physical interpretation of the experiment for the binomial trial. Instead of the experiment being the flip of a coin with probability of heads equal to p, let it be the drawing of a ball from an urn with N balls, R of which are white and $N - R$ red. Let the drawing of a white ball be a "success" and the drawing of a red ball a "failure." The probabilities of success and failure are, therefore, $p = R/N$ and $q = (N - R)/N$, respectively. Suppose that we draw n balls from the urn *with replacement* (i.e., the ball is returned to the urn after each drawing). These n drawings obviously constitute n binomial trials since they are independent and the probability of success remains constant at $p = R/N$ throughout each trial.

Suppose, however, that we draw the n balls *without replacement*. The trials in this case are no longer binomial because the probability of success changes from one drawing to the next. In fact, the probability that the first x $(x < R)$ drawings yield successes is

$$\frac{R}{N} \cdot \frac{R-1}{N-1} \cdots \frac{R-(x-1)}{N-(x-1)} = \frac{R!}{(R-x)!} \Big/ \frac{N!}{(N-x)!}$$

and the probability that the last $n - x$ yield failures is

$$\frac{N-R}{N-x} \cdot \frac{N-R-1}{N-x-1} \cdots \frac{N-R-(n-x-1)}{N-(n-1)}$$

$$= \frac{(N-R)!}{(N-R-n+x)!} \Big/ \frac{(N-x)!}{(N-n)!}.$$

But x successes and $n - x$ failures can be obtained in $\binom{n}{x}$ ways, so that the probability of x successes in n trials *without replacement* is

$$
\begin{aligned}
P(X = x) &= \binom{n}{x}\left[\frac{R!}{(R-x)!}\bigg/\frac{N!}{(N-x)!}\right]\cdot\left[\frac{(N-R)!}{(N-R-n+x)!}\bigg/\frac{(N-x)!}{(N-n)!}\right] \\
&= \frac{n!}{x!\,(n-x)!}\cdot\frac{R!(N-x)!}{(R-x)!\,N!}\cdot\frac{(N-R)!\,(N-n)!}{(N-R-n+x)!(N-x)!} \\
&= \left[\frac{R!}{x!\,(R-x)!}\cdot\frac{(N-R)!}{(n-x)!\,(N-R-n+x)!}\right]\bigg/\frac{N!}{n!\,(N-n)!} \\
&= \binom{R}{x}\binom{N-R}{n-x}\bigg/\binom{N}{n}. \qquad (7.18)
\end{aligned}
$$

Thus, once again we have arrived at the density function for the hypergeometric distribution.

Consequently, we see that if we have a finite dichotomous population and sample *without replacement*, the distribution of successes is given by the hypergeometric distribution, while if we sample *with replacement* the distribution of successes is given by the binomial.

7.5 MOMENTS OF THE HYPERGEOMETRIC DISTRIBUTION

By using the following identity,[2] where a, b, n are positive integers,

$$\binom{a}{0}\binom{b}{n} + \binom{a}{1}\binom{b}{n-1} + \cdots + \binom{a}{n}\binom{b}{0} = \binom{a+b}{n}, \qquad (7.19)$$

it is easy to show that

$$\sum_{x=0}^{n}\binom{R}{x}\binom{N-R}{n-x}\bigg/\binom{N}{n} = \binom{N}{n}\bigg/\binom{N}{n} = 1. \qquad (7.20)$$

Mean. The easiest way to obtain the mean is through its definition. We have

$$E(X) = \sum_{x=0}^{n} x\binom{R}{x}\binom{N-R}{n-x}\bigg/\binom{N}{n}. \qquad (7.21)$$

[2] See Feller (1968, p. 64).

Again we note that the term for $x = 0$ is also 0, so that we can sum from 1 to n. Next, cancelling x from the numerator,

$$\sum_{x=1}^{n} \frac{xR!}{x!(R-x)!} \cdot \binom{N-R}{n-x} \Big/ \binom{N}{n} = \sum_{x=1}^{n} \frac{R!}{(x-1)!(R-x)!} \binom{N-R}{n-x} \Big/ \binom{N}{n},$$

and then multiplying and dividing by R/N, we obtain after some manipulation,

$$\frac{R}{N} \sum_{x=1}^{n} \frac{R!}{(x-1)!(R-x)!} \binom{N-R}{n-x} \Big/ \frac{N!}{n!(N-n)!} \frac{R}{N}$$

$$= \frac{nR}{N} \sum_{x=1}^{n} \frac{(R-1)!}{(x-1)!(R-x)!} \binom{N-R}{n-x} \Big/ \frac{(N-1)!}{(n-1)!(N-n)!}. \tag{7.22}$$

Since $R - 1 - (x - 1) = R - x$ and $N - 1 - (n - 1) = N - n$, the sum on the right-hand side is simply the sum of the densities for a population of $N - 1$ with $R - 1$ successes, and we know this to be equal to 1. Therefore,

$$E(X) = n \frac{R}{N} = np. \tag{7.23}$$

In other words, the expected number of successes with the hypergeometric distribution, as with the binomial distribution, is equal to the sample size times the proportion of successes in the initial population.

Variance. We go through a similar set of manipulations to obtain the variance. By definition,

$$E(X^2) = \sum_{x=0}^{n} x^2 \binom{R}{x} \binom{N-R}{n-x} \Big/ \binom{N}{n}$$

$$= \sum_{x=1}^{n} \frac{xR!}{(x-1)!(R-x)!} \binom{N-R}{n-x} \Big/ \binom{N}{n}$$

$$= n \frac{R}{N} \sum_{x=1}^{n} \frac{x(R-1)!}{(x-1)!(R-x)!} \binom{N-R}{n-x} \Big/ \frac{(N-1)!}{(n-1)!(N-n)!}. \tag{7.24}$$

Next, we subtract and add

$$n \frac{R}{N} \sum_{x=1}^{n} \frac{(R-1)!}{(x-1)!(R-x)!} \binom{N-R}{n-x} \Big/ \frac{(N-1)!}{(n-1)!(N-n)!}$$

so that

$$E(X^2) = n \frac{R}{N} \sum_{x=1}^{n} \frac{(x-1)(R-1)!}{(x-1)!(R-x)!} \binom{N-R}{n-x} \Big/ \binom{N-1}{n-1}$$

$$+ n \frac{R}{N} \sum_{x=1}^{n} \frac{(R-1)!}{(x-1)!(R-x)!} \binom{N-R}{n-x} \Big/ \binom{N-1}{n-1}. \tag{7.25}$$

From the argument used in deriving the mean, we know that the second sum on the right-hand side of (7.25) is 1, while the first sum on the right, after some manipulation, yields

$$(n-1)\frac{R-1}{N-1}.$$

Therefore,

$$E(X^2) = \frac{n(n-1)R(R-1)}{N(N-1)} + n\frac{R}{N}, \tag{7.26}$$

whence it follows that the variance is equal to

$$E[(X-\mu)^2] = \frac{n(n-1)R(R-1)}{N(N-1)} + n\frac{R}{N} - \left(n\frac{R}{N}\right)^2, \tag{7.27}$$

which after simplification becomes

$$E[(X-\mu)^2] = npq\left(\frac{N-n}{N-1}\right), \tag{7.28}$$

where $p = R/N$ and $q = 1 - p$.

Thus, the variance of the hypergeometric distribution is equal to the variance of the binomial distribution multiplied by $(N-n)/(N-1)$. For $n = 1$, the two distributions are identical, of course, but for $n > 1$, $(N-n)/(N-1)$ is less than 1, so that the variance of the hypergeometric is less than the variance of the binomial distribution. In other words, sampling with replacement from a finite population has a higher variance than sampling without replacement.

As N and R become large and if n is small relative to N, $(R-1)/(N-1)$ tends to R/N and the hypergeometric distribution approaches the binomial. In the limit, that is, for an infinite dichotomous population, they are, of course, identical. As for the binomial, the hypergeometric distribution has been extensively tabulated, and there is less need now than in the past to use the binomial as an approximation.

7.6 THE POISSON DISTRIBUTION

The third discrete distribution to be discussed is the *Poisson distribution.* Named after its discoverer, N. Poisson, the Poisson distribution can be viewed as an approximation to the binomial or as a distribution that arises naturally from a very simple random process. Historically, it has been referred to as the distribution of the occurrence of rare events, such as the number of deaths resulting from horsekicks in the Prussian cavalry in the nineteenth century or the number of wrong numbers dialed over a short period of time.

The Poisson Distribution as an Approximation to the Binomial

In the binomial density function

$$f(x)_0 = \binom{n}{x} p^x q^{n-x}, \tag{7.29}$$

let $n \to \infty$ and $p \to 0$ in such a way that np remains finite and approaches a positive constant λ. Under these conditions, we can write the binomial density as

$$\begin{aligned} f(x) &= \frac{n!}{x!(n-x)!} \left(\frac{\lambda}{n}\right)^x \frac{[1-\lambda/n]^n}{[1-\lambda/n]^x} \\ &= \frac{\lambda^x}{x!} \cdot \frac{n(n-1)\cdots(n-x+1)}{n^x} \cdot \frac{[1-\lambda/n]^n}{[1-\lambda/n]^x} \\ &= \frac{\lambda^x}{x!} \left(1 - \frac{\lambda}{n}\right)^n \cdot \left\{\frac{[1-1/n](1-(x-1)/n]}{[1-\lambda/n]^x}\right\}. \end{aligned} \tag{7.30}$$

As $n \to \infty$, the product within the braces tends to 1 and

$$\lim_{n \to \infty} \left(1 - \frac{\lambda}{n}\right)^n = e^{-\lambda}. \tag{7.31}$$

Therefore, under the conditions postulated, $f(x)$ reduces to

$$f(x) = \frac{e^{-\lambda}\lambda^x}{x!}, \qquad x = 0, 1, 2, \ldots, \tag{7.32}$$

which is a single-parameter density function of discrete x called the *Poisson distribution*. Its distribution function is obviously

$$F(x) = \sum_{k=0}^{x} \frac{e^{-\lambda}\lambda^k}{k!}. \tag{7.33}$$

As an approximation to the binomial, the Poisson distribution is very good even for moderate values of n, say of the order of 50, if λ is of the order of 1 or 2.

The Poisson Distribution as a Random Process

The Poisson distribution can be derived not only as an approximation to the binomial distribution, but also directly from a simple and useful model of random behavior of events distributed randomly over time or space.

Indeed, this is an area of rapidly expanding interest in probability theory, and the Poisson, therefore, has importance far beyond that of a mere approximation to the binomial.

Let us suppose that we have a process that triggers events through time and we wish to know at some fixed but arbitrary point in time, say $t = 0$, the probability that the number of events that have occurred up to that time is x. We shall make the following assumptions:

1. The occurrence of events in the current interval of time is independent of their occurrence or nonoccurrence in past intervals.
2. The probability of exactly one occurrence of the event during the short interval $(t, t + \Delta t)$ is $b\Delta t$ (we neglect higher-order terms), where b is a positive constant. If Δt is doubled, the probability of one event is also doubled. Moreover, by assumption this probability depends only on the length of the interval and not on its starting point. The probability is the same for all intervals of length Δt.
3. The probability of more than one event in the short interval Δt is assumed to be of a lower order of magnitude than Δt and is neglected; that is, the probability of two or more events in the interval Δt approaches 0 faster than does Δt.

From these assumptions, it is clear that if we take $(0, t)$ and $(t, t + \Delta t)$ as nonoverlapping intervals, x events up to time $t + \Delta t$ can occur in only two ways: x events during the interval $(0, t)$ and 0 during $(t, t + \Delta t)$ or $x - 1$ during $(0, t)$ and 1 during $(t, t + \Delta t)$. If we denote by $f(x, t)$ the probability that x events occur up to time t, then we have

$$f(x, t + \Delta t) = f(x, t) \cdot (1 - b\Delta t) + f(x - 1, t) \cdot b\Delta t, \tag{7.34}$$

which, as $\Delta t \to 0$, defines the differential equation,

$$\lim_{t \to 0} \frac{f(x, t + \Delta t) - f(x, t)}{\Delta t} = \frac{df(x, t)}{dt} = -bf(x, t) + bf(x - 1, t). \tag{7.35}$$

We need to determine $f(x, t)$. If we begin with $x = 0$, since $f(-1, t) = 0$, (7.35) reduces to

$$\frac{df(0, t)}{dt} = -bf(0, t), \tag{7.36}$$

a solution being

$$f(0, t) = e^{-bt}, \tag{7.37}$$

the constant of integration being determined by the reasonable assumption that $f(0, 0) = 0$.

Next, put $x = 1$. Then using (7.37) in (7.35),

$$\frac{df(1, t)}{dt} = -bf(1, t) + be^{-bt}, \tag{7.38}$$

a solution being

$$f(1, t) = bte^{-bt}, \tag{7.39}$$

the constant of integration being determined by the again reasonable assumption that $f(1, 0) = 0$. Finally x repetitions of this argument [always assuming $f(x, 0) = 0$] yields

$$f(x, t) = \frac{e^{-bt}(bt)^x}{x!}, \qquad x = 0, 1, 2, \ldots, \tag{7.40}$$

which is the density function for the Poisson distribution with $\lambda = bt$ and t a continuous function of time.

7.7 MOMENTS OF THE POISSON DISTRIBUTION

It is easy to check that the Poisson density function in fact sums to 1, for we have

$$\begin{aligned} \sum_{x=0}^{\infty} e^{-\lambda} \frac{\lambda^x}{x!} &= e^{-\lambda}\left(1 + \lambda + \frac{\lambda^2}{2!} + \frac{\lambda^3}{3!} + \cdots\right) \\ &= e^{-\lambda} e^{\lambda} = 1. \end{aligned} \tag{7.41}$$

We shall derive the mean and variance first by their definitions and then by use of the Poisson moment-generating function.

Mean. By definition, we have

$$\begin{aligned} E(X) &= \sum_{x=0}^{\infty} x e^{-\lambda} \frac{\lambda^x}{x!} \\ &= \sum_{x=1}^{\infty} \frac{e^{-\lambda}\lambda^x}{(x-1)!} \\ &= \lambda \sum_{x=1}^{\infty} \frac{e^{-\lambda}\lambda^{x-1}}{(x-1)!}. \end{aligned} \tag{7.42}$$

Replacing $x - 1$ by y, it is evident that the sum is 1, so that

$$\mu = \lambda. \tag{7.43}$$

Variance. Again, by definition,

$$
\begin{aligned}
E(X^2) &= \sum_{x=0}^{\infty} x^2 \frac{e^{-\lambda}\lambda^x}{x!} \\
&= \sum_{x=1}^{\infty} x \frac{e^{-\lambda}\lambda^x}{(x-1)!} \\
&= \lambda \sum_{x=1}^{\infty} \frac{xe^{-\lambda}\lambda^{x-1}}{(x-1)!} \\
&= \lambda \sum_{x=1}^{\infty} \frac{(x-1)e^{-\lambda}\lambda^{x-1}}{(x-1)!} + \lambda \sum_{x=1}^{\infty} \frac{e^{-\lambda}\lambda^{x-1}}{(x-1)!}. \qquad (7.44)
\end{aligned}
$$

Replacing $x - 1$ by y, we see that the first sum is simply the mean, and therefore equal to λ, while the second sum is 1. Consequently,

$$E(X^2) = \lambda^2 + \lambda, \qquad (7.45)$$

hence the variance is

$$E[(X - \mu)^2] = \lambda^2 + \lambda - \lambda^2 = \lambda. \qquad (7.46)$$

Thus, with the Poisson distribution, both the mean and the variance are equal to λ.

The moment-generating function is also very easily obtained. By definition, it is equal to

$$
\begin{aligned}
E(e^{tX}) &= \sum_{x=0}^{\infty} e^{tx} e^{-\lambda} \frac{\lambda^x}{x!} \\
&= e^{-\lambda} \sum_{x=0}^{\infty} \frac{(\lambda e^t)^x}{x!} \\
&= e^{-\lambda}\left[1 + \lambda e^t + \frac{(\lambda e^t)^2}{2!} + \frac{(\lambda e^t)^3}{3!} + \cdots\right]. \qquad (7.47)
\end{aligned}
$$

The sum is simply the series expansion for $e^{\lambda e^t}$, so that

$$M_X(t) = e^{-\lambda} e^{\lambda e^t}. \qquad (7.48)$$

Differentiation of (7.47) with respect to t gives

$$\frac{dM_X(t)}{dt} = e^{-\lambda} e^{\lambda e^t} \cdot \lambda e^t, \qquad (7.49)$$

which when evaluated at $t = 0$ yields the mean

$$\left.\frac{dM_X(t)}{dt}\right|_{t=0} = \lambda. \qquad (7.50)$$

To obtain the variance, we differentiate (7.48) twice with respect to t:

$$\frac{d^2M_X(t)}{dt^2} = e^{-\lambda}e^{\lambda e^t} \cdot \lambda^2 e^{2t} + e^{-\lambda}e^{\lambda e^t} \cdot \lambda e^t, \tag{7.51}$$

evaluate at $t = 0$:

$$\left.\frac{d^2M_X(t)}{dt^2}\right|_{t=0} = \lambda^2 + \lambda, \tag{7.52}$$

and then subtract λ^2. Hence,

$$E[(X - \mu)^2] = \lambda, \tag{7.53}$$

in agreement with (7.46).

7.8 AN EXAMPLE USING THE POISSON DISTRIBUTION

Table 7.1 gives observations on telephone connections to the wrong number. A total of $N = 267$ numbers were observed over a period of time. The column headed N_x indicates how many numbers had exactly x bad connections, while the column headed $N_p(x, 8.74)$ gives how many numbers would have exactly x bad connections as predicted by the Poisson distribution with $\lambda = 8.74$. (8.74 is the mean number of wrong connections per phone.) The agreement between the actual and predicted values, it will be noted, is very good.

TABLE 7.1 Connections to Wrong Number

X	N_x	$N_p(x, 8.74)$
≤ 2	1	2.05
3	5	4.76
4	11	10.39
5	14	18.16
6	22	26.45
7	43	33.03
8	31	36.09
9	40	35.04
10	35	30.63
11	20	24.34
12	18	17.72
13	12	11.92
14	7	7.44
15	6	4.33
≥ 16	2	4.65
	267	267.00

SOURCE: Taken from F. Thorndyke, "Applications of Poisson's Probability Summation," *The Bell System Technical Journal*, vol. 5 (1926), pp. 604–624.

7.9 THE NEGATIVE BINOMIAL DISTRIBUTION

The final discrete distribution to be discussed in this chapter is one which arises in the context of waiting times. Suppose that we have a succession of n Bernoulli trials and that we are interested in how long it will take for the rth success to occur, r being a fixed positive integer. In particular, let us inquire into the probability that the rth success occurs at the mth trial, where $m \leq n$. It is clear that this probability will not depend on n, but rather only on p (the probability of success in any trial), r, and m. Since $m \geq r$, let us write $m = r + k$, where $k \geq 0$.

Let us denote the probability that the rth success occurs at the trial number $r + k$ by $f(k;\, p,\, r)$. It equals the probability that exactly k failures precede the rth success. The rth success can occur at trial $r + k$ if and only if among the preceding $r + k - 1$ trials there are k failures, and the $(r + k)$th trial results in success. The probabilities of these events are $\binom{r+k-1}{k} \cdot p^{r-1}q^k$ (where $q = 1 - p$) and p, so that

$$f(k;\, p,\, r) = \binom{r+k-1}{k} \cdot p^r q^k. \tag{7.54}$$

However, in accordance with the identity,[3]

$$\binom{-r}{k} = \binom{r+k-1}{k} \cdot (-1)^k, \tag{7.55}$$

this probability can be rewritten as

$$f(k;\, p,\, r) = \binom{-r}{k} p^r(-q)^k, \qquad k = 0, 1, 2, \ldots. \tag{7.56}$$

[3] Derivation of this identity is as follows:
In the formula

$$\binom{r}{k} = \frac{r(r-1)\cdots(r-k+1)}{k!},$$

replace r by $-r$:

$$\binom{-r}{k} = \frac{-r(-r-1)\cdots(-r-k+1)}{k!}.$$

Factoring $(-1)^k$ from the right-hand side and rearranging then yields

$$\binom{-r}{k} = \frac{(-1)^k(r+k-1)\cdots(r+1)r}{k!}$$

or

$$\binom{-r}{k} = \frac{(-1)^k(r+k-1)!}{k!(r-1)!} = \binom{r+k-1}{k}(-1)^k.$$

Suppose now that the Bernoulli trials are continued as long as necessary for r successes to occur. We now must ask whether it is possible that the trials never end, that is, whether an infinite sequence of trials may produce fewer than r successes. In order for the rth success to turn up after finitely many trials, it is necessary that $f(k; p, r)$ go to zero as k goes to infinity. This will be the case only if $\sum_{k=0}^{\infty} f(k; p, r)$ is equal to 1.

However, from Newton's binomial formula,[4]

$$\sum_{k=0}^{\infty} \binom{-r}{k}(-q)^k = (1-q)^{-r}, \tag{7.57}$$

so that

$$\begin{aligned} \sum_{k=0}^{\infty} f(k; p, r) &= p^r(1-q)^{-r} \\ &= p^r p^{-r} \\ &= 1, \end{aligned} \tag{7.58}$$

as required. The possibility that the rth success is never attained can thus be safely discounted.

In a waiting time problem, r is necessarily a positive integer. However, the quantity defined by (7.54) or (7.55) is nonnegative and holds for any positive r and positive integer k. For arbitrary fixed $r > 0$ and $0 < p < 1$, the sequence $f(k; p, r)$, $k = 0, 1, 2, \ldots$, defines the *negative binomial distribution*, the name arising from the relation used in (7.57). When r is a positive integer, $f(k; p, r)$, as we have already seen, may be interpreted as the density function of the *probability distribution for the waiting time to the rth success*. In some contexts, it is also referred to as the *Pascal distribution*. For the waiting time to the first success ($r = 1$), $f(k; r, p)$ reduces to pq^k, which is the density function for the *geometric distribution*.

The moments of the negative binomial distribution are easily obtained through the moment-generating function for k:

$$\begin{aligned} M_K(t) &= \sum_{k=0}^{\infty} e^{kt}\binom{-r}{k} p^r(-q)^k \\ &= \sum_{k=0}^{\infty} \binom{-r}{k} p^r(-qe^t)^k \\ &= p^r(1-qe^t)^{-r}. \end{aligned} \tag{7.59}$$

Differentiating once,

$$\frac{dM_K(t)}{dt} = rp^r qe^t(1-qe^t)^{-r-1} \tag{7.60}$$

[4] See Theorem A.2.3 of Appendix 2.

and then again,

$$\frac{d^2M_K(t)}{dt^2} = rp^r qe^t(1 - qe^t)^{-r-1} + (r + 1)rp^r q^2 e^{2t}(1 - qe^t)^{-r-2}. \tag{7.61}$$

Evaluating (7.60) at $t = 0$, we have for the mean,

$$\begin{aligned} E(K) &= rp^r q(1 - q)^{-r-1} \\ &= rp^r qp^{-r-1} \\ &= \frac{rq}{p}, \end{aligned} \tag{7.62}$$

while from (7.61), we have for the variance,

$$\begin{aligned} E(K^2) - [E(K)]^2 &= \frac{rq}{p} + \frac{r(r + 1)q^2}{p^2} - \frac{r^2q^2}{p^2} \\ &= \frac{rq}{p} + \frac{rq^2}{p^2} \\ &= \frac{rq}{p^2}. \end{aligned} \tag{7.63}$$

REFERENCES

Feller, W., *An Introduction to Probability Theory and its Applications*, vol. I, 3rd ed., Wiley, 1968, chap. VI.

Freeman, H., *Introduction to Statistical Inference*, Addison-Wesley, 1963, chaps. 10–13.

CHAPTER 8

SPECIAL CONTINUOUS DISTRIBUTIONS

In this chapter, we begin the study of distributions which involve a continuous random variable. Because of its fundamental importance, the central focus will be on the normal distribution, but several other distributions will be discussed as well. These include the exponential, Laplace, Gamma, Weibull, and Pareto.

8.1 THE NORMAL DISTRIBUTION

Nearly everyone is familiar in some way with the normal distribution, for its density function is the familiar bell-shaped curve that comes to mind when someone mentions "statistics." The normal distribution (or the Gaussian distribution, as it is sometimes called) is the most frequently encountered and the most widely used distribution in statistics. Many natural, demographic, and social phenomena appear to be distributed according to normal law, and it is the distribution that forms the foundation for nearly all statistical inference in scientific research. In addition to its frequent empirical appearance, the normal is also one of the most mathematically tractable theoretical distributions known. However, it perhaps receives its deepest theoretical justification from several rather remarkable limit theorems, one of which will be discussed and proven later in this chapter.

The density function for the normal distribution is given by

$$f(x; a, b) = \frac{1}{b\sqrt{2\pi}} e^{-(x-a)^2/2b^2}, \qquad -\infty < x < \infty, \tag{8.1}$$

where a and b are parameters. The normal distribution is thus a function of two parameters, a and b, which, it will be shown shortly, turn out to be the mean and standard deviation of the distribution, respectively. The density function is graphed in Figure 8.1 for three different values of b. Changing a merely shifts the curves to the left or right without changing their shape. The distribution is seen to be symmetrical about a, and to be more concentrated the smaller the value of b is. The distribution function for the normal distribution is obviously given by

$$F(x) = \frac{1}{b\sqrt{2\pi}} \int_{-\infty}^{x} e^{-(z-a)^2/2b^2}\, dz. \tag{8.2}$$

We shall begin our study of the normal distribution by verifying that the function in (8.1) integrates over x to one, thus establishing that it, in fact, defines a density. Unfortunately, this verification is not particularly tidy because of the fact that the function in (8.1) does not integrate to a simple closed expression. Proceeding, nevertheless, let A represent the area under the curve; thus,

$$A = \frac{1}{b\sqrt{2\pi}} \int_{-\infty}^{\infty} e^{-(x-a)^2/2b^2}\, dx, \tag{8.3}$$

which, upon making the substitution $y = (x - a)/b$, becomes

$$A = \frac{1}{\sqrt{2\pi}} \int_{-\infty}^{\infty} e^{-y^2/2}\, dy. \tag{8.4}$$

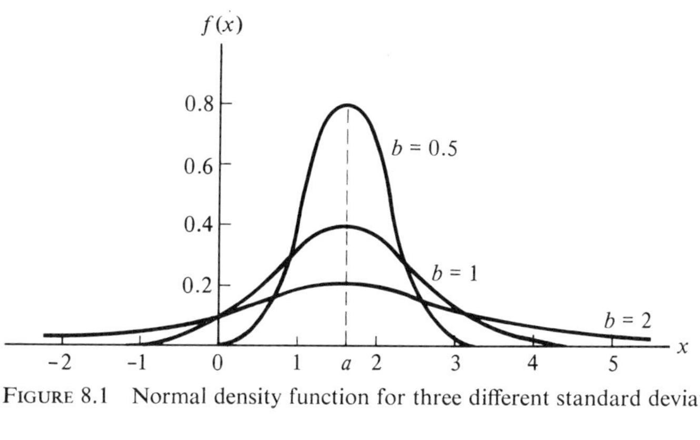

FIGURE 8.1 Normal density function for three different standard deviations.

We wish to show that A is equal to one. This is most easily done by showing that $A^2 = 1$, and then concluding, since $f(x)$ is never negative, that $A = 1$. Thus,

$$A^2 = \frac{1}{\sqrt{2\pi}} \int_{-\infty}^{\infty} e^{-y^2/2}\, dy \frac{1}{\sqrt{2\pi}} \int_{-\infty}^{\infty} e^{-z^2/2}\, dz$$

$$= \frac{1}{\sqrt{2\pi}} \int_{-\infty}^{\infty} \int_{-\infty}^{\infty} e^{-(y^2+z^2)/2}\, dy\, dz. \tag{8.5}$$

To evaluate this double integral, we make a translation to polar coordinates through the substitutions $y = r \sin \theta$ and $z = r \cos \theta$; hence,

$$A^2 = \frac{1}{2\pi} \int_0^{\infty} \int_0^{2\pi} re^{-r^2/2}\, d\theta\, dr$$

$$= \int_0^{\infty} re^{-r^2/2}\, dr$$

$$= 1. \tag{8.6}$$

Consequently, $A = 1$, as was to be shown.

8.2 MEAN AND VARIANCE OF THE NORMAL DISTRIBUTION

We shall derive the mean of the normal distribution from its definition. Thus,

$$E(X) = k \int_x xe^{-(x-a)^2/2b^2}\, dx, \tag{8.7}$$

where for convenience we have let $k = 1/b\sqrt{2\pi}$. If we add and subtract

$$ak \int_x e^{-(x-a)^2/2b^2}\, dx,$$

(8.7) becomes

$$E(X) = k \int_x (x-a)e^{-(x-a)^2/2b^2}\, dx + ak \int_x e^{-(x-a)^2/2b^2}\, dx$$

$$= b^2 k \int_x \frac{(x-a)}{b^2} e^{-(x-a)^2/2b^2}\, dx + ak \int_x e^{-(x-a)^2/2b^2}\, dx, \tag{8.8}$$

where we have multiplied and divided the first integral by b^2 in order to make $(x - a)/b^2$ an integrating factor. The second term is simply a because

k times the integral is equal to the area under the density function and therefore equal to one. Consequently,

$$E(X) = -b^2ke^{-(x-a)^2/2b^2}\Big]_{-\infty}^{\infty} + a$$

$$= a, \tag{8.9}$$

because $e^{-(x-a)^2/2b^2}$ is equal to zero at both $-\infty$ and $+\infty$. Thus, as stated earlier, the mean of the normal distribution is seen to be a.

The variance can also be derived from its definition, but it is easier to use the moment-generating function. Since we have just established that the mean is equal to a, we shall derive the moment-generating function for the moments about the point $\mu = a$. From the definition, we have

$$M_{X-\mu}(t) = k\int_x e^{t(x-\mu)}e^{-(x-\mu)^2/2b^2}\,dx$$

$$= k\int_x e^{[t(x-\mu)-(x-\mu)^2]/2b^2}dx, \tag{8.10}$$

where $k = 1/b\sqrt{2\pi}$ as defined above. We now complete the square in the exponent:

$$\frac{2b^2t(x-\mu)-(x-\mu)^2}{2b^2} = \frac{-[(x-\mu)^2 - 2b^2t(x-\mu) + b^4t^2]}{2b^2} + \frac{b^4t^2}{2b^2},$$

so that

$$M_{X-\mu}(t) = e^{b^2t^2/2}k\int_x e^{-(x-\mu-b^2t^2)^2/2b^2}\,dx, \tag{8.11}$$

which, since

$$k\int_x e^{-(x-\mu-b^2t^2)^2/2b^2}\,dx$$

is the area under the density function for a normal random variable with mean equal to $\mu + b^2t^2$ and hence equal to one, equals

$$M_{X-\mu}(t) = e^{b^2t^2/2}. \tag{8.12}$$

Therefore, differentiating (8.12) twice with respect to t,

$$\frac{d^2M_{X-\mu}(t)}{dt^2} = b^2e^{b^2t^2/2} + b^4t^2e^{b^2t^2/2}, \tag{8.13}$$

and then evaluating at $t = 0$, we see that (as stated earlier) the variance σ^2 of the normal distribution is b^2.

Since it will be needed later, it will be useful at this point to note that from Equation (6.27) the generating function for moments about the origin for the normal distribution is

$$M_X(t) = e^{t\mu + t^2\sigma^2/2}. \tag{8.14}$$

8.3 STANDARDIZED RANDOM VARIABLES

Consider the following random variable,

$$Y = \frac{X - \mu}{\sigma}, \tag{8.15}$$

where μ is the mean of X and σ is its standard deviation. Then

$$\begin{aligned} E(Y) &= E\left(\frac{X-\mu}{\sigma}\right) \\ &= \frac{E(X) - E(\mu)}{\sigma} \\ &= \frac{\mu - \mu}{\sigma} \\ &= 0, \end{aligned} \tag{8.16}$$

and

$$\begin{aligned} \operatorname{var} Y &= \operatorname{var}\left(\frac{X-\mu}{\sigma}\right) \\ &= E\left[\frac{X-\mu}{\sigma}\right]^2 \\ &= \frac{1}{\sigma^2} E(X-\mu)^2 \\ &= \frac{1}{\sigma^2} \cdot \sigma^2 \\ &= 1. \end{aligned} \tag{8.17}$$

In other words, by subtracting the mean and dividing by the standard deviation, we transform a random variable into a new random variable that has zero mean and unit standard deviation. This new random variable is called a *standardized* random variable, and the transformation is called *standardization*.

Although standardization is valid for any random variable, it leads to a particularly happy result for the normal random variable, for, if X is normal with mean μ and variance σ^2, then $Y = (X - \mu)/\sigma$ has the density function

$$f(y) = \frac{1}{\sqrt{2\pi}} e^{-y^2/2}. \tag{8.18}$$

(A full justification of this transformation is given in Chapter 10.) In other words, the standardized normal variate is normal with mean 0 and variance 1. Since this is the case, it follows that the normal distribution with mean 0 and variance 1 is all that needs to be tabulated since the distribution for any other mean and variance can be found by the transformation

$$X = \mu + \sigma Y. \tag{8.19}$$

8.4 THE NORMAL DISTRIBUTION AS THE LIMIT DISTRIBUTION OF THE BINOMIAL

We shall now show how the normal distribution arises as the limit form for the binomial distribution. The fact made use of in this derivation is that two distributions are identical if they have the same moment-generating function. Proof of this theorem is given in Appendix 3.

We begin with the standardized binomial variable,

$$Y = \frac{X - np}{\sqrt{npq}}, \tag{8.20}$$

which we know has mean 0 and variance 1. The moment-generating function of Y is

$$\begin{aligned} M_Y(t) &= M_{X-\mu}\left(\frac{t}{\sigma}\right) = e^{-t\mu/\sigma} M_X\left(\frac{t}{\sigma}\right) \\ &= e^{-t\mu/\sigma}(q + pe^{t/\sigma})^n, \end{aligned} \tag{8.21}$$

so that

$$\ln M_Y(t) = -\frac{t\mu}{\sigma} + n \ln(q + pe^{t/\sigma}). \tag{8.22}$$

Then expanding $e^{t/\sigma}$ in its power series and making use of the fact that $p + q = 1$,

$$\ln M_Y(t) = -\frac{t\mu}{\sigma} + n \ln\left\{1 + p\left[\frac{t}{\sigma} + \frac{1}{2!}\frac{t^2}{\sigma^2} + \frac{1}{3!}\frac{t^3}{\sigma^3} + \cdots\right]\right\}. \tag{8.23}$$

Next, denote by zp multiplied by the expression within the square brackets. For $|z| < 1$, the expansion of $\ln(1 + z)$ is

$$\ln(1 + z) = z - \frac{z^2}{2} + \frac{z^3}{3} - \cdots,$$

so that for sufficiently small (but arbitrary) z, we have

$$\ln M_Y(t) = -\frac{t\mu}{\sigma} + n\left(z - \frac{z^2}{2} + \frac{z^3}{3} - \cdots\right). \tag{8.24}$$

Substituting back for z, collecting terms separately in t, t^2, and so on and using $\mu = np$ and $\sigma = \sqrt{npq}$, we find

$$\ln M_Y(t) = \frac{t^2}{2} + \text{terms in } t^k, \qquad k \geq 3. \tag{8.25}$$

However, all the terms in t^k involve n only through terms of the form

$$\frac{n}{\sigma^k} = \frac{n}{(npq)^{k/2}}, \qquad k \geq 3,$$

which vanish as $n \to \infty$. Therefore,

$$\lim_{n\to\infty} \ln M_Y(t) = \frac{t^2}{2} \tag{8.26}$$

and, since for a continuous function f, $f(\lim g)$ is equal to $\lim f(g)$,

$$\lim_{n\to\infty} M_Y(t) = e^{t^2/2}, \tag{8.27}$$

which as we saw above is the moment-generating function for a normal variable with mean 0 and variance 1. Thus, we conclude that the binomial tends to the normal distribution as the number of trials becomes large.

8.5 OTHER CONTINUOUS DISTRIBUTIONS

Exponential Distribution

The density function for the exponential distribution is

$$f(x; \alpha) = \alpha e^{-\alpha x}, \tag{8.28}$$

defined for $x \geq 0$ and α a positive constant. Its graph is given in Figure 8.2. Verification that the area beneath this function is equal to one involves

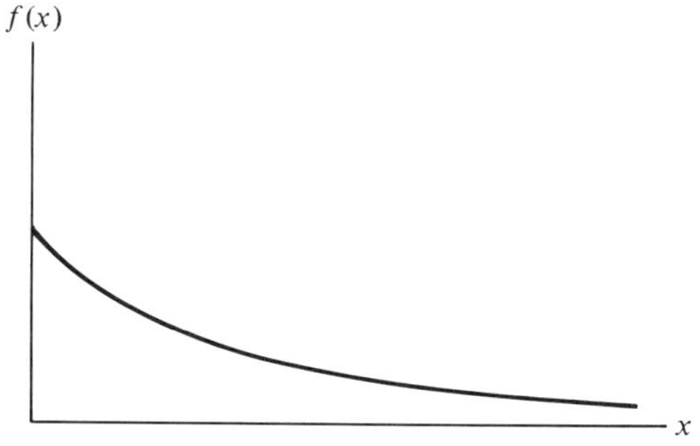

FIGURE 8.2 Exponential distribution.

a straightforward integration and is left as an exercise for the reader. We shall obtain the mean and variance of the distribution from their definitions. In each case, we proceed via integration by parts.

For the mean,

$$E(X) = \alpha \int_0^\infty x e^{-\alpha x}\, dx. \tag{8.29}$$

Let $u = x$ and $dv = \alpha e^{-\alpha x}\, dx$; then $v = -e^{-\alpha x}$ and

$$\begin{aligned} E(X) &= -xe^{-\alpha x}\Big]_0^\infty + \int_0^\infty e^{-\alpha x}\, dx \\ &= -xe^{-\alpha x} - \frac{1}{\alpha} e^{-\alpha x}\Big]_0^\infty \\ &= \frac{1}{\alpha} \qquad \text{(using l'Hôpital's rule on } \lim_{x \to \infty} xe^{-\alpha x}). \end{aligned} \tag{8.30}$$

Thus, we find the mean to be equal to the reciprocal of α, the distribution's parameter.

Turning to the variance, we have for $E(X^2)$,

$$\begin{aligned} E(X^2) &= \alpha \int_0^\infty x^2 e^{-\alpha x}\, dx \\ &= -x^2 e^{-\alpha x}\Big]_0^\infty - 2 \int_0^\infty x e^{-\alpha x}\, dx \\ &= 0 - \frac{2}{\alpha} x e^{-\alpha x}\Big]_0^\infty + \frac{2}{\alpha} \int_0^\infty e^{-\alpha x}\, dx \\ &= 0 - 0 - \frac{2}{\alpha^2} e^{-\alpha x}\Big]_0^\infty \\ &= \frac{2}{\alpha^2}. \end{aligned} \tag{8.31}$$

Consequently, subtracting $1/\alpha^2$ from $E(X^2)$ we have for the variance of the exponential distribution,

$$\sigma^2 = \frac{1}{\alpha^2}. \tag{8.32}$$

The exponential distribution plays an important role in describing a large class of phenomena, especially in the area of reliability theory. In particular, the distribution has the following interesting property. Assume that we have a light bulb which burns continuously and whose probability of having a life of t hours is given by

$$P(t < X < t + dt) = \alpha e^{-\alpha t}\, dt, \tag{8.33}$$

where X is a random variable representing the life of the bulb. The probability that the light bulb will still be working after t hours will then be equal to

$$\begin{aligned} P(X > t) &= \alpha \int_t^\infty e^{-\alpha t}\, dt \\ &= e^{-\alpha t}. \end{aligned} \tag{8.34}$$

Suppose that the bulb is burning at time t, and let us seek the probability that the bulb is also burning at time $t + s$ $(s > 0)$. By definition, this conditional probability will be given by

$$\begin{aligned} P(X > t + s \mid X > t) &= \frac{P(X > t + s)}{P(X > t)} \\ &= \frac{e^{-\alpha(t+s)}}{e^{-\alpha t}} \\ &= e^{-\alpha s}. \end{aligned} \tag{8.35}$$

Thus, we find this probability depends only on s, and not on t. This is an important property of the exponential distribution, and is frequently referred to as "absence of memory." The reader should convince himself that the geometric distribution, alluded to briefly near the end of the last chapter, also has this property.

Laplace Distribution

The Laplace (or double exponential) random variable is continuous over the entire real line and has the density function

$$f(x; \alpha, \theta) = \frac{\alpha}{2} e^{-\alpha|x-\theta|}, \qquad -\infty < x < \infty, \tag{8.36}$$

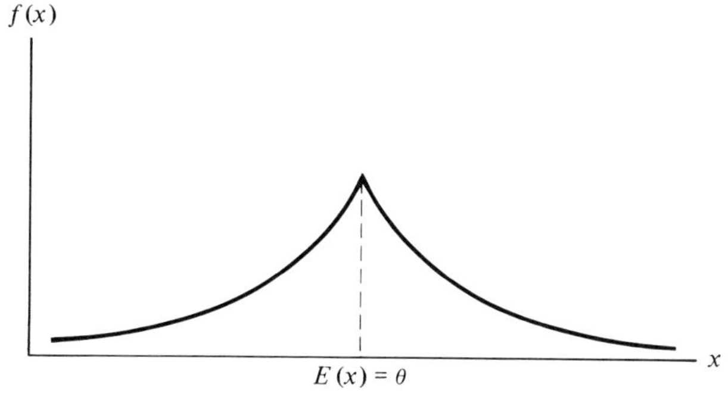

FIGURE 8.3 Laplace distribution.

where α and θ are constants, $\alpha > 0$. The general shape of the distribution is as given in Figure 8.3. Since $|x - \theta|$ is symmetrical about zero, it is evident that the distribution is symmetrical about θ.

For the mean, we have

$$E(X) = \frac{\alpha}{2} \int_{-\infty}^{\infty} x e^{-\alpha|x-\theta|} \, dx$$

$$= \frac{\alpha}{2} \int_{-\infty}^{\theta} x e^{\alpha(x-\theta)} \, dx + \frac{\alpha}{2} \int_{\theta}^{\infty} x e^{\alpha(\theta - x)} \, dx, \tag{8.37}$$

which upon integration by parts, yields

$$E(X) = \frac{1}{2} e^{-\alpha\theta} \left[x e^{\alpha x} - \frac{1}{\alpha} e^{\alpha x} \right]_{-\infty}^{\theta} + \frac{1}{2} e^{\alpha\theta} \left[-x e^{-\alpha x} - \frac{1}{\alpha} e^{-\alpha x} \right]_{\theta}^{\infty}$$

$$= \frac{1}{2}\left(\theta - \frac{1}{\alpha}\right) + \frac{1}{2}\left(\theta + \frac{1}{\alpha}\right)$$

$$= \theta. \tag{8.38}$$

For the second moment, we have

$$E(X^2) = \frac{\alpha}{2} \int_{-\infty}^{\infty} x^2 e^{-\alpha|x-\theta|} \, dx \tag{8.39}$$

$$= \frac{\alpha}{2} \int_{-\infty}^{\theta} x^2 e^{\alpha(x-\theta)} \, dx + \frac{\alpha}{2} \int_{\theta}^{\infty} x^2 e^{\alpha(\theta-x)} \, dx$$

$$= \theta^2 + \frac{2}{\alpha},$$

the details of twice integrating by parts being left to the reader as an exercise. Subtracting $[E(X)]^2$, we then obtain for the variance,

$$\sigma^2 = \frac{2}{\alpha}. \tag{8.40}$$

Gamma Distribution

A distribution which plays an important role in statistics is the gamma distribution, which has a density function defined as follows:

$$f(x; \alpha, \beta) = \frac{1}{\alpha!\,\beta^{\alpha+1}} x^{\alpha} e^{-x/\beta}, \qquad 0 < x < \infty. \tag{8.41}$$

This is a two-parameter family of distributions, the parameters being α and β. Restrictions on the parameters are that β must be positive, while α must be greater than -1. The function is plotted in Figure 8.4 for $\beta = 1$ and three values of α. As is clear from examining the form of the function, varying the value of β merely changes the scale of the two axes.

Our first task will be to show that the function has unit area and therefore defines a density. Accordingly, we shall evaluate the integral

$$A = \int_0^{\infty} \frac{1}{\beta^{\alpha+1}} x^{\alpha} e^{-x/\beta}\, dx, \tag{8.42}$$

and show that it has area equal to $\alpha!$. We begin with a substitution of y for x/β, which transforms A to a function of α only:

$$A(\alpha) = \int_0^{\infty} y^{\alpha} e^{-y}\, dy. \tag{8.43}$$

However, this integral is a well-known form, defining the *gamma function* for $\alpha + 1$, that is,

$$\Gamma(\alpha + 1) = \int_0^{\infty} y^{\alpha} e^{-y}\, dy. \tag{8.44}$$

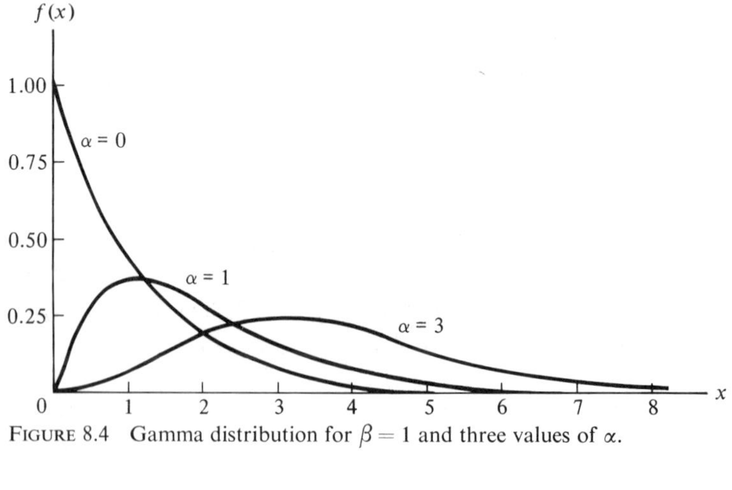

FIGURE 8.4 Gamma distribution for $\beta = 1$ and three values of α.

This function has the property that[1]

$$\begin{aligned}\Gamma(\alpha+1) &= \alpha\Gamma(\alpha)\\ &= \alpha(\alpha-1)\Gamma(\alpha-1)\\ &= \alpha(\alpha-1)\cdots(\alpha-k)\Gamma(\alpha-k),\end{aligned} \tag{8.45}$$

and, therefore, can serve as the definition of $\alpha!$ for *any* positive α, not just for α a positive integer. Thus, since $A(\alpha) = \Gamma(\alpha+1)$ and $\alpha! \equiv \Gamma(\alpha+1)$, we have $A = \alpha!$ by definition. Thus, it is evident from where the distribution derives its name.

The cumulative distribution for the gamma random variable is

$$F(x) = \int_0^x \frac{1}{\alpha!\beta^{\alpha+1}}\, t^\alpha e^{-t/\beta}\, dt. \tag{8.46}$$

For α a positive integer, $F(x)$ can be found by successive integration by parts to be equal to

$$F(x) = 1 - \left[1 + \frac{x}{\beta} + \frac{1}{2!}\left(\frac{x}{\beta}\right)^2 + \frac{1}{3!}\left(\frac{x}{\beta}\right)^3 + \cdots + \frac{1}{\alpha!}\left(\frac{x}{\beta}\right)^\alpha\right] e^{-x/\beta}. \tag{8.47}$$

However, for α not an integer, the function must be evaluated by numerical methods. $F(x)$ has been extensively tabulated and is referred to as the *incomplete gamma function.*

The mean and variance of the distribution are easily obtained from the moment-generating function. For the latter, we have

$$M_X(t) = \int_0^\infty e^{tx} \frac{1}{\alpha!\beta^{\alpha+1}}\, x^\alpha e^{-x/\beta}\, dx. \tag{8.48}$$

On substituting $y = x/\beta$, this becomes

$$\begin{aligned}M_X(t) &= \frac{1}{\alpha!}\int_0^\infty y^\alpha e^{-y(1-\beta t)}\, dy\\ &= \frac{1}{(1-\beta t)^{\alpha+1}}\int_0^\infty \frac{(1-\beta t)^{\alpha+1}}{\alpha!}\, y^\alpha e^{-y(1-\beta t)}\, dt\\ &= \frac{1}{(1-\beta t)^{\alpha+1}}\end{aligned} \tag{8.49}$$

[1] See Taylor (1955, pp. 648–649).

provided that $t < 1/\beta$, since the last integral represents the area under a gamma distribution with parameters α and $\beta' = 1/(1 - \beta t)$ and is therefore equal to one.

On differentiating $M_X(t)$ twice with respect to t and then evaluating the results at $t = 0$, we will eventually obtain

$$\mu = \beta(\alpha + 1), \tag{8.50}$$

$$\sigma^2 = \beta^2(\alpha + 1), \tag{8.51}$$

for the mean and variance, respectively.

Weibull Distribution

In the example discussed in connection with the exponential distribution, it was assumed that the reliability of the light bulb decayed at a constant exponential rate. We now turn to a distribution which has even wider application than the exponential in the study of the reliability of working systems. However, before introducing this distribution, we require some preliminary definitions.

Reliability

The *reliability* of a component (or system) at time t, say $R(t)$, is defined as $R(t) = P(T > t)$, where T is the life length of the component. R is called the *reliability function.*

Failure Rate

The (instantaneous) *failure rate* Z (sometimes referred to as the *hazard function*) associated with the random variable T is defined by

$$Z(t) = \frac{f(t)}{1 - F(t)} = \frac{f(t)}{R(t)}, \tag{8.52}$$

where $f(t)$ and $F(t)$ are the density and distribution functions of T, respectively.

In order to interpret the failure rate, let us consider the conditional probability, $P(t < T \leq t + \Delta T | T > t)$, which is the probability that the component will fail during the next Δt time units, given that it is functioning properly at time t. However, by definition, this conditional probability is equal to

$$\begin{aligned} P(t < T \leq t + \Delta t \mid T > t) &= \frac{P(t < T < t + \Delta t)}{P(T > t)} \\ &= \frac{\int_t^{t+\Delta t} f(\tau)\, d\tau}{P(T > t)} \\ &= \frac{\Delta t f(\xi)}{R(t)}, \end{aligned} \tag{8.53}$$

where $t \leq \xi \leq t + \Delta t$.

This last expression is for Δt small and (assuming f to be continuous at 0^+) approximately equal to $\Delta t Z(t)$. Thus, ignoring higher-order terms, $\Delta t Z(t)$ can be taken as representing the probability of the component failing between t and $t + \Delta t$, given that it is functioning properly at time t.

From the above, we see that the failure rate $Z(t)$ is determined uniquely by the density function of T, $f(t)$. We shall now show that the converse also holds; that is, that Z uniquely determines f.

THEOREM 8.1

If T, the time to failure, is a continuous random variable with density function $f(t)$ and such that $F(0) = 0$, then $f(t)$ may be expressed in terms of the failure rate $Z(t)$ as follows:

$$f(t) = Z(t) e^{-\int_0^t Z(\tau)\, d\tau}. \tag{8.54}$$

PROOF. Since $R(t) = 1 - F(t)$, we have $R'(t) = -F'(t) = -f(t)$, so that

$$\begin{aligned} Z(t) &= \frac{f(t)}{R(t)} \\ &= \frac{-R'(t)}{R(t)}. \end{aligned}$$

Therefore,

$$\begin{aligned} \int_0^t Z(\tau)\, d\tau &= -\int_0^t \frac{R'(\tau)}{R(\tau)}\, d\tau \\ &= -\ln R(\tau)\Big]_0^t \\ &= -\ln R(t) + \ln R(0) \\ &= -\ln R(t) \end{aligned}$$

provided that ln $R(0)$ is equal to zero. This will hold if and only if $R(0) = 1$, which is guaranteed by the assumption that $F(0) = 0$. (This assumption simply says that the probability of initial failure is equal to zero.) Hence,

$$R(t) = e^{-\int_0^t Z(\tau)\,d\tau},$$

and consequently,

$$\begin{aligned} f(t) &= F'(t) \\ &= \frac{d}{dt}[1 - R(t)] \\ &= Z(t)e^{-\int_0^t Z(\tau)\,d\tau} \end{aligned}$$

as was to be shown.

Turning now to the problem at hand, suppose that the failure rate Z, associated with T, the life span of a component, has the following form:

$$Z(t) = (\alpha\beta)t^{\beta-1}, \tag{8.55}$$

where α and β are positive constants. From (8.54), we will then obtain for the density function for T,

$$f(t; \alpha, \beta) = (\alpha\beta)t^{\beta-1}e^{-\alpha t^\beta}, \tag{8.56}$$

for $t > 0$. The random variable having the density function given by this expression is said to have the *Weibull distribution*.

The Weibull distribution represents an appropriate model for describing the reliability and failure functions for situations in which there is a system composed of a number of components and failure is essentially due to the "most severe" flaw among a large number of flaws in the system. In particular, it has been found to describe well the malfunctioning characteristics of B-52 aircraft.

Figure 8.5 shows the density function for the Weibull distribution for $\alpha = 1$ and $\beta = 1, 2, 3$. Note that for $\beta = 1$, the density in (8.56) reduces to the density function for the exponential distribution [cf. (8.28)]. The exponential is thus a special case of the Weibull. The reliability function for the distribution is given by

$$\begin{aligned} R(t) &= \int_t^\infty (\alpha\beta)\tau^{\beta-1}e^{-\alpha\tau^\beta}\,d\tau \\ &= e^{-\alpha t^\beta}, \end{aligned} \tag{8.57}$$

obtained by straightforward integration.

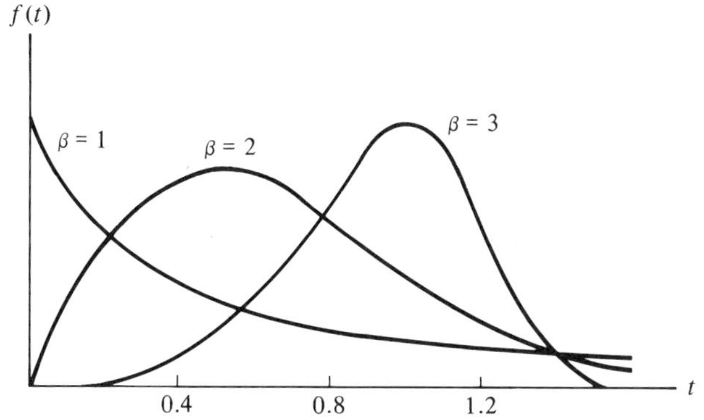

FIGURE 8.5 Weibull distribution with $\alpha = 1$ and $\beta = 1, 2, 3$.

Neither the mean nor the variance of the Weibull distribution can be written in closed form, but only in terms of the gamma function. For the mean,

$$E(T) = \int_0^\infty (\alpha\beta)t^\beta e^{-\alpha t^\beta}\, dt. \tag{8.58}$$

We want to show that this expression can be reduced to

$$E(T) = \alpha^{-1/\beta}\Gamma\left(\frac{1}{\beta + 1}\right). \tag{8.59}$$

Begin by letting

$$y = \alpha t^\beta, \tag{8.60}$$

so that

$$dy = \alpha\beta t^{\beta-1}\, dt \tag{8.61}$$

and

$$dt = \frac{dy}{\alpha\beta t^{\beta-1}}. \tag{8.62}$$

From (8.60),

$$t^{-1} = \left(\frac{y}{\alpha}\right)^{-1/\beta} \tag{8.63}$$

and

$$t^{\beta-1} = \left(\frac{y}{a}\right)^{1-1/\beta} \tag{8.64}$$

so that

$$dt = \frac{dy}{\alpha\beta\left(\dfrac{y}{\alpha}\right)^{1-1/\beta}}. \tag{8.65}$$

Consequently, returning to (8.58),

$$\begin{aligned} E(T) &= \int_0^\infty \beta y e^{-y}(\alpha\beta)^{-1}\left(\frac{y}{\alpha}\right)^{1/\beta-1} dy \\ &= \alpha^{-1/\beta}\int_0^\infty e^{-y}y^{1/\beta}\,dy \\ &= \alpha^{-1/\beta}\Gamma\left(\frac{1}{\beta+1}\right), \end{aligned} \tag{8.66}$$

as was to be shown.

The derivation of the variance is along similar lines. For $E(T^2)$, we have by definition,

$$E(T^2) = \int_0^\infty \alpha\beta t^{\beta+1}e^{-\alpha t^\beta}\,dt. \tag{8.67}$$

As before, let $y = \alpha t^\beta$, so that $dy = \alpha\beta t^{\beta-1}\,dt$ and [from (8.64)] $dt = dy/\alpha\beta(y/\alpha)^{1-1/\beta}$. Hence,

$$\begin{aligned} E(T^2) &= \int_0^\infty \beta y\left(\frac{y}{\alpha}\right)^{1/\beta} e^{-y}\left(\frac{1}{\alpha\beta(y/\alpha)^{1-1/\beta}}\right) dy \\ &= \alpha^{-2/\beta}\int_0^\infty e^{-y}y^{2/\beta}\,dy \\ &= \alpha^{-2/\beta}\Gamma\left(\frac{2}{\beta+1}\right). \end{aligned} \tag{8.68}$$

Subtracting the square of (8.66), we consequently find for the variance of the Weibull distribution,

$$\sigma^2 = \alpha^{-2/\beta}\left\{\Gamma\left(\frac{2}{\beta+1}\right) - \left[\Gamma\left(\frac{1}{\beta+1}\right)\right]^2\right\}. \tag{8.69}$$

Pareto Distribution

Around the turn of the century, it was observed by the Italian economist, Vilfredo Pareto, that for high-income households, the proportion of households with incomes greater than the amount x is given by

$$P(X > x) = Ax^{-\alpha}, \qquad x \geq x_0 > 0, \tag{8.70}$$

where α is a positive constant. A is also a constant, which we shall evaluate in a moment, depending on α and x_0, the lower bound of x.

Interpreting X as a probability (the probability that a household has a level of income greater than x), the expression in (8.70) gives the upper tail of the cumulative distribution of X:

$$1 - F(x) = Ax^{-\alpha}. \tag{8.71}$$

We then have for the density function of X,

$$F'(x) = \alpha A x^{-(\alpha+1)}, \qquad x \geq x_0 > 0. \tag{8.72}$$

However, for (8.72) to be a legitimate density, its integral must have unit area, that is, we must have

$$1 = \int_{x_0}^{\infty} \alpha A x^{-(\alpha+1)}\, dx. \tag{8.73}$$

But

$$\begin{aligned} \int_{x_0}^{\infty} \alpha A x^{-(\alpha+1)}\, dx &= A \int_{x_0}^{\infty} \alpha x^{(-\alpha+1)}\, dx \\ &= -Ax^{-\alpha}\Big]_{x_0}^{\infty} \\ &= A{x_0}^{-\alpha} \qquad (\text{since } \alpha > 0). \end{aligned} \tag{8.74}$$

Consequently,

$$A = {x_0}^{\alpha}, \tag{8.75}$$

and hence for the density function, $f(x; \alpha, x_0)$,

$$f(x; \alpha, x_0) = \alpha {x_0}^{\alpha} x^{-(\alpha+1)}, \qquad x \geq x_0 > 0. \tag{8.76}$$

A random variable with this density function is said to have the *Pareto distribution.*

Taking the logarithm of (8.76),

$$\ln f(x; \alpha, x_0) = a - (\alpha + 1) \ln x, \tag{8.77}$$

where $a = \ln \alpha + \alpha \ln x_0$, we see that the Pareto density is linear in logarithms, with (negative) slope equal to $-(\alpha + 1)$. In addition to the upper tail of the distribution of income, the Pareto distribution has been found to describe a wide variety of economic, social, and physical phenomena. The distribution of firms by size, changes in short-term speculative prices, word lengths, cities by size, bodies of water, and intervals between errors in transmitting high-frequency electronic communications are just a few instances of where the Pareto distribution has been found applicable.

We shall derive the mean and variance of the Pareto distribution through their definitions. For the mean,

$$
\begin{aligned}
E(X) &= \int_{x_0}^{\infty} x\alpha x_0^{\alpha} x^{(-\alpha+1)}\,dx \\
&= \frac{\alpha x_0^{\alpha}}{1-\alpha}\int_{x_0}^{\infty}(1-\alpha)x^{-\alpha}\,dx \\
&= \frac{\alpha x_0^{\alpha}}{1-\alpha}\,x^{1-\alpha}\Big]_{x_0}^{\infty} \\
&= \begin{cases} \dfrac{\alpha x_0}{\alpha-1}, & \text{if } \alpha > 1 \\ \infty, & \text{if } \alpha \le 1. \end{cases}
\end{aligned}
\tag{8.78}
$$

Thus, we see that the mean of the Pareto distribution exists only for $\alpha > 1$.

For $E(X^2)$,

$$
\begin{aligned}
E(X^2) &= \int_{x_0}^{\infty} x^2\alpha x_0^{\alpha} x^{-(\alpha+1)}\,dx \\
&= \frac{\alpha x_0^{\alpha}}{2-\alpha}\int_{x_0}^{\infty}(2-\alpha)x^{1-\alpha}\,dx \\
&= \frac{\alpha x_0}{2-\alpha}\,x^{2-\alpha}\Big]_{x_0}^{\infty} \\
&= \begin{cases} \dfrac{\alpha x_0^2}{\alpha-2}, & \text{if } \alpha > 2 \\ \infty, & \text{if } \alpha \le 2. \end{cases}
\end{aligned}
\tag{8.79}
$$

For $\alpha > 2$, the variance will therefore be equal to

$$
\begin{aligned}
\sigma^2 &= \frac{\alpha x_0^2}{\alpha-2} - \left(\frac{\alpha x_0}{\alpha-1}\right)^2 \\
&= \frac{\alpha x_0^2}{(\alpha-1)^2(\alpha-2)}.
\end{aligned}
\tag{8.80}
$$

The variance of the Pareto random variable does not exist (i.e., is infinite) for α less than or equal to 2. This fact is of particular interest, for in the cases where the Pareto distribution has been found to be most applicable,

α is usually estimated to be less than 2. For the upper tail of the distribution of income, for example, α is typically about 1.7.

REFERENCES

Freeman, H., *Introduction to Statistical Inference*, Addison-Wesley, 1963, chaps. 15, 16.

Mandelbrot, B., "New Methods in Statistical Economics," *Journal of Political Economy*, October 1963.

Meyer, P. L., *Introductory Probability and Statistical Applications*, Addison-Wesley, 1965, chap. 11.

Mood, A. M. and Graybill, F. A., *Introduction to the Theory of Statistics*, 2nd ed., McGraw-Hill, 1963, chap. 6.

Taylor, A. E., *Advanced Calculus*, Ginn, 1955.

CHAPTER 9

RANDOM VARIABLES IN *N* DIMENSIONS

Thus far our discussion has been entirely in terms of probability distributions for a single random variable. In this chapter we will extend the basic notions of probability to a random variable in several dimensions. We shall begin with random variables in two dimensions.

9.1 THE JOINT DISTRIBUTION OF TWO RANDOM VARIABLES

Although we did not take note of it at the time, we have already been in contact with the joint distribution of two random variables, namely, in Chapter 3 when we were considering the tossing of two dice. The two random variables are, respectively, the faces of the two dice. The joint distribution of the two random variables is given by the 36 possible outcomes with their associated probabilities. If we denote the face of the first die by X and that of the second by Y, then the density function for the *joint occurrence* of $X = x$ and $Y = y$ is (assuming that the dice are fair):

$$P(X = x,\ Y = y) = f(x, y) = \frac{1}{36}. \tag{9.1}$$

Moreover, it is obvious that $f(x, y) \geq 0$ for all x and y, that

$$\sum_{x=1}^{6} \sum_{y=1}^{6} f(x, y) = 1, \tag{9.2}$$

and that

$$P(2 \le X \le 4, 3 \le Y \le 5) = \sum_{x=2}^{4} \sum_{y=3}^{5} f(x, y). \tag{9.3}$$

Thus, $f(x, y)$ satisfies the axioms of Chapter 3 for a probability density function.

More generally, we write the density function for the *joint occurrence* of two discrete random events X and Y as

$$P(X = x, Y = y) = f(x, y), \tag{9.4}$$

and the distribution function as

$$P(X \le x, Y \le y) = F(x, y) = \sum_{z_1 \le x} \sum_{z_2 \le y} f(z_1, z_2). \tag{9.5}$$

For two continuous random variables, the probability that X is within the interval x to $x + dx$ and Y is within the interval y to $y + dy$ is

$$P(x < X < x + dx, y < Y < y + dy) = f(x, y)\, dx\, dy \tag{9.6}$$

and the distribution function is

$$P(X < x, Y < y) = F(x, y) = \int_{-\infty}^{x} \int_{-\infty}^{y} f(w, v)\, dv\, dw \tag{9.7}$$

where at the continuity points of F,

$$\frac{\partial^2 F(x, y)}{\partial x\, \partial y} = f(x, y). \tag{9.8}$$

Moreover, for $f(x, y)$ to define a true density, we must have

$$\int_{-\infty}^{\infty} \int_{-\infty}^{\infty} f(x, y)\, dx\, dy = 1. \tag{9.9}$$

Often we are interested in the probabilities taken by one of the random variables for *all values of the other*. By integrating (or summing) out the variable that we are not interested in, we define the two *marginal distributions*, namely,

$$f_x(x) = \int_y f(x, y)\, dy \tag{9.10}$$

$$f_y(y) = \int_x f(x, y)\, dx, \tag{9.11}$$

(9.10) being the marginal density function for X and (9.11) the marginal density function for Y. For discrete random variables, the marginal density functions are defined as

$$f_x(x) = \sum_y f(x, y), \tag{9.12}$$

$$f_y(y) = \sum_x f(x, y). \tag{9.13}$$

We also have the marginal distribution functions for X and Y,

$$F_x(x) = \int_{-\infty}^{x} \int_{-\infty}^{\infty} f(z, y)\, dy\, dz, \tag{9.14}$$

$$F_y(y) = \int_{-}^{y} \int_{-}^{\infty} f(x, z)\, dx\, dz, \tag{9.15}$$

for continuous X, and

$$F_x(x) = \sum_{z \le x} \sum_y f(z, y), \tag{9.16}$$

$$F_y(y) = \sum_{z \le y} \sum_x f(x, z), \tag{9.17}$$

for discrete X.

Since the marginal distributions for a bivariate distribution are distributions in a single variable they are used in defining the moments—means, variances, and so on—for the two variables taken separately. The mean for X is given by

$$E(X) = \int_x \int_y x f(x, y)\, dy\, dx, \tag{9.18}$$

and the mean for Y by

$$E(Y) = \int_y \int_x y f(x, y)\, dx\, dy. \tag{9.19}$$

Similarly, for the variances,

$$E[(X - \mu_x)^2] = \int_x \int_y (x - \mu_x)^2 f(x, y)\, dy\, dx, \tag{9.20}$$

$$E[(Y - \mu_y)^2] = \int_y \int_x (y - \mu_y)^2 f(x, y)\, dx\, dy. \tag{9.21}$$

The Covariance

It is easy to generalize the notion of a moment to include that of a cross moment for a joint distribution in two random variables. The most useful cross moment is the *covariance* and is defined as

$$E[(X - \mu_x)(Y - \mu_y)] = \int_x \int_y (x - \mu_x)(y - \mu_y) f(x, y)\, dy\, dx \tag{9.22}$$

and is usually denoted by the symbol σ_{xy} or simply by cov (X, Y). The covariance is often used as a measure of *linear* association. It is evident from the definition that there is no restriction on the sign of the covariance in that it can be either positive or negative (or, for that matter, zero). When cov (X, Y) is positive, $(x - \mu_x)$ and $(y - \mu_y)$, loosely speaking, have the same sign more often than they have opposite signs, while the reverse is true when cov (X, Y) is negative. In the first case, we say that there is a *positive* association between X and Y, and a *negative* association in the second case.

The Correlation Coefficient

From the Schwartz inequality, it is true that

$$\left\{\int_x \int_y (x - \mu_x)(y - \mu_y) f(x, y)\, dy\, dx\right\}^2$$

$$\leq \int_x \int_y (x - \mu_x)^2 f(x, y)\, dy\, dx \cdot \int_y \int_x (y - \mu_y)^2 f(x, y)\, dx\, dy, \qquad (9.23)$$

or in a more convenient notation

$$[\text{cov}\,(X, Y)]^2 \leq \text{var}\,(X)\,\text{var}\,(Y). \qquad (9.24)$$

Hence, it follows that

$$0 \leq \frac{[\text{cov}\,(X, Y)]^2}{\text{var}\,(X)\,\text{var}\,(Y)} \leq 1. \qquad (9.25)$$

Moreover, since cov (X, Y) can be either positive or negative, we also have

$$-1 \leq \frac{\text{cov}\,(X, Y)}{[\text{var}\,(X)\,\text{var}\,(Y)]^{1/2}} \leq 1. \qquad (9.26)$$

Expression (9.26) is one of the most frequently encountered relationships in statistics and is true for *any* joint distribution of two random variables so long as both variables possess finite variances. The quantity

$$\frac{\text{cov}\,(X, Y)}{[\text{var}\,(X)\,\text{var}\,(Y)]^{1/2}} = \frac{\sigma_{xy}}{\sigma_x \sigma_y} \qquad (9.27)$$

is called the *correlation coefficient* between X and Y and is often denoted by the symbol ρ. A correlation of 1 indicates a perfect positive linear association, while -1 indicates perfect negative linear association. We see that if X and Y are in standard form (that is, if each has mean 0 and variance 1), the covariance and correlation are identical.

9.2 INDEPENDENCE OF RANDOM VARIABLES

In Chapter 3, it was stated that two events A and B are independent of one another if $P(AB)$ is equal to $P(A) \cdot P(B)$. We now extend the definition of independence to random variables and probability distributions.

Given a joint distribution in the random variables X and Y, $f(x, y)$, we say that X and Y are *independent* if and only if the following is true:

$$f(x, y) = f_x(x) \cdot f_y(y) \tag{9.28}$$

—that is, if and only if the joint density function can be factored into two density functions, the first involving only x and the second involving only y.

If X and Y are independent, then

$$\begin{aligned} E(XY) &= \int_x \int_y xyf(x, y)\, dy\, dx \\ &= \int_x xf_x(x)\, dx \cdot \int_y yf_y(y)\, dy \\ &= \mu_x \cdot \mu_y . \end{aligned} \tag{9.29}$$

Moreover,

$$\begin{aligned} \operatorname{cov}(X, Y) &= \int_x \int_y (x - \mu_x)(y - \mu_y) f(x, y)\, dy\, dx \\ &= \int_x (x - \mu_x) f_x(x)\, dx \cdot \int_y (y - \mu_y) f_y(y)\, dy \\ &= 0. \end{aligned} \tag{9.30}$$

Thus, we see that if two jointly distributed random variables are distributed independently, then the mean of their products is equal to the product of their means and their covariance is zero. From formula (9.27), it is obvious that a zero covariance implies a zero correlation, hence independent random variables are uncorrelated. However, as Figure 9.1 illustrates,

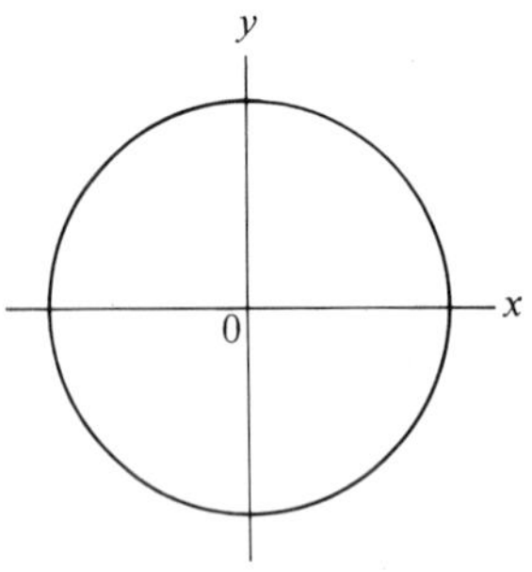

FIGURE 9.1 x and y uncorrelated and dependent.

variables that are uncorrelated (as X and Y in this figure are) are not necessarily independent.

Also, we should note that it follows from Equations (9.10) and (9.11) that $f_x(x)$ and $f_y(y)$ in Equation (9.28) are the marginal density functions for X and Y, respectively.

We shall illustrate the foregoing with an experiment with which we are very familiar, the rolling of two dice. Assuming that the dice are fair, we know from the classical definition of probability that the joint density function is equal to

$$f(x, y) = \begin{cases} \dfrac{1}{36}, & x, y = 1, 2, \ldots, 6 \\ 0, & \text{otherwise.} \end{cases} \tag{9.31}$$

By summing over the 6 values of y for each value of x, the marginal distribution of X, $f_x(x)$, is seen to be

$$f_x(x) = \frac{1}{6}, \qquad x = 1, 2, \ldots, 6, \tag{9.32}$$

and similarly for the marginal distribution of Y:

$$f_y(y) = \frac{1}{6}, \qquad x = 1, 2, \ldots, 6. \tag{9.33}$$

The mean of X is then

$$\mu_x = \sum_{x=1}^{6} x f_x(x)$$

$$= \frac{(1 + 2 + 3 + 4 + 5 + 6)}{6} = 3.5, \tag{9.34}$$

which is also the mean of Y, while the variance is

$$\sigma_x{}^2 = \sum_{x=1}^{6} (x - \mu_x)^2 f_x(x)$$

$$= \frac{[(1 - 3.5)^2 + (2 - 3.5)^2 + (3 - 3.5)^2 + (4 - 3.5)^2 + (5 - 3.5)^2 + (6 - 3.5)^2]}{6}$$

$$= \frac{17.50}{6}. \tag{9.35}$$

and similarly for the variance of Y. For the covariance of X and Y, formula (9.22) yields

$$\begin{aligned}\operatorname{cov}(X, Y) &= \sum_{x=1}^{6} \sum_{y=1}^{6} (x - \mu_x)(y - \mu_y) f(x, y) \\ &= [2.5(2.5 + 1.5 + 0.5 - 0.5 - 1.5 - 2.5) \\ &\quad + 1.5(2.5 + 1.5 + 0.5 - 0.5 - 1.5 - 2.5) \\ &\quad + 0.5(2.5 + 1.5 + 0.5 - 0.5 - 1.5 - 2.5) \\ &\quad - 0.5(2.5 + 1.5 + 0.5 - 0.5 - 1.5 - 2.5) \\ &\quad - 1.5(2.5 + 1.5 + 0.5 - 0.5 - 1.5 - 2.5) \\ &\quad - 2.5(2.5 + 1.5 + 0.5 - 0.5 - 1.5 - 2.5)]/36 \\ &= 0, \end{aligned} \tag{9.36}$$

which confirms what we already knew beforehand, namely, that the outcomes of the two dice are uncorrelated. That they are also independent is confirmed by the fact that the joint density function is equal to the product of the two marginal densities, that is,

$$f(x, y) = \frac{1}{36} = \frac{1}{6} \cdot \frac{1}{6} = f_x(x) \cdot f_y(y). \tag{9.37}$$

9.3 CONDITIONAL DISTRIBUTIONS

As with the notion of independence, we now extend the notion of conditional probability to probability distributions. If (X, Y) is a two-dimensional random variable with density function $f(x, y)$, then the *conditional* distribution of X given y is defined as the ratio of the joint density function of (X, Y) to the marginal density of Y—that is,

$$f(x \mid y) = \frac{f(x, y)}{f_y(y)} \qquad \text{for } f_y(y) \neq 0. \tag{9.38}$$

And similarly the conditional distribution of Y is

$$f(y \mid x) = \frac{f(x, y)}{f_x(x)} \qquad \text{for } f_x(x) \neq 0. \tag{9.39}$$

By multiplying (9.38) through by $f_y(y)$, we obtain a formula that is of frequent use in statistics, namely,

$$f(x, y) = f(x \mid y) \cdot f_y(y), \tag{9.40}$$

which says that the joint density is equal to the product of the conditional density of X and the marginal density of Y. From (9.39), it is evident that a similar formula obtains with the roles of X and Y interchanged.

If X and Y are independent, then, since the joint density is equal to the product of the two marginal densities, it follows that

$$\begin{aligned} f(x \mid y) &= \frac{f_x(x) \cdot f_y(y)}{f_y(y)} \\ &= f_x(x). \end{aligned} \tag{9.41}$$

In other words, the conditional and marginal distributions are identical when X and Y are independent.

Conditional Expectation

We now define conditional expectation. Let (X, Y) be a two-dimensional random variable with density function $f(x, y)$ and let $f(x|y)$ be the conditional distribution of X given y. Then the *conditional expectation* of X given y is

$$\begin{aligned} E(X \mid y) &= \int_x x f(x \mid y)\, dx \\ &= \int_x x \frac{f(x, y)}{f_y(y)}\, dx \qquad \text{for continuous } f(x, y) \\ &= \sum_x x f(x \mid y) \\ &= \sum_x x \frac{f(x, y)}{f_y(y)} \qquad \text{for discrete } f(x, y). \end{aligned} \tag{9.42}$$

Similar formulas are obtained, for the conditional expectation of Y given x. It should be noted that $E(X \mid y)$ is *not* a random variable since it is a function of a *particular* value of Y, namely, $Y = y$.

Formula (9.42) can be generalized to the conditional expectation of functions of X; that is, if $H = g(x)$, then

$$E(H \mid y) = \int_x g(x) f(x \mid y)\, dx. \tag{9.43}$$

In particular, the conditional variance of X is defined as follows:

$$E[X - E(X \mid y)]^2 = \int_x [x - E(X \mid y)]^2 f(x \mid y)\, dx. \tag{9.44}$$

Finally, if we take the expected value of the conditional expectation of X given y, we end up with the *unconditional* mean of X, that is,

$$\mu_x = E_y[E_x(X \mid Y)] = \int_y \int_x x f(x \mid y) f_y(y)\, dx\, dy, \tag{9.45}$$

where the symbols in the middle term mean that we first take the expectation with respect to X for given Y and then take the expectation with respect to Y.

9.4 THE BIVARIATE NORMAL DISTRIBUTION

Of all the distributions in two variables, the bivariate normal is the most frequently encountered. Its density function is defined as

$$f(x, y) = \frac{1}{2\pi\sigma_x \sigma_y(1-p^2)^{1/2}} \exp\left\{-\frac{1}{2}(1-\rho^2)^{-1}\left\{\left[\frac{(x-\mu_x)}{\sigma_x}\right]^2 - 2\rho(x-\mu_x)\frac{(y-\mu_y)}{\sigma_x\sigma_y} + \left[\frac{(y-\mu_y)}{\sigma_y}\right]^2\right\}\right., \tag{9.46}$$

where ρ is the correlation coefficient between X and Y. The density in (9.46) is a bell-shaped surface in three-dimensional space, and has the following properties:

1. A plane parallel to the xy-plane and intersecting $f(x, y)$ intersects $f(x, y)$ in an ellipse.
2. A plane perpendicular to the xy-plane will intersect $f(x, y)$ in a curve of the normal density form.

Two special cases of the second property are the marginal distributions of X and Y, and we shall now demonstrate that these two distributions are normal. Beginning with the definition

$$f_x(x) = \int_y f(x, y)\, dy, \tag{9.47}$$

we first make the substitution $v = (y - \mu_y)/\sigma_y$ and $dv = dy/\sigma_y$. The exponent in (9.46) then becomes

$$-\frac{1}{2(1-\rho^2)}\left[\left(\frac{x-\mu_x}{\sigma_x}\right)^2 - 2\rho v\left(\frac{x-\mu_x}{\sigma_x}\right) + v^2\right]. \tag{9.48}$$

Next, we complete the square in v, which transforms (9.48) to

$$\begin{aligned}
-\frac{1}{2(1-\rho^2)}&\left[\left(\frac{x-\mu_x}{\sigma_x}\right)^2 + v^2 - 2\rho v\left(\frac{x-\mu_x}{\sigma_x}\right)\right.\\
&\left. + \rho^2\left(\frac{x-\mu_x}{\sigma_x}\right)^2 - \rho^2\left(\frac{x-\mu_x}{\sigma_x}\right)^2\right]\\
&= -\frac{1}{2(1-\rho^2)}\left[(1-\rho^2)\left(\frac{x-\mu_x}{\sigma_x}\right)^2 + \left(v - \rho\cdot\frac{x-\mu_x}{\sigma_x}\right)^2\right]\\
&= -\frac{1}{2}\left(\frac{x-\mu_x}{\sigma_x}\right)^2 - \frac{1}{2(1-\rho^2)}\left(v-\rho\cdot\frac{x-\mu_x}{\sigma_x}\right)^2.
\end{aligned} \tag{9.49}$$

Therefore, $f_x(x)$ becomes

$$f_x(x) = \int_v \frac{1}{2\pi\sigma_x(1-\rho^2)^{1/2}} \exp\left\{-\frac{1}{2}\left[\frac{(x-\mu_x)}{\sigma_x}\right]^2 - \frac{1}{2(1-\rho^2)} \cdot \left[v - \frac{\rho(X-\mu_x)}{\sigma_x}\right]^2\right\} dv. \tag{9.50}$$

We now make the substitution

$$\omega = \frac{v - \rho[(x-\mu_x)/\sigma_x]}{(1-\rho^2)^{1/2}}, \qquad d\omega = \frac{dv}{(1-\rho^2)^{1/2}},$$

which yields

$$\begin{aligned} f_x(x) &= \frac{1}{2\pi\sigma_x} \int_\omega \exp\left\{-\frac{1}{2}\left[\frac{(x-\mu_x)}{\sigma_x}\right]^2 - \frac{1}{2}\omega^2\right\} d\omega \\ &= \frac{1}{\sigma_x\sqrt{2\pi}} \exp\left\{-\frac{1}{2}\left[\frac{(x-\mu_x)}{\sigma_x}\right]^2\right\} \int_\omega \frac{1}{\sqrt{2\pi}} \exp\left(-\frac{1}{2}\omega^2\right) d\omega, \end{aligned} \tag{9.51}$$

which, since the integral on the right is equal to the area under the density function for a standard normal variable,[1] becomes

$$f_x(x) = \frac{1}{\sqrt{2\pi}\sigma_x} \exp\left\{-\frac{1}{2}\left[\frac{(x-\mu_x)}{\sigma_x}\right]^2\right\}. \tag{9.52}$$

However, this is the density function for a normal variable with mean μ_x and variance σ_x^2. A similar argument will yield the marginal distribution for Y:

$$f_y(y) = \frac{1}{\sqrt{2\pi}\sigma_y} \exp\left\{-\frac{1}{2}\left[\frac{(y-\mu_y)}{\sigma_y}\right]^2\right\}. \tag{9.53}$$

Hence we see that the marginal distributions for the bivariate normal distribution are also normal.

From our previous knowledge of the normal distribution for a single random variable, it is immediately evident from the two marginal distributions that the mean of X is μ_x, its variance σ_x^2, and the mean of Y is μ_y and its variance σ_y^2.

[1]

$$E(\omega) = \frac{E(v) - E\{\rho[(x-\mu_x)/\sigma_x]\}}{(1-\rho^2)^{1/2}} = 0.$$

$$\text{var } \omega = \frac{1}{1-\rho^2}\left\{\text{var } v - \rho^2 \text{ var}\left[\frac{(x-\mu_x)}{\sigma_x}\right]\right\} = \frac{1}{1-\rho^2}[1-\rho^2] = 1.$$

If ρ is zero—that is, if X and Y are uncorrelated—the joint density function (9.46) collapses to

$$f(x, y) = \frac{1}{2\pi\sigma_x\sigma_y} \exp\left\{-\frac{1}{2}\left\{\left[\frac{(x-\mu_x)}{\sigma_x}\right]^2 + \left[\frac{(y-\mu_y)}{\sigma_y}\right]^2\right\}\right\}, \tag{9.54}$$

which factors into the two marginal densities:

$$f(x, y) = \frac{1}{\sqrt{2\pi}\sigma_x} \exp\left\{-\frac{1}{2}\left[\frac{(x-\mu_x)}{\sigma_x}\right]^2\right\} \cdot \frac{1}{\sqrt{2\pi}\sigma_y} \exp\left\{-\frac{1}{2}\left[\frac{(y-\mu_y)}{\sigma_y}\right]^2\right\}. \tag{9.55}$$

In view of (9.28), we find that, in this case, uncorrelatedness implies independence.

Having the marginal distributions, it is an easy step to obtain the conditional distributions. For the conditional distribution of X, we have

$$f(x \mid y) = \frac{f(x, y)}{f_x(x)}, \tag{9.56}$$

which after some manipulation can be reduced to the form

$$f(x \mid y) = \frac{1}{[2\pi\sigma_x^2(1-\rho^2)]^{1/2}} \exp\left\{-\frac{\frac{1}{2}}{\sigma_x^2(1-\rho^2)}\left\{x - \mu_x - \rho\frac{\sigma_x}{\sigma_y}(y-\mu_y)\right\}^2\right\}. \tag{9.57}$$

The marginal distribution of X is thus seen to be normal with mean $\mu_x + \rho(\sigma_x/\sigma_y)(y - \mu_y)$ and variance $\sigma_x^2(1 - \rho^2)$.

A similar expression is obtained for the conditional distribution of Y.

The Regression Function

Equation (9.57) shows that the mean value of the conditional distribution of X is equal to

$$E(X \mid y) = \mu_x + \rho\frac{\sigma_x}{\sigma_y}(y - \mu_y). \tag{9.58}$$

The quantity $E(X \mid y)$ is called the *regression function of X on y*, and, in the case of the bivariate normal distribution; Equation (9.58) shows the regression function to be linear. This is a rather remarkable result, and provides the theoretical foundation for much of statistical inference in scientific research.

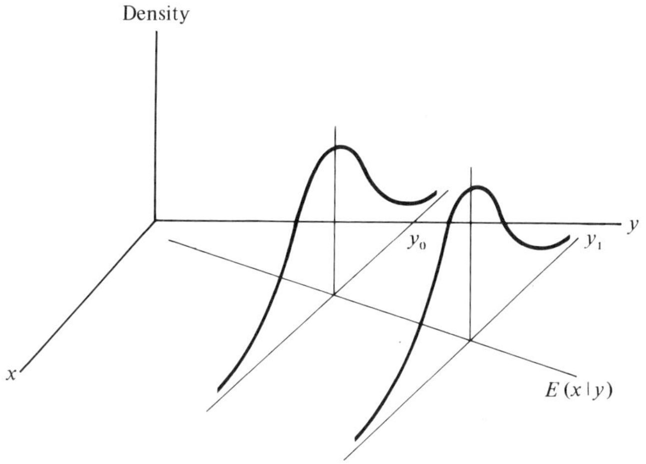

FIGURE 9.2 Regression function of x on y.

Figure 9.2 shows the regression function of X on y for a particular set of parameters, μ_x, μ_y, σ_x, σ_y, and ρ. It also shows the conditional density function of X for two values of y, y_0, and y_1.

9.5 MULTIVARIATE DISTRIBUTIONS

Let us now extend the foregoing to random variables in n dimensions. Let $\mathbf{X}' = (X_1, X_2, \ldots, X_n)$ be an n-dimensional (row) vector representing a point in an n-dimensional sample space,[2] and consider the event, assuming the X_i's to take on discrete values only, that $X_1 = x_1, X_2 = x_2, \ldots, X_n = x_n$ simultaneously. Let this probability be given by $f(x_1, x_2, \ldots, x_n)$,

$$P(X_1 = x_1, X_2 = x_2, \ldots, X_n = x_n) = f(x_1, x_2, \ldots, x_n). \qquad (9.59)$$

The function $f(x_1, x_2, \ldots, x_n)$ defines the joint density function for the event $(X_1 = x_1, X_2 = x_2, \ldots, X_n = x_n)$. If, for example, X_i represents the face showing of the ith die of the simultaneous rolling of n identical (and fair) dice, then the sample space in question will consist of 6^n points (consisting of the 6^n possible outcomes arising from the roll of n dice—or, alternatively, of the rolling of the same die n times), and

$$f(x_1, x_2, \ldots, x_n) = \frac{1}{6^n}. \qquad (9.60)$$

[2] The convention to be followed is that a vector is a column vector unless primed. Also vectors in this section will be represented in boldface, italic if random.

The vector $X' = (X_1, X_2, \ldots, X_n)$ is referred to as an n-dimensional random variable (or vector). When the X_i's can take on a continuum of values, our interest is in the event $(x_1 < X_1 < x_1 + dx_1,\ x_2 < X_2 < x_2 + dx_2, \ldots, x_n < X_n < x_n + dx_n)$, and its probability will be given by

$$P(x_1 < X_1 < x_1 + dx_1, \ldots, x_n < X_n < x_n + dx_n) = f(x_1, x_2, \ldots, x_n)\, dx_1\, dx_2 \cdots dx_n. \qquad (9.61)$$

For $f(x_1, \ldots, x_n)$ to be a legitimate density, we must of course have

$$\int_{x_1} \cdots \int_{x_n} f(x_1, \ldots, x_n)\, dx_1 \cdots dx_n = 1. \qquad (9.62)$$

The cumulative distribution function is given by

$$F(x_1, x_2, \ldots, x_n) = \int_{-\infty}^{x_1} \int_{-\infty}^{x_2} \cdots \int_{-\infty}^{x_n} f(z_1, z_2, \ldots, z_n)\, dz_1\, dz_2 \cdots dz_n \qquad (9.63)$$

for continuous X_i and by

$$F(x_1, x_2, \ldots, x_n) = \sum_{z_1 \le x_1} \sum_{z_2 \le x_2} \cdots \sum_{z_n \le x_n} f(z_1, z_2, \ldots, z_n) \qquad (9.64)$$

for discrete X_i. The relationship between F and f is, of course,

$$f(x_1, x_2, \ldots, x_n) = \frac{\partial^n F(x_1, x_2, \ldots, x_n)}{\partial x_1\, \partial x_2 \cdots \partial x_n} \qquad (9.65)$$

at the continuity points of F.

Marginal and conditional distributions are defined as for the bivariate case, only there are now many more to contend with. The marginal distribution for X_1, for example, is obtained by integrating (or summing) out $x_2, x_3, \ldots, x_n$:

$$f_{x_1}(x_1) = \int_{x_2} \int_{x_3} \cdots \int_{x_n} f(x_1, x_2, \ldots, x_n)\, dx_2 \cdots dx_n. \qquad (9.66)$$

However, we may on some occasions be interested in the joint distribution of just a subset of variables of $\mathbf{X}$, say $X_1, X_2, \ldots, X_k$, $k < n$, obtained by integrating (or summing) out $x_{k+1}, \ldots, x_n$:

$$f_{x_1 x_2 \cdots x_k}(x_1, x_2, \ldots, x_k) = \int_{x_{k+1}} \cdots \int_{x_n} f(x_1, x_2, \ldots, x_n)\, dx_{k+1} \cdots dx_n. \qquad (9.67)$$

Conditional distributions are defined analogously. For example, the conditional distribution of X_1 given $X_2, \ldots, X_n, f(x_1 \mid x_2, x_3, \ldots, x_n)$, is defined as

$$f(x_1 \mid x_2, x_3, \ldots, x_n) = \frac{f(x_1, x_2, \ldots, x_n)}{f_{x_2 x_3 \cdots x_n}(x_2, \ldots, x_n)}. \tag{9.68}$$

Similarly, for the conditional distribution of $x_1, \ldots, x_k$ given $x_{k+1}, \ldots, x_n$, $f(x_1, \ldots, x_k \mid x_{k+1}, \ldots, x_n)$,

$$f(x_1, \ldots, x_k \mid x_{k+1}, \ldots, x_n) = \frac{f(x_1, \ldots, x_n)}{f_{x_{k+1} \cdots x_n}(x_{k+1}, \ldots, x_n)}. \tag{9.69}$$

Finally, if we are interested in the conditional distribution of x_1 given $x_2, f(x_1 \mid x_2)$, we would have

$$f(x_1 \mid x_2) = \frac{f_{x_1 x_2}(x_1, x_2)}{f_{x_2}(x_2)}. \tag{9.70}$$

With regard to means, variances, and covariances, we will have

$$\begin{aligned} E(X_i) &= \int_{x_i} x_i f_{x_i}(x_i)\, dx \qquad (9.71)\\ &= \int_{x_1} \cdots \int_{x_n} x_i f(x_1, \ldots, x_n)\, dx_1 \cdots dx_n. \end{aligned}$$

$$E[(X_i - \mu_i)^2] = \int_{x_i} (x_i - \mu_i)^2 f_{x_i}(x_i)\, dx_i, \tag{9.72}$$

and for the covariance of X_i and X_j,

$$\begin{aligned} &E[(X_i - \mu_i)(X_j - \mu_j)] \\ &\quad = \int_{x_i} \int_{x_j} (x_i - \mu_i)(x_j - \mu_j) f_{x_i}(x_i) f_{x_j}(x_j)\, dx_i\, dx_j. \end{aligned} \tag{9.73}$$

For later reference, it will be useful to combine (9.72) and (9.73) for all i and j into a single expression, the *covariance matrix* of X,

$$E[(X - \mu)(X - \mu)'] = \{E[(X_i - \mu_i)(X_j - \mu_j)\}, \qquad i, j = 1, n. \tag{9.74}$$

In this expression, $(X - \mu)$ is a column vector,

$$(X - \mu) = \begin{pmatrix} X_1 - \mu_1 \\ X_2 - \mu_2 \\ \vdots \\ X_n - \mu_n \end{pmatrix} \tag{9.75}$$

and $(X - \mu)'$ is its transpose. The covariance matrix is thus $n \times n$ in dimension consisting of variances on the diagonal and covariances on the off-diagonal. Since $E[(X_i - \mu_i)(X_j - \mu_j)] = E[(X_j - \mu_j)(X_i - \mu_i)]$, the covariance matrix is, of course, symmetric.

Finally, we will note that the n variables X_i will be distributed independently if their joint density function can be factored into a product of n functions,

$$f(x_1, x_2, \ldots, x_n) = f_1(x_1) \cdot f_2(x_2) \cdots f_n(x_n), \tag{9.76}$$

where $f_1(x_1)$ is the marginal density function of X_1, and so on.

9.6 MULTIVARIATE MOMENT-GENERATING FUNCTIONS

As with univariate distributions, the moment-generating function often provides the most convenient method for obtaining the moments of a multivariate distribution. By definition, the moment-generating function for a distribution of an n-dimensional random variable is given by

$$E(e^{t_1X_1+t_2X_2+\cdots+t_nX_n}) = \int_{x_1}\int_{x_2}\cdots\int_{x_n} e^{t_1x_1+t_2x_2+\cdots+t_nx_n} \cdot f(x_1, x_2, \ldots, x_n)\, dx_1\, dx_2 \cdots dx_n, \tag{9.77}$$

assuming that the n-fold integral on the right exists in a neighborhood of $t'[=(t_1, t_2, \ldots, t_n)] = 0$. In order to convince the reader that this expression does in fact provide a moment-generating function in the sense that this term has been used previously, let us proceed by assuming $n = 2$. In this case,

$$E(e^{t_1X_1+t_2X_2}) = \int_{x_1}\int_{x_2} e^{t_1x_1+t_2x_2} f(x_1, x_2)\, dx_1\, dx_2. \tag{9.78}$$

Expanding $e^{t_1x_1}$ and $e^{t_2x_2}$ in their power series, this expression becomes

$$E(e^{t_1X_1+t_2X_2}) = \int_{x_1}\int_{x_2} [1 + t_1x_1 + \tfrac{1}{2}(t_1x_1)^2 + \cdots][1 + t_2x_2 + \tfrac{1}{2}(t_2x_2)^2 + \cdots] f(x_1, x_2)\, dx_1\, dx_2. \tag{9.79}$$

Differentiation of this expression with respect to t_1 and evaluation at $t_1, t_2 = 0$ then yields

$$\left.\frac{\partial E(e^{t_1X_1+t_2X_2})}{\partial t_1}\right|_{t_1,t_2=0} = \int_{x_1}\int_{x_1} x_1 f(x_1, x_2)\, dx_1\, dx_2 = E(X_1). \tag{9.80}$$

Differentiation with respect to t_1 twice and again evaluation at t_1, $t_2 = 0$ yields

$$\left.\frac{\partial^2 E(e^{t_1X_1+t_1X_1})}{\partial t_1^2}\right|_{t_1,t_2=0} = \int_{x_1}\int_{x_2} x_1{}^2 f(x_1, x_2)\, dx_1\, dx_2$$
$$= E(X_1^2), \tag{9.81}$$

and so on. Since similar formulas are obviously obtained for X_2, it is clear that (9.78) does in fact provide a generating function for the moments of $\mathbf{X}$ and, consequently, that expression (9.77) is valid as a moment-generating function for the general case.

The second-order cross moments of X_1 and X_2 are obtained by differentiation first with respect to t_1, then with respect to t_2, and subsequent evaluation at t_1, $t_2 = 0$:

$$\left.\frac{\partial^2 E(e^{t_1X_1+t_2X_2})}{\partial t_1 \partial t_2}\right|_{t_1,t_2=0} = \int_{x_1}\int_{x_2} x_1 x_2 f(x_1, x_2)\, dx_1\, dx_2$$
$$= E(X_1X_2). \tag{9.82}$$

Since

$$E[(X_1 - \mu_1)(X_2 - \mu_2)] = E(X_1X_2) - \mu_1 E(X_2) - E(X_1)\mu_2 + \mu_1\mu_2$$
$$= E(X_1X_2) - \mu_1\mu_2, \tag{9.83}$$

the covariance of X_1 and X_2 is obtained simply by subtraction of $E(X_1)E(X_2)$ from $E(X_1X_2)$.

In general, therefore, with reference to (9.77), it is clear that

$$E(X_iX_j) = \left.\frac{\partial^2 E(e^{t_1X_1+\cdots+t_nX_n})}{\partial t_i \partial t_j}\right|_{t=0} \tag{9.84}$$

and, moreover, that

$$E[(X_i - \mu_i)(X_j - \mu_j)] = E(X_iX_j) - E(X_i)E(X_j). \tag{9.85}$$

Before leaving this section, we shall prove an important theorem which states that if the moment-generating function (assuming that it exists) for the joint distribution of n random variables $X_1, \ldots, X_n$ can be factored into the moment-generating functions for the marginal distributions of the X_i, then the X_i are distributed independently. We shall prove the theorem for the case of continuous X and $n = 2$.

THEOREM 9.1
Let X_1 and X_2 represent random variables that have the joint density function $f(x_1, x_2)$ and marginal density functions $f_1(x_1)$ and $f_2(x_2)$, respectively. Let $M(t_1, t_2)$ denote the moment-generating function for $f(x_1, x_2)$. Then X_1 and X_2 are distributed independently if and only if

$$M(t_1, t_2) = M(t_1)M(t_2), \tag{9.86}$$

where $M(t_1)$ is the moment-generating function for $f_1(x_1)$, and similarly for $M(t_2)$.

PROOF. For sufficiency: If X_1 and X_2 are independent, then

$$\begin{aligned} M(t_1, t_2) &= \int_{x_1}\int_{x_2} e^{t_1x_1+t_2x_2} f(x_1, x_2)\, dx_1\, dx_2 \\ &= \int_{x_1} e^{t_1x_1} f_1(x_1)\, dx_1 \int_{x_2} e^{t_2x_2} f_2(x_2)\, dx_2 \\ &= M(t_1)M(t_2). \end{aligned} \tag{9.87}$$

For necessity: By definition,

$$M(t_1) = \int_{x_1} e^{t_1x_1} f_1(x_1)\, dx_1, \tag{9.88}$$

$$M(t_2) = \int_{x_2} e^{t_2x_2} f_2(x_2)\, dx_2. \tag{9.89}$$

Then

$$\begin{aligned} M(t_1)M(t_2) &= \int_{x_1} e^{t_1x_1} f_1(x_1)\, dx_1 \int_{x_2} e^{t_2x_2} f_2(x_2)\, dx_2 \\ &= \int_{x_1}\int_{x_2} e^{t_1x_1+t_2x_2} f_1(x_1) f_2(x_2)\, dx_1\, dx_2 \\ &= M(t_1, t_2) \end{aligned} \tag{9.90}$$

by hypothesis. However, by definition,

$$M(t_1, t_2) = \int_{x_1}\int_{x_2} e^{t_1x_1+t_2x_2} f(x_1, x_2)\, dx_1\, dx_2, \tag{9.91}$$

whence, since the moment-generating function is unique,

$$f(x_1, x_2) = f(x_1)f(x_2), \tag{9.92}$$

as was to be shown.

9.7 MULTINOMIAL DISTRIBUTION

The first specific multivariate distribution that we shall discuss is the multinomial distribution, which is a straightforward generalization of the binomial distribution. With the binomial distribution, the mutually exclusive and exhaustive outcomes were termed success and failure with probabilities p and q $(p + q = 1)$, respectively. Let us now assume that the number of possible outcomes is increased to m; designate the first outcome by E_1, the second by E_2, and so on, and assume that these occur with probabilities $p_1, p_2, \ldots, p_m$, respectively, where $p_1 + p_2 + \cdots + p_m = 1$.

As in the binomial, let n independent trials be made, and designate $X_1, X_2, \ldots, X_m$ as random variables, where x_1 refers to the number of occurrences of E_1, x_2 the number of occurrences of E_2, and so on. Obviously, $x_1, x_2, \ldots, x_n$ are nonnegative integers subject to the condition $x_1 + x_2 + \cdots + x_m = n$. Our task is to determine the joint density function $f(x_1, x_2, \ldots, x_m)$.

Now, the first x_1 trials could yield E_1, the second x_2 trials E_2, and so forth. The probability of this *ordered* result is ${p_1}^{x_1}{p_2}^{x_2} \cdots {p_m}^{x_m}$. However, x_1 alike events E_1, x_2 alike events E_2, ..., and x_m alike events E_m can be obtained in

$$\frac{n!}{x_1!\,x_2! \cdots x_m!}$$

mutually exclusive ways.[3] Moreover, it is to be noted that each of these ways is equally likely. Consequently, the joint density we seek will be given by

$$f(x_1, x_2, \ldots, x_n) = \frac{n!}{x_1!\, x_2! \cdots x_m!}\, {p_1}^{x_1}{p_2}^{x_2} \cdots {p_m}^{x_m}, \tag{9.93}$$

for $x_1, x_2, \ldots, x_m$ nonnegative integers whose sum is n. This is the *multinomial* density function, and depends on m continuous parameters $p_1, p_2, \ldots, p_m$ subject to the constraints that they sum to one, one discrete parameter n, and m random variables $X_1, X_2, \ldots, X_m$. The name of the distribution arises from the fact that (9.93) represents the general term in the

[3] See Section 4.2.

expansion of the multinomial $(p_1 + p_2 + \cdots + p_m)^n$. This being the case, it follows immediately that[4]

$$\sum \frac{n!}{x_1!\, x_2! \ldots x_m!} p_1^{x_1} p_2^{x_2} \cdots p_m^{x_m} = (p_1 + p_2 + \cdots + p_m)^n$$

$$= 1, \tag{9.94}$$

where the sum is to be understood as being over all possible values of $x_1, x_2, \ldots, x_m$ which satisfy $x_1 + x_2 + \cdots + x_m = n$.

As an example of the multinomial distribution, suppose that a true die is to be rolled a total of 10 times. There will thus be six events in all, E_1 the event of a 1 showing, E_2 the event of a 2 showing, and so on, with probabilities $p_1 = p_2 \cdots = p_6 = \frac{1}{6}$. The random variable X_1 will refer to the number of 1's which occur in the 10 trials, X_2 to the number of 2's which occur, and so on, with $x_1 + x_2 + \cdots + x_6 = 10$.

The moments of the multinomial distribution are easily obtained through the moment-generating function. For the latter, we have in view of expression (9.77)

$$M(t_1, t_2, \ldots, t_m) = \sum \frac{n!}{x_1!\, x_2! \cdots x_m!} e^{t_1 x_1 + \cdots + t_m x_m} p_1^{x_1} \cdots p_m^{x_m}$$

$$= \sum \frac{n!}{x_1!\, x_2! \ldots x_m!} (p_1 e^{t_1})^{x_1} \cdots (p_m e^{t_m})^{x_m}$$

$$= (p_1 e^{t_1} + \cdots + p_m e^{t_m})^n. \tag{9.95}$$

Once again, it is to be understood that the sum is taken over all possible values of $x_1, x_2, \ldots, x_m$ subject to $x_1 + x_2 + \cdots + x_m = n$.

Differentiation of (9.95) with respect to t_i and then evaluation at $t_1, t_2, \ldots, t_m = 0$ yields for the mean of X_i,

$$E(X_i) = np_i. \tag{9.96}$$

Differentiation of (9.95) twice with respect to t_i, evaluation at $t = 0$, and then subtraction of $[E(X_i)]^2$ gives for the variance of X_i,

$$\sigma^2(X_i) = np_i(1 - p_i). \tag{9.97}$$

Finally, the covariance of X_i with X_j is obtained by differentiating of (9.95) first with respect to t_i, then with respect to t_j, evaluation at $t = 0$, and then subtraction of $E(X_i)E(X_j)$. The result is

$$E[(X_i - \mu_i)(X_j - \mu_j)] = -np_i p_j. \tag{9.98}$$

[4] See Theorem A.2.2 of Appendix 2.

The negative sign in this expression thus indicates that X_i and X_j are negatively correlated.

REFERENCES

Freeman, H., *An Introduction to Statistical Inference*, Addison-Wesley, 1963, chap. 14, 17.

Hogg, R. V. and Craig, A. T., *Introduction to Mathematical Statistics*, 3rd ed., Macmillan, 1970, chap. 2.

Mood, A. M. and Graybill, F. A., *Introduction to the Theory of Statistics*, 2nd ed., McGraw-Hill, 1963, chap. 4 .

CHAPTER 10

CHANGE OF VARIABLES

We interrupt our study of multivariate distributions in order to present a topic, change of variables, which is of crucial importance to the sequel. In particular, a change of variables is instrumental in obtaining the closed form of several distributions associated with the normal distribution and, we have already seen in the case of univariate distributions, a change of variable frequently leads to simplification in the evaluation of certain integrals.

The theory underlying a change of variable is very straightforward for univariate distributions, but is quite complicated for multivariate distributions. Indeed, the theorem on which the multivariate change of variables rests is frequently shied away from even in texts on advanced calculus. Our procedure in this chapter will be to demonstrate the change of variable in detail for one and two variables; proof for the general case of n variables, however, is omitted.

10.1 CHANGE OF VARIABLE FOR A SINGLE RANDOM VARIABLE

Discrete Random Variables

A change of variable for a discrete random variable is straightforward, and we shall illustrate what is involved with the binomial distribution

$$f(x) = \binom{n}{x} p^x q^{n-x}, \qquad x = 0, 1, \ldots, n. \tag{10.1}$$

Consider the change of variable defined by the transformation $y = x^2$. Since the inverse function $x = \pm\sqrt{y}$ is not single-valued, we shall only consider the positive square root, which then restricts the relationship between x and y to be one-to-one.

We are interested in the density function of y. Because there is a one-to-one relationship between the occurrence of x and the occurrence of y, namely,

$$x\colon \quad 0, 1, 2, 3, 4, \ldots,$$

$$Y\colon \quad 0, 1, 4, 9, 16, \ldots,$$

it is evident that $g(y) = f(x)$ where $y = x^2$. However, because f is known and g is not, we must express $g(y)$ in terms of $f(x)$ with x replaced by y. Therefore,

$$g(y) = f(x) = f(\sqrt{y}) = \binom{n}{\sqrt{y}} p^{\sqrt{y}} q^{n-\sqrt{y}}, \qquad y = 0, 1, 4, 9, 16, \ldots, \tag{10.2}$$

and the problem is solved.

Continuous Random Variables

For continuous random variables, a one-to-one correspondence implies that $y = \phi(x)$ is either an increasing or a decreasing function of x over its entire domain. Either case implies that $y = \phi(x)$ has a unique inverse function $x = \psi(y)$, which itself is either increasing or decreasing over the entire domain of y. We shall consider first the case where the function $\phi(x)$ is increasing.

Let $f(x)$ be the density function of X and, as above, let $y = \phi(x)$ and $x = \psi(y)$. We seek the density function $g(y)$ of Y. From these definitions, it is evident that the events

$$c < Y < d \qquad \text{and} \qquad \psi(c) < X < \psi(d)$$

are equivalent in the sense that they must occur simultaneously for arbitrary c and d, $c < d$. For these equivalent events, we have

$$P(c < Y < d) = P[\psi(c) < X < \psi(d)], \tag{10.3}$$

or, for continuous X and Y,

$$\int_c^d g(y)\, dy = \int_{\psi(c)}^{\psi(d)} f(x)\, dx. \tag{10.4}$$

From (10.4), we can readily calculate the probabilities beneath the density $g(y)$, even if $g(y)$ is unknown, so long as $f(x)$ and its limits are known. For example, if

$$f(x) = e^{-x}, \qquad x \geq 0, \tag{10.5}$$

where $y = x^2$ with $x = \sqrt{y}$ as the unique inverse, we would have

$$\int_2^4 g(y)\, dy = \int_{\sqrt{2}}^2 e^{-x}\, dx = 0.2325 \tag{10.6}$$

without ever having to calculate the explicit form of $g(y)$.

To determine the form of $g(y)$, we proceed by expressing all x quantities in the right-hand integral of (10.4) in terms of their equivalent y quantities, namely,

$$f(x) = f[\psi(y)] \qquad \text{and} \qquad dx = \frac{d\psi(y)}{dy}\, dy, \tag{10.7}$$

so that (10.4) becomes

$$\int_c^d g(y)\, dy = \int_c^d f[\psi(y)] \frac{d\psi(y)}{dy}\, dy. \tag{10.8}$$

Hence, it follows that the density function of y is equal to

$$g(y) = f[\psi(y)] \frac{d\psi(y)}{dy}. \tag{10.9}$$

For the example given above,

$$f(x) = e^{-x}, \qquad x \geq 0, \qquad y = x^2, \tag{10.10}$$

we will have, since $d\sqrt{y}/dy = \frac{1}{2}y^{-1/2}$,

$$g(y) = \tfrac{1}{2} e^{-\sqrt{y}} y^{-1/2}. \tag{10.11}$$

As a second example, we will take one with a great deal of general importance. Let X be a continuous random variable with density function $f(x)$, and let the transformation from x to y be given by

$$y = \phi(x) = \int_{-\infty}^x f(z)\, dz, \qquad -\infty \leq x \leq \infty. \tag{10.12}$$

In other words, $\phi(x)$ is the distribution function of X. Note that for

$x = -\infty$, $y = 0$, while for $x = \infty$, $y = 1$, so that $0 \le y \le 1$. Then, differentiating the inverse function $\psi(y)$, we have

$$\begin{aligned}\frac{d\psi(y)}{dy} &= \frac{dx}{dy} \\ &= \frac{1}{dy/dx} \\ &= \frac{1}{f(x)} \\ &= \frac{1}{f[\psi(y)]}. \end{aligned} \tag{10.13}$$

Hence,

$$\begin{aligned} g(y) &= f[\psi(y)] \cdot \frac{d\psi(y)}{dy} \\ &= \frac{f[\psi(y)]}{f[\psi(y)]} \\ &= 1, \qquad 0 \le y \le 1. \end{aligned} \tag{10.14}$$

This result shows that (10.12) transfoms a continuous random variable X with *any* density function into the continuous random variable Y with *uniform* density function $g(y) = 1$ with $0 \le y \le 1$.

The implication of this is far reaching, for it means that any continuous distribution can be transformed into any other continuous distribution by first transforming the first distribution to the uniform distribution and then taking the inverse of the transformation that takes the second distribution into the uniform distribution.

If $y = \phi(x)$ is a decreasing function of x, then the events

$$c < Y < d \qquad \text{and} \qquad \psi(d) < X < \psi(c)$$

are equivalent for arbitrary c and d, $c < d$, so that

$$\int_c^d g(y)\, dy = \int_{\psi(d)}^{\psi(c)} f(x)\, dx. \tag{10.15}$$

As before, areas under $g(y)$ can be computed by evaluating the (known) right-hand side of (10.15), and, also as before, we can determine the form of $g(y)$ by transforming all x quantities in the right-hand integral to equivalent y quantities so that,

$$\int_c^d g(y)\, dy = \int_c^d -f[\psi(y)] \frac{d\psi(y)}{dy}\, dy. \tag{10.16}$$

The minus sign is needed since $d\psi(y)/dy$ is negative.

As an example, consider

$$f(x) = e^{-x}, \qquad x \geq 0,$$
$$y = -4x + 3. \tag{10.17}$$

We have the (unique) inverse $x = (3 - y)/4$, so that

$$g(y) = -\tfrac{1}{4}e^{-(3-y)/4}, \tag{10.18}$$

because $d\psi(y)/dy = -\frac{1}{4}$.

Because $d\psi(y)/dy > 0$, when y is an increasing function of x and $d\psi(y)/dy < 0$ when decreasing, the two results (10.7) and (10.8) can be combined to form

$$g(y) = f[\psi(y)]\left|\frac{d\psi(y)}{dy}\right| \tag{10.19}$$

for any monotonic continuous transformation, increasing or decreasing.

10.2 THE CAUCHY DISTRIBUTION

As a second illustration of a change of variable for a single random variable, let us consider the physical model illustrated in Figure 10.1. The practical problem is to estimate from the distribution of particles on the screen the location of the source of radioactive energy. On the assumption that the source is a known distance (say unity) from the screen, this problem is equivalent to estimating θ. Here, however, we are concerned with the form of the distribution of the particles and its moments and we shall assume that θ is known.

To begin with, let us assume that all angles of omission α are equally likely, which is to say that the density function of α is uniform over interval $-\pi/2$ to $\pi/2$, so that

$$f(\alpha) = \frac{1}{\pi}. \tag{10.20}$$

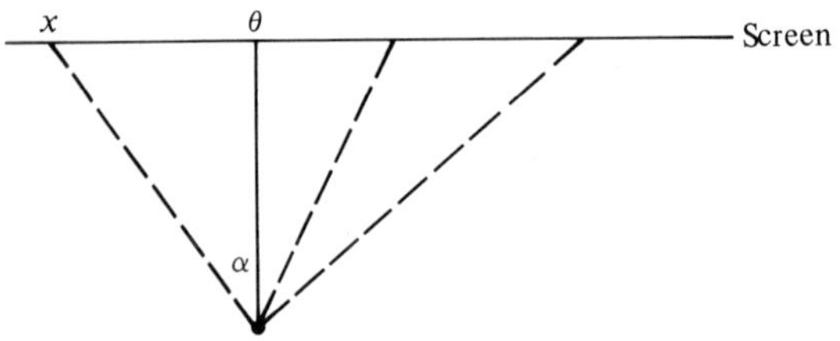

FIGURE 10.1 Radioactive energy source.

Suppose, however, that we are interested in the density function for X, the distance that the particle strikes the screen from θ. From elementary trigonometry, we have $\tan \alpha = x - \theta$, so that $dx = \sec^2 \alpha \, d\alpha$ and

$$\begin{aligned} d\alpha &= \frac{dx}{\sec^2 \alpha} \\ &= \frac{dx}{1 + \tan^2 \alpha} \\ &= \frac{dx}{1 + (x - \theta)^2}. \end{aligned} \tag{10.21}$$

Consequently, since $f(\alpha) = f[\psi(x)] = 1/\pi$, we have for the probability element for X.

$$g(x)\,dx = \frac{1}{\pi} \frac{dx}{1 + (x - \theta)^2}, \qquad -\infty \leq x \leq \infty. \tag{10.22}$$

The random variable having this probability element is said to have the *Cauchy* distribution. Its density function is illustrated in Figure 10.2.

In addition to the physical phenomenon that it describes, the Cauchy distribution is also of interest because it is an example of a distribution which possesses no moments beyond the 0th. That this is so follows directly from the fact that the integral defining the first moment

$$E(X) = \int_{-\infty}^{\infty} \frac{x}{\pi} \frac{1}{1 + (x - \theta)^2} \, dx \tag{10.23}$$

is not absolutely convergent.

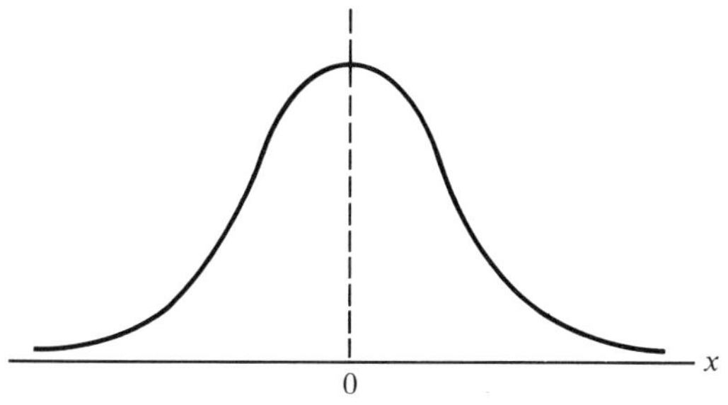

FIGURE 10.2 Cauchy distribution.

10.3 CHANGE OF VARIABLES FOR TWO RANDOM VARIABLES[1]

A change of variable for two or more random variables involves a mathematical apparatus that is considerably more complicated than that for a single random variable, and may be something with which the typical reader is only casually acquainted. With this in mind, the present section develops the underlying mathematical theory in some detail, for if the two-variable case is understood, an understanding of the n-variable case is reasonably assured. The theorem of ultimate interest relates to the transformation of variables in a multiple integral (double integral in the case of two variables and an n-fold integral in the case of n variables); readers familiar with this theorem will accordingly find much of this section essentially a review.

Let $f(x, y)$ be the density function for the joint distribution of X and Y and suppose that X and Y are transformed to new variables U and V according to the transformation

$$\begin{aligned} u &= \phi(x, y), \\ v &= \psi(x, y). \end{aligned} \tag{10.24}$$

We assume ϕ and ψ to be single-valued and continuously differentiable. Analogous to the single-variable case, we seek the density $g(u, v)$ of the "new" random variables U and V. Now, it is clear that if this function is to exist, we must, as a starter, be able to express x and y in terms of u and v, and moreover, do this uniquely. Consequently, our first task it to establish the conditions under which there exist functions h and k.

$$\begin{aligned} x &= h(u, v), \\ y &= k(u, v), \end{aligned} \tag{10.25}$$

which define the transformation which is inverse to the transformation defined by ϕ and ψ in (10.24).

The theorem which establishes these conditions is essentially an application of the implicit function theorem for simultaneous equations, and so to understand the former it is necessary to understand the latter. However, the implicit function theorem for simultaneous equations requires knowledge of the implicit function theorem for a single equation,

[1] Proofs of the theorems appearing in this section are taken from Taylor (1955 chaps. 8, 9, and 13).

so let us begin with a statement and proof of this. Our proof shall be for three variables.

THEOREM 10.1 (Implicit Function Theorem)
Let $F(x, y, z)$ be a function defined on an open set S containing the point (x_0, y_0, z_0). Suppose that F has continuous first partial derivatives on S. In addition, assume that $F(x_0, y_0, z_0) = 0$ and $F_3(x_0, y_0, z_0) = \partial F/\partial z \neq 0$. Under these conditions, there exists a box-like region defined by certain inequalities

$$|x - x_0| < a, \qquad |y - y_0| < b, \qquad |z - z_0| < c,$$

lying in the region defined by S and such that the following assertions are true:

Let R be the rectangular region defined by $|x - x_0| < a$ and $|y - y_0| < b$ in the xy-plane. Then:

1. For any (x, y) in R, there is a unique z such that $|z - z_0| < c$ and $F(x, y, z) = 0$. We shall express this dependence of z on (x, y) as

$$z = f(x, y). \tag{10.26}$$

2. The function f is continuous in R.
3. The function f has continuous first partial derivatives given by

$$f_1(x, y) = -\frac{F_1(x, y, z)}{F_3(x, y, z)}$$

and

$$f_2(x, y) = -\frac{F_2(x, y, z)}{F_3(x, y, z)},$$

where z is given by (10.26).

PROOF. The first part of the proof is concerned with finding suitable values for the positive constants a, b, and c mentioned in the theorem. Let A be a rectangular parallelopiped (box) with center at (x_0, y_0, z_0) such that the whole of A is contained in S, and such that $F_3(x, y, z)$ has everywhere in A the sign that it has at (x_0, y_0, z_0). This choice of A is possible since S is open and F_3 is continuous.

For definiteness, let us assume $F_3 > 0$ in A. Consider the top and bottom faces of the box A. If we denote the height of the box by $2c$, these faces will lie in the planes $z = z_0 \pm c$. Because F_3 is positive, the value of F increases as we go upward along any line in A parallel to the z-axis, and because $F(x_0, y_0, z_0) = 0$, it follows that

$$F(x_0, y_0, z_0 + c) > 0 \qquad \text{and} \qquad F(x_0, y_0, z_0 - c) < 0.$$

Because of the continuity of F, F will be positive in a small rectangle in the plane $z = z_0 + c$ with center at $(x_0, y_0, z_0 + c)$ and negative in a small rectangle in the plane $z = z_0 - c$ with center at $(x_0, y_0, z_0 - c)$. We choose positive numbers a and b so that these rectangles are determined by the inequalities

$$|x - x_0| < a \qquad \text{and} \qquad |y - y_0| < b.$$

In choosing a and b, we take care that a and b are such that the box B defined by the inequalities

$$|x - x_0| < a, \qquad |y - y_0| < b, \qquad |z - z_0| < c$$

is no larger than the box A.

Now consider the value of F along the segment in which a line parallel to the z-axis intersects the box B. As we move upward along this segment, the value of F increases. At the lower end of the segment, $F < 0$, while at the upper end $F > 0$. F being continuous, it follows that there is *just one* point on the segment at which $F = 0$. For a given pair (x, y), the z-coordinate of this point can be denoted by $z = f(x, y)$. This proves the first assertion of the theorem.

Let us now prove that the function $f(x, y)$ just obtained is continuous. To prove continuity at (x_0, y_0), we must show that, given an $\varepsilon > 0$, there exists a $\delta > 0$ such that

$$|f(x, y) - f(x_0, y_0)| < \varepsilon \quad \text{whenever}$$

$$|x - x_0| < \delta \quad \text{and} \quad |y - y_0| < \delta. \tag{10.27}$$

We may assume that $\varepsilon \leq c$. Now $f(x_0, y_0) = z_0$, and because $F(x_0, y_0, z_0) = 0$,

$$F(x_0, y_0, z_0 + \varepsilon) > 0 \quad \text{and} \quad F(x_0, y_0, z_0 - \varepsilon) < 0.$$

Consequently, by the same argument used in proving assertion 1 of the theorem, we see that if we choose $\delta > 0$ so that

$$F(x, y, z_0 + \varepsilon) > 0 \qquad \text{and} \qquad F(x, y, z_0 - \varepsilon) < 0$$

for

$$|x - x_0| < \delta \qquad \text{and} \qquad |y - y_0| < \delta, \tag{10.28}$$

then to each (x, y) satisfying (10.28) there corresponds a unique z such that $|z - z_0| < \varepsilon$ and $F(x, y, z) = 0$. This z must be equal to $f(x, y)$ by assertion 1. Hence (10.27) holds, thus establishing the continuity of $f(x, y)$ at (x_0, y_0).

To prove continuity of f at any other point (x, y) of R, let $z_1 = f(x_1, y_1)$, and assume that F satisfies at (x_1, y_1, z_1) the same hypotheses as are stated in the theorem relative to the point (x_0, y_0, z_0). Hence, by what has already been proven (as applied to this new situation), all the points (x, y, z) in the vicinity of (x_1, y_1, z_1) such that $F(x, y, z) = 0$ are furnished by a single-valued function, say $g(x_1, y_1)$, which is continuous at (x_1, y_1). All of these points are in box B however, and in view of this, the uniqueness part of assertion 1 of the theorem assures us that $f(x, y) = g(x, y)$ whenever (x, y) is near (x_1, y_1).

There remains only assertion 3 of the theorem. We shall only deal with $\partial f/\partial x$, since the treatment of $\partial f/\partial y$ differs only in the letters used. Let (x, y) be a point of R and let $z = f(x, y)$. We wish to show that

$$\lim_{\Delta x \to 0} \frac{f(x + \Delta x, y) - f(x, y)}{\Delta x} = -\frac{F_1(x, y, z)}{F_3(x, y, z)}. \tag{10.29}$$

It is assumed that Δx is sufficiently small that $x + \Delta x$ remains in R. Let

$$\Delta z = f(x + \Delta x, y) - f(x, y),$$

and consider F as a function of x and z only. Then, by the Law of the Mean,

$$F(x + \Delta x, y, z + \Delta z) - F(x, y, z) = \Delta x F_1(\tilde{x}, y, \tilde{z}) + \Delta z F_3(\tilde{x}, y, \tilde{z}), \tag{10.30}$$

where $\tilde{x} = x + \theta\, \Delta x$ and $\tilde{z} = z + \theta\, \Delta z$, with $0 < \theta < 1$. Because

$$\begin{aligned} F(x + \Delta x, y, z + \Delta z) &= F[x + \Delta x, y, f(x + \Delta x, y)] \\ &= F(x, y, z) \end{aligned}$$

by the definition of the function f, the left-hand side of (10.30) is zero, so that

$$\frac{\Delta z}{\Delta x} = -\frac{F_1(\tilde{x}, y, \tilde{z})}{F_3(\tilde{x}, y, \tilde{z})}. \tag{10.31}$$

Now, as $\Delta x \to 0$, $\Delta z \to 0$ by the continuity of f; hence, $\tilde{x} \to x$ and $\tilde{z} \to z$. Since F_1 and F_3 are continuous, the truth of (10.25) is thus seen to follow from (10.31), and accordingly establishes the formula

$$f_1(x, y) = -\frac{F_1(x, y, z)}{F_3(x, y, z)}, \qquad (10.32)$$

Finally, to complete the proof of the theorem, the continuity of f_1 follows from the continuity of F_1 and F_3.

Although we have only proved the implicit function theorem for the case of three variables, there is nothing essential in the theorem or its proof which is dependent on the number three. The analytical details and geometrical language employed in the proof of the theorem can all be easily modified to accommodate as many variables as desired. In view of this, we shall simply state the theorem for the general case of $n + 1$ variables, omitting its proof.

THEOREM 10.2 (Implicit Function Theorem: General Case)
Let $F(x_1, \ldots, x_n, z)$ be defined in an $(n + 1)$-dimensional neighborhood of the point $(a_1, \ldots, a_n, c)$. Suppose that F has continuous partial derivatives in this neighborhood, and in addition, assume that $F(a_1, \ldots, a_n, c) = 0$ and $F_{n+1}(a_1, \ldots, a_n, c) = \partial F/\partial z \neq 0$. Under these conditions, there exists a box-like region defined by certain inequalities

$$|x_1 - a_1| < \alpha_1, \ldots, |x_n - a_n| < \alpha_n, \qquad |z - c| < \beta,$$

lying in the above neighborhood, and such that the following assertions are true:

Let R be the n-dimensional region defined by

$$|x_1 - a_1| < \alpha_1, \ldots, |x_n - a_n| < \alpha_n.$$

Then:

1. For any $(x_1, \ldots, x_n)$ in R, there exists a unique z such that $|z - c| < \beta$ and $F(x_1, \ldots, x_n, z) = 0$. We express this dependence of z on $(x_1, \ldots, x_n)$ by writing

$$z = f(x_1, \ldots, x_n). \qquad (10.33)$$

2. The function f is continuous in R.

3. The function f has continuous first partial derivatives given by

$$\frac{\partial}{\partial x_i} f(x_1, \ldots, x_n) = -\frac{\dfrac{\partial}{\partial x_i} F(x_1, \ldots, x_n, z)}{\dfrac{\partial}{\partial z} F(x_1, \ldots, x_n, z)}, \quad i = 1, n,$$

where $z = f(x_1, \ldots, x_n)$.

Next on the agenda is to extend the implicit function theorem to simultaneous equations. For definiteness, suppose that we have four variables x, y, u, v connected by the two relationships:

$$F(x, y, u, v) = 0, \tag{10.34}$$

$$G(x, y, u, v) = 0. \tag{10.35}$$

We are interested in the conditions under which we can express u and v, say, in terms of x and y. If F and G were linear, that is, if

$$F(x, y, u, v) = a_{11}x + a_{12}y + b_{11}u + b_{12}v = 0, \tag{10.36}$$

$$G(x, y, u, v) = a_{21}x + a_{22}y + b_{21}u + b_{22}v = 0, \tag{10.37}$$

we know that this is possible if and only if the determinant of the matrix

$$\mathbf{B} = \begin{bmatrix} b_{11} & b_{12} \\ b_{21} & b_{22} \end{bmatrix} \tag{10.38}$$

$$= \begin{bmatrix} \dfrac{\partial F}{\partial u} & \dfrac{\partial F}{\partial v} \\ \dfrac{\partial G}{\partial u} & \dfrac{\partial G}{\partial v} \end{bmatrix},$$

is not equal to zero. This determinant is called a *Jacobian* determinant and is usually denoted by the symbol $\partial(F, G)/\partial(u, v)$, that is,

$$\frac{\partial(F, G)}{\partial(u, v)} = \left\| \begin{bmatrix} \dfrac{\partial F}{\partial u} & \dfrac{\partial F}{\partial v} \\ \dfrac{\partial G}{\partial u} & \dfrac{\partial G}{\partial v} \end{bmatrix} \right\|. \tag{10.39}$$

Let us take up the case where F and G are not linear.

THEOREM 10.3 (Implicit Function Theorem: Simultaneous Equations) Let S be a neighborhood of the point P_0: (x_0, y_0, u_0, v_0) in the 4-dimensional space of the coordinates x, y, u, v. Suppose that the functions F and G occurring in the system (10.34)–(10.35) are continuous and have continuous first partial derivatives in S. Assume also that both functions vanish at the point P_0, but that the Jacobian given by (10.39) does not. Under these conditions, there exists a box-like region lying in S, defined by the inequalities

$$|x - x_0| < a, \qquad |y - y_0| < b, \tag{10.40}$$

$$|u - u_0| < \alpha, \qquad |v - v_0| < \beta, \tag{10.41}$$

such that the following assertions are true:

Let R be the region defined in xy-space by the inequalities in (10.40). Then:

1. To any (x, y) in R there corresponds a unique pair of values u, v such that the inequalities (10.41) are satisfied and the functions F and G vanish. This correspondence defines u and v as functions of x and y, say

$$\begin{aligned} u &= f(x, y), \\ v &= g(x, y). \end{aligned} \tag{10.42}$$

2. The functions f, g are continuous in R.
3. The functions f, g have continuous partial derivatives given by

$$\begin{aligned} \frac{\partial f}{\partial x} &= -\frac{1}{J}\frac{\partial(F, G)}{\partial(x, v)}, \\ \frac{\partial g}{\partial x} &= -\frac{1}{J}\frac{\partial(F, G)}{\partial(u, x)}, \end{aligned} \tag{10.43}$$

and similar formulas with x replaced by y, where

$$J = \frac{\partial(F, G)}{\partial(u, v)}. \tag{10.44}$$

PROOF. To begin with, note that the two partial derivatives $\partial F/\partial u$, $\partial F/\partial v$ cannot both be zero at the point P_0, for if they were, the Jacobian J would vanish there, contrary to hypothesis. Accordingly, for definiteness, assume that $\partial F/\partial v$ is different from zero at P_0. We can now apply Theorem 10.2 to the equation, $F(x, y, u, v) = 0$, taking $n = 3$, and $x_1 = x$, $x_2 = y$, $x_3 = u$, and $z_4 = v$. As a result, we obtain a function

$$v = h(x, y, u), \tag{10.45}$$

defined for (x, y, u) in a neighborhood of (x_0, y_0, u_0), and furnishing a solution of the equation $F(x, y, u, v) = 0$ for v. Next, we substitute in G, obtaining

$$H(x, y, u) = G[x, y, u, h(x, y, u)]. \tag{10.46}$$

The equation $G(x, y, u, v) = 0$ is thus replaced by the equation $H(x, y, u) = 0$, and we next seek to solve this equation for u. For this to be done, it is necessary that $\partial H/\partial u \neq 0$. However,

$$\frac{\partial H}{\partial u} = \frac{\partial G}{\partial u} + \frac{\partial v}{\partial u}\frac{\partial G}{\partial v}, \tag{10.47}$$

from the rule for composite functions, and from Theorem 10.2 we have that

$$\begin{aligned}\frac{\partial v}{\partial u} &= \frac{\partial h}{\partial u}\\ &= -\frac{\partial F/\partial u}{\partial F/\partial v}.\end{aligned} \tag{10.48}$$

Consequently,

$$\begin{aligned}\frac{\partial H}{\partial u} &= \frac{\partial G}{\partial u} - \frac{\partial G}{\partial v}\frac{\partial F/\partial u}{\partial F/\partial v}\\ &= \frac{\dfrac{\partial G}{\partial u}\dfrac{\partial F}{\partial v} - \dfrac{\partial G}{\partial v}\dfrac{\partial F}{\partial u}}{\dfrac{\partial F}{\partial v}}\\ &= -\frac{J}{\partial F/\partial v}.\end{aligned} \tag{10.49}$$

Whence $\partial H/\partial u \neq 0$, since neither the Jacobian J nor $\partial F/\partial v$ vanish (by hypothesis) at P_0, and in view of the continuity of F and G, neither do they vanish in a small neighborhood of P_0. Accordingly, we can once again employ Theorem 10.2, this time to the equation $H(x, y, u) = 0$, to obtain a solution for u,

$$u = f(x, y). \tag{10.50}$$

Finally, substituting (10.50) in (10.45) for u, we obtain v as a function of x, y:

$$\begin{aligned}v &= h[x, y, f(x, y)]\\ &= g(x, y).\end{aligned} \tag{10.51}$$

Assertion 1 of the theorem is thus established.

Turning now to the second and third assertions, in applying Theorem 10.2, we are assured that the functions h and f are continuous and possess continuous first partial derivatives. The function g, since it is a composite function of h and f, will also be continuous and have continuous first partial derivatives. It only remains, therefore, to derive the expressions for $\partial g/\partial x$ and $\partial g/\partial x$.

From

$$F[x, y, f(x, y), g(x, y)] = 0, \tag{10.52}$$

$$G[x, y, f(x, y), g(x, y)] = 0, \tag{10.53}$$

we have [since $f(x, y) = u$ and $g(x, y) = v$]

$$\frac{\partial F}{\partial x} + \frac{\partial F}{\partial u}\frac{\partial f}{\partial x} + \frac{\partial F}{\partial v}\frac{\partial g}{\partial x} = 0, \tag{10.54}$$

$$\frac{\partial G}{\partial x} + \frac{\partial G}{\partial u}\frac{\partial f}{\partial x} + \frac{\partial G}{\partial v}\frac{\partial g}{\partial x} = 0, \tag{10.55}$$

or

$$\begin{bmatrix} \dfrac{\partial F}{\partial u} & \dfrac{\partial F}{\partial v} \\ \\ \dfrac{\partial G}{\partial u} & \dfrac{\partial G}{\partial v} \end{bmatrix} \begin{pmatrix} \dfrac{\partial f}{\partial x} \\ \\ \dfrac{\partial g}{\partial x} \end{pmatrix} = - \begin{pmatrix} \dfrac{\partial F}{\partial x} \\ \\ \dfrac{\partial G}{\partial x} \end{pmatrix}. \tag{10.56}$$

Consequently,

$$\begin{pmatrix} \dfrac{\partial f}{\partial x} \\ \\ \dfrac{\partial g}{\partial x} \end{pmatrix} = -\frac{1}{J} \begin{bmatrix} \dfrac{\partial G}{\partial v} & -\dfrac{\partial F}{\partial v} \\ \\ -\dfrac{\partial G}{\partial u} & \dfrac{\partial F}{\partial u} \end{bmatrix} \begin{pmatrix} \dfrac{\partial F}{\partial x} \\ \\ \dfrac{\partial G}{\partial x} \end{pmatrix}, \tag{10.57}$$

or

$$\begin{aligned} \frac{\partial f}{\partial x} &= -\frac{1}{J}\left(\frac{\partial G}{\partial v}\frac{\partial F}{\partial x} - \frac{\partial F}{\partial v}\frac{\partial G}{\partial x}\right) \\ &= -\frac{1}{J}\frac{\partial(F, G)}{\partial(x, v)} \end{aligned} \tag{10.58}$$

and

$$\begin{aligned} \frac{\partial g}{\partial x} &= -\frac{1}{J}\left(\frac{\partial F}{\partial u}\frac{\partial G}{\partial x} - \frac{\partial F}{\partial x}\frac{\partial G}{\partial u}\right) \\ &= -\frac{1}{J}\frac{\partial(F, G)}{\partial(u, x)}, \end{aligned} \tag{10.59}$$

as was to be shown. It is clear that similar formulas hold for $\partial f/\partial y$ and $\partial g/\partial y$ with x replaced by y. Proof of the theorem is thus complete.

Let us now return to the problem at hand, namely, to establish the conditions under which the transformation defined by ϕ and ψ,

$$u = \phi(x, y),\quad v = \psi(x, y), \tag{10.24}$$

admits to an inverse transformation which allows x and y to be expressed in terms of u and v.

$$x = h(u, v),\quad y = k(u, v). \tag{10.25}$$

The following theorem provides the necessary conditions.

THEOREM 10.4

Let $u = \phi(x, y)$ and $v = \psi(x, y)$ define a continuously differentiable transformation for all pairs (x, y) in some neighborhood of a point (x_0, y_0). Let $u_0 = \phi(x_0, y_0)$ and $v_0 = \psi(x_0, y_0)$, and suppose that the Jacobian $\partial(\phi, \psi)/\partial(x, y)$ is not zero at (x_0, y_0). Then there exist positive numbers a, b, α, β and functions $h(u, v)$ and $k(u, v)$ defined when $|u - u_0| < a$ and $|v - v_0| < b$ such that the following assertions are true:

Let R be the rectangular region in the uv-plane defined by the inequalities $|u - u_0| < a$, $|v - v_0| < b$, and let S be the rectangular region in the xy-plane defined by the inequalities $|x - x_0| < \alpha$, $|y - y_0| < \beta$. Then:

1. To any (u, v) in R there corresponds a unique (x, y) in S such that $u = \phi(x, y)$, $v = \psi(x, y)$ and this unique pair is given by

$$x = h(u, v),\quad y = k(u, v). \tag{10.60}$$

2. The functions h, k are continuous and have continuous partial derivatives given by

$$\frac{\partial h}{\partial u} = \frac{1}{J}\frac{\partial \psi}{\partial y},\qquad \frac{\partial k}{\partial u} = -\frac{1}{J}\frac{\partial \psi}{\partial x},$$
$$\frac{\partial h}{\partial v} = -\frac{1}{J}\frac{\partial \phi}{\partial y},\qquad \frac{\partial k}{\partial v} = \frac{1}{J}\frac{\partial \phi}{\partial x}, \tag{10.61}$$

where J is the Jacobian of the transformation from (x, y) to (u, v),

$$J = \frac{\partial(\phi, \psi)}{\partial(x, y)}. \tag{10.62}$$

In the formulas in (10.61), it is understood that x, y are to be expressed in terms of u, v by (10.60).

PROOF. This theorem is essentially just a special case of Theorem 10.3, for the system of equations

$$\begin{aligned} \phi(x, y) - u &= 0, \\ \psi(x, y) - v &= 0. \end{aligned} \tag{10.63}$$

With the appropriate shifts in notation, (10.44) becomes (10.62), and the equations in (10.43) become two of the equations in (10.61). There is nothing more to the proof; we just apply Theorem 10.3 to the equations in (10.63).

To summarize the foregoing, we have just found that in order to be able to express x and y in terms of u and v, the functions ϕ and ψ must: (1) be continuous, (2) have continuous first partial derivatives, and (3) have Jacobian $[\equiv \partial(\phi, \psi)/\partial(x, y)]$ different from zero in the region of the xy-plane over which the functions ϕ, ψ are defined. In view of the fact that the Jacobian is a continuous function of functions which are themselves continuous, the third condition also implies that the Jacobian cannot change sign in the region of interest. Although these conditions are somewhat restrictive, we shall nevertheless assume that they are satisfied.

We are now in a position to take up the central task of this section, which is to deduce the form of the density function of the joint distribution of U and V. In particular, our problem is as follows: Let R be the region in the xy-plane over which $f(x, y)$ is defined, and let S be the region in the uv-plane which is the image of R under the transformation defined by $u = \phi(x, y)$, $v = \psi(x, y)$. Then we must have

$$\iint_S g(u, v)\, du\, dv = \iint_R f(x, y)\, dx\, dy. \tag{10.64}$$

What is the form of the function g?

However, before we can answer this question, we must place some restrictions on the region R, and in addition, provide one further result. In particular, R must be compact (i.e., closed and bounded), and its boundary must consist of a finite number of simple closed curves which are non-intersecting and each of which is sectionally smooth.[2] Such a region will be referred to as *regular*. In view of our assumptions on the functions ϕ and ψ, S will be regular also.

[2] A curve is sectionally smooth if it is composed of a finite number of smooth arcs joined end to end. (An arc is *smooth* if it does not intersect itself and if each point on it possesses a tangent whose direction varies continuously as the point moves along the arc.) Such a curve may have corners at the junction points.

THEOREM 10.5
Let A be the area of R. Then, under the foregoing assumptions on ϕ, ψ, and R,

$$A = \iint_S |J(u, v)| \, du \, dv, \tag{10.65}$$

where

$$J(u, v) = \left| \begin{bmatrix} \dfrac{\partial x}{\partial u} & \dfrac{\partial y}{\partial u} \\ \dfrac{\partial x}{\partial v} & \dfrac{\partial y}{\partial v} \end{bmatrix} \right|. \tag{10.66}$$

PROOF. Proof of this theorem requires two results, for which proofs are provided in Appendix 4. The first relates to the fact that

$$A = \int_C x \, dy, \tag{10.67}$$

where C denotes the complete boundary of the region R. R is assumed to be oriented in a positive sense, which is to say that one is assumed to advance counterclockwise along C. Let C' denote the boundary of S. We orient C' by taking the positive direction along C' to be that which corresponds, under the transformation defined by ϕ and ψ, to the positive direction along C; that is, as (x, y) moves along C in the positive direction, so does its image point (u, v) along C'. (That C is mapped into C' is guaranteed by the assumption that the Jacobian of the transformation does not change sign in R.) With this convention, we have

$$\int_C x \, dy = \int_{C'} h(u, v)\left[\frac{\partial k}{\partial u} \, du + \frac{\partial k}{\partial v} \, dv\right] \tag{10.68}$$

since $x = h(u, v)$, $y = k(u, v)$, and $dy = (\partial k/\partial u) \, du + (\partial k/\partial v) \, dv$ holds for corresponding points of C and C'.

Next, we apply Green's theorem in the uv-plane, which is proven in Appendix 4, to the line integral in (10.68). Green's theorem states that

$$\int_C [P(x, y) \, dx + Q(x, y)] \, dy = \iint_R \left[\frac{\partial Q}{\partial x} - \frac{\partial P}{\partial y}\right] dx \, dy, \tag{10.69}$$

where P and Q are continuous and have continuous first partial derivatives in R. Instead of $P\,dx + Q\,dy$, we have

$$h\frac{\partial k}{\partial u}\,du + h\frac{\partial k}{\partial v}\,dv,$$

so that corresponding to $\partial Q/\partial x - \partial P/\partial y$, we have

$$\frac{\partial}{\partial u}\left(h\frac{\partial k}{\partial u}\right) - \frac{\partial}{\partial v}\left(h\frac{\partial k}{\partial v}\right).$$

On carrying out the indicated differentiations, this latter expression is found to be equal to $J(u, v)$. Consequently,

$$\int_{C'}\left[h\frac{\partial k}{\partial u}\,du + \frac{\partial k}{\partial v}\,dv\right] = \pm\iint_S J(u, v)\,du\,dv. \tag{10.70}$$

The choice of sign on the right is determined by the orientation of C'. If the orientation that we have given to C' coincides with the usual positive orientation of the boundary of S, the plus sign is the one wanted, while in the contrary case, the minus sign is correct. Combining (10.67), (10.68), (10.69), and (10.70), we find that

$$A = \iint_S \pm J(u, v)\,du\,dv. \tag{10.71}$$

Since A is positive and J is always of the same sign, it follows that the sign in (10.71) must be the same as the sign of J. Hence, whichever the sign, formula (10.65) is correct.

If the regions R, S are connected (if one is connected, then so is the other),[3] the Law of the Mean can be applied to (10.65) with the following result: Let A' be the area of S. *Then there is some point* (u, v) *in* S *such that*

$$A = |J(u, v)|\,A'. \tag{10.72}$$

The magnitude of the Jacobian can therefore be interpreted as a measure of the distortion of areas induced by the mapping of R into S. If $|J| = 1$, the mapping is said to be *equiareal.*

Let us now turn to the double integral in (10.64).

[3] R is *connected* if any two points of R can be joined by a path consisting of a finite number of line segments, the whole path lying in R. The continuity of ϕ and ψ guarantees the connectedness of S.

THEOREM 10.6 (Change of Variables with Double Integrals) If $f(x, y)$ is continuous in R and if the mapping $u = \phi(x, y)$, $v = \psi(x, y)$ meets the conditons stated prior to Theorem 10.5, then

$$\iint_R f(x, y)\, dx\, dy = \iint_S f[h(u, v), k(u, v)]\,|J(u, v)|\, du\, dv. \tag{10.73}$$

PROOF. Let R be divided into a finite number of connected regular subregions $R_1, \ldots, R_n$ of areas $\Delta A_1, \ldots, \Delta A_n$, respectively. Let the corresponding subregions in the uv-plane be $S_1, \ldots, S_n$, with areas $\Delta A_1', \ldots, \Delta A_n'$. Choose a point (u_i, v_i) in S_i such that by (10.72),

$$\Delta A_i = |J(u_i, v_i)|\,\Delta A_i', \tag{10.74}$$

and let (x_i, y_i) be the corresponding point of R. For convenience, let

$$H(u, v) = f[h(u, v), k(u, v)]. \tag{10.75}$$

Then

$$\sum_{i=1}^{n} f(x_i, y_i)\Delta A_i = \sum_{i=1}^{n} H(u_i, v_i)\,|J(u_i, v_i)|\,\Delta A_i'. \tag{10.76}$$

We now pass to the limit as $n \to \infty$ and the maximum size of the subregions approaches zero. By Theorem A.4.3 of Appendix 4, we obtain formula (10.73) as the result.

Thus, we have finally arrived at the form of $g(u, v)$, the density function of the joint distribution of U and V, which from (10.73) is equal to

$$g(u, v) = f[h(u, v), k(u, v)]\,|J(u, v)|. \tag{10.77}$$

As an illustration of a change of variable with two random variables, suppose that

$$f(x, y) = e^{-(x+y)}, \qquad 0 \le x, y < \infty, \tag{10.78}$$

and let

$$\begin{aligned} u &= x + y, \\ v &= x - y, \end{aligned} \tag{10.79}$$

Then

$$\begin{aligned} x &= \frac{u + v}{2}, \\ y &= \frac{u - v}{2}, \end{aligned} \tag{10.80}$$

and

$$J(u, v) = \frac{\partial x}{\partial u}\frac{\partial y}{\partial v} - \frac{\partial x}{\partial v}\frac{\partial y}{\partial u} \tag{10.81}$$
$$= -\tfrac{1}{2}.$$

Hence,

$$g(u, v) = \tfrac{1}{2}e^{-u}. \tag{10.82}$$

These transformations have the interesting property that, for this case, a bivariate distribution is transformed into a univariate one.

10.4 CHANGE OF VARIABLES IN THE GENERAL CASE

Let us now take up the change of variables in the general case. Assume that we have n random variables $X_1, \ldots, X_n$ with joint density function $f(x_1, \ldots, x_n)$, and suppose that $X_1, \ldots, X_n$ are transformed to "new" variables $U_1, \ldots, U_n$ via the n transformations,

$$\begin{aligned} u_1 &= \phi_1(x_1, \ldots, x_n), \\ &\vdots \\ u_n &= \phi_n(x_1, \ldots, x_n). \end{aligned} \tag{10.83}$$

Let R be the region in the n-dimensional space of $x_1, \ldots, x_n$ for which the transformations in (10.83) are defined, and let S be the region in the n-dimensional space of $u_1, \ldots, u_n$ which is the image of R under these transformations. Assume that the region R is regular as defined in connection with Theorem 10.6; assume that the functions $\phi_1, \ldots, \phi_n$ are continuous and have continuous first partial derivatives. Finally, assume that the Jacobian of the transformation from u to x,

$$J(u_1, \ldots, u_n) = \frac{\partial(x_1, \ldots, x_n)}{\partial(u_1, \ldots, u_n)}, \tag{10.84}$$

is nonzero and has the same sign everywhere in R. Then, under these conditions,

$$\begin{aligned} \int_R f(x_1, \ldots, x_n)\, dx_1 \ldots dx_n = \int_S f[h_1(u_1, \ldots, u_n), \ldots, h_n(u_1, \ldots, u_n)] \\ \cdot |J(u_1, \ldots, u_n)|\, du_1 \ldots du_n, \end{aligned} \tag{10.85}$$

where $h_1, \ldots, h_n$ define the inverse transformations from $u_1, \ldots, u_n$ to $x_1, \ldots, x_n$. It is to be understood that the single integrals in (10.85) in fact represent n-fold integrals over the regions R and S, respectively.

We shall not provide a proof of the validity of this formula, as this would require knowledge of certain results in analysis with which readers are unlikely to be familiar and which are too delicate to attempt to develop here.[4] However, if the reader understands Theorem 10.6, then formula (10.85) ought not present any difficulty as to plausibility, since it is a natural generalization of the formula for double integrals in (10.73).

If $g(u_1, \ldots, u_n)$ is the density function for the joint distribution of $U_1, \ldots, U_n$, then from the right-hand integral in (10.85), it follows that

$$g(u_1, \ldots, u_n) = f[h_1(u_1, \ldots, u_n), \ldots, h_n(u_1, \ldots, u_n)]\,|J(u_1, \ldots, u_n)|. \tag{10.86}$$

Examples will be provided in connection with the multivariate normal distribution in Chapter 12.

REFERENCES

Crowell, R. H. and Williamson, R. E., *Calculus of Vector Functions*, Prentice Hall, 1962, Appendix II.

Freeman, H., *An Introduction to Statistical Inference*, Addison-Wesley, 1963, chap. 8.

Taylor, A. E., *Advanced Calculus*, Ginn, 1955, chaps. 8, 9, 13.

[4] For a proof of the theorem on the change of variable in multiple integrals, see Crowell and Williamson (1962, Appendix II).

CHAPTER 11

CONVOLUTION

Not infrequently in statistics, we find ourselves with a random variable which is a sum of other random variables and need to know its distribution. To be specific, suppose that X and Y are two random variables that are independently distributed with density functions $f(x)$ and $g(y)$. Let $U = X + Y$. What is the density function of U, say $h(u)$?

11.1 SOLUTION BY CHANGE OF VARIABLE

Let us obtain the solution first by a simple change of variable. Let

$$\begin{aligned} u &= x + y, \\ v &= y. \end{aligned} \tag{11.1}$$

Then

$$\begin{aligned} x &= u - v, \\ y &= v, \end{aligned} \tag{11.2}$$

and

$$J(u, v) = \frac{\partial x}{\partial u}\frac{\partial y}{\partial v} - \frac{\partial y}{\partial u}\frac{\partial x}{\partial v}. \tag{11.3}$$

The joint density function for U and V, $g(u, v)$, is therefore

$$\begin{aligned} \phi(u, v) &= f[x(u, v)]g[y(u, v)]\,|J(u, v)| \\ &= f(u - v)g(v). \end{aligned} \tag{11.4}$$

Finally, the density funtion for U is obtained by integrating out v, namely,

$$h(u) = \int_{-\infty}^{\infty} f(u - v)g(v)\, dv. \tag{11.5}$$

Alternately, if we had taken $v = x$ as the dummy variable instead of $v = y$, we would have found

$$h(u) = \int_{-\infty}^{\infty} f(v)g(u - v)\, dv, \tag{11.6}$$

as the reader can easily verify.

11.2 SOLUTION BY CONVOLUTION

Let us reach the solution in (11.5) by a different path. Let $H(u)$ be the distribution function of U. Then

$$\begin{aligned} H(u) &= P(U \leq u) \\ &= P(X + Y \leq u) \\ &= \iint_{x+y<u} f(x)\, g(y)\, dx\, dy. \end{aligned} \tag{11.7}$$

From $X + Y \leq u$, the upper limit of Y is $u - x$; hence,

$$\begin{aligned} H(u) &= \int_{-\infty}^{\infty} \int_{-\infty}^{u-x} f(x)g(y)\, dy\, dx \\ &= \int_{-\infty}^{\infty} f(x) \int_{-\infty}^{u-x} g(y)\, dy\, dx \\ &= \int_{-\infty}^{\infty} f(x)G(u - x)\, dx. \end{aligned} \tag{11.8}$$

If, for known G and f, the integration can be carried out, the problem is solved. Alternatively, with $u - y$ as the upper limit to X, we would have

$$\begin{aligned} H(u) &= \int_{-\infty}^{\infty} g(y) \int_{-\infty}^{u-y} f(x)\, dx\, dy \\ &= \int_{-\infty}^{\infty} g(y)F(u - y)\, dy. \end{aligned} \tag{11.9}$$

If $H(u)$ can be differentiated beneath the integral with respect to u, then from (11.8) and (11.9), we have

$$\begin{aligned} h(u) &= \frac{dH(u)}{du} \\ &= \int_{-\infty}^{\infty} g(u-x) f(x)\, dx \\ &= \int_{-\infty}^{\infty} f(u-y)\, g(y)\, dy, \end{aligned} \tag{11.10}$$

as already obtained in (11.5) and (11.6).

The operations shown in (11.8) and (11.9) or in (11.10) are referred to as the *convolution of* $f(x)$ *and* $g(y)$. We describe (11.10) by saying that the density function of the sum of two independent random variables is the convolution of the density functions of the random variables. While we have only treated continuous X and Y, it should be clear that formulas parallel to (11.8)–(11.10) with integrals replaced by sums hold for discrete X and Y.

For n variables and $U = X_1 + X_2 + \cdots + X_n$, $H(u)$ and $h(u)$ can be obtained by repeated application of (11.8) and (11.9), first to $X_1 + X_2$, then to $(X_1 + X_2) + X_3$, and so on. For the general problem, we write

$$\begin{aligned} U &= (X_1 + X_2 + \cdots + X_{n-1}) + X_n \\ &= Y + X_n. \end{aligned} \tag{11.11}$$

Then, with subscripts on F, G, f, and g indicating the number of random variables involved, (11.8) and (11.9) become

$$\begin{aligned} H_n(u) &= \int_{-\infty}^{\infty} f_1(x_n) G_{n-1}(u - x_n)\, dx_n \\ &= \int_{-\infty}^{\infty} g_{n-1}(y) F_1(u-y)\, dy, \end{aligned} \tag{11.12}$$

and if $H_n(u)$ is differentiable beneath the integral, (11.10) becomes

$$\begin{aligned} h_n(u) &= \int_{-\infty}^{\infty} f_1(x_n) g_{n-1}(u - x_n)\, dx_n \\ &= \int_{-\infty}^{\infty} f_1(u-y) g_{n-1}(y)\, dy. \end{aligned} \tag{11.13}$$

11.3 APPLICATION TO UNIFORM RANDOM VARIABLES

Let us assume that the X_i in the sum $U = X_1 + \cdots + X_n$ are each distributed uniformly and independently over the interval 0 to 1. As above, let $Y = X_1 + X_2 + \cdots + X_{n-1}$, so that $U = Y + X_n$. Then, in formula (11.13),

$$f_1(x_n) = \begin{cases} 1, & 0 \leq x_n \leq 1, \\ 0, & \text{otherwise}, \end{cases} \tag{11.14}$$

or

$$f_1(u - y) = \begin{cases} 1, & 0 \leq u - y \leq 1, \\ 0, & \text{otherwise}. \end{cases} \tag{11.15}$$

This being the case, the second integral in (11.13) accordingly reduces to

$$h_n(u) = \int_{u-1}^{u} g_{n-1}(y)\, dy, \tag{11.16}$$

the range of y being from $u - 1$ to u. Thus, for the sum of n independently distributed uniform variables, we find the density function of U from the density function of $X_1 + \cdots + X_{n-1}$, the density function of the latter being found from an earlier convolution, and so on.

We shall illustrate what is involved for $n = 2$. With $U = X_1 + X_2$, the range of U is from 0 to 2, and the lower and upper limits of the integral in (11.16) are -1 and 2, respectively. However, it is to be noted that, since with $n = 2$, $g_1 = f_1$, the functional form of $g_1(y)$ varies within the range $u = 0$ and $u = 2$. The integral in (11.16) must accordingly be split into parts such that the functional form of $g_1(y)$ is constant within each part. In this example, $g_1(y)$ must be either 1 or 0. We consider first (11.16) for $0 \leq u \leq 1$, and then for $1 \leq u \leq 2$.

When $0 \leq u \leq 1$, the lower limit of the integral satisfies $-1 \leq u - 1 \leq 0$. Within this range, only $u - 1 = 0$ need be considered, since $g_1(y) = 0$ for all values of y less than zero. When $0 \leq u \leq 1$, the upper limit of the integral satisfies $0 \leq u \leq 1$, for all values of which $g_1(y)$ is nonvanishing. Pulling these results together, we consequently have

$$\begin{aligned} \int_{u-1}^{u} g_1(y)\, dy &= \int_0^u dy \\ &= u \end{aligned} \tag{11.17}$$

for $0 \leq u \leq 1$.

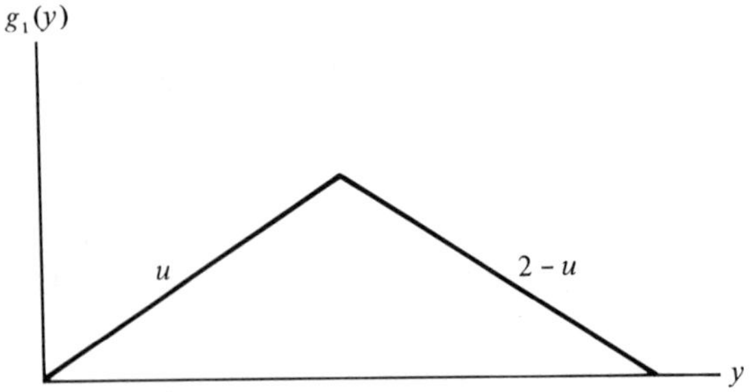

FIGURE 11.1 Density function for 2 independent rectangular random variables.

When $1 \le u \le 2$, the lower limit of the integral in (11.16) satisfies $0 \le u - 1 \le 1$, for all values of which $g_1(y)$ does not vanish. When $1 \le u \le 2$, the upper limit of the integral satisfies $1 \le u \le 2$, but of the ranges, only $u = 1$ need be considered, since $g_1(y) = 0$ for all values of y greater than one. Consequently,

$$\int_{u-1}^{u} g_1(y)\,dy = \int_{u-1}^{1} dy$$
$$= 2 - u \qquad (11.18)$$

for $1 \le u \le 2$. Combining (11.17) and (11.18), the density function for U is then as shown in Figure 11.1. For values of u outside the range $0 \le u \le 1$, $h_2(u) = 0$.

REFERENCES

Feller, W., *An Introduction to Probability and Its Applications*, vol. I, 3rd ed., Wiley, 1968, pp. 266–270.

Freeman, H., *An Introduction to Statistical Inference*, Addison-Wesley, 1963, chap. 20.

CHAPTER 12

MULTIVARIATE NORMAL DISTRIBUTION

In this chapter, we discuss the n-variable generalization of the bivariate normal distribution that was studied in Chapter 9. Since this distribution provides much of the theoretical underpinning of statistical inference in science, it is important that its properties be carefully examined.[1]

12.1 DENSITY FUNCTION OF THE MULTIVARIATE NORMAL DISTRIBUTION

The univariate normal density function can be written

$$f(y) = ke^{-1/2(y-\beta)\alpha(y-\beta)}, \qquad -\infty < y < \infty, \tag{12.1}$$

where α is positive and k is chosen so that the area beneath (12.1) over the entire y-axis is equal to one. The density function of the multivariate normal distribution has an analogous form. The single variable y is replaced by a vector

$$\mathbf{x} = \begin{pmatrix} x_1 \\ \vdots \\ x_n \end{pmatrix}, \tag{12.2}$$

[1] This chapter makes extensive use of linear algebra. The results required are presented in Appendix 5. In this chapter vectors representing random variables will be written in upper case boldface italics, while vectors representing constants or particular values of random variables will be written in lower case boldface nonitalics. Matrices will appear in boldface upper case nonitalics.

the scalar β by a vector of constants,

$$\mathbf{b} = \begin{pmatrix} b_1 \\ \vdots \\ b_n \end{pmatrix}, \tag{12.3}$$

and the positive scalar α by a matrix $\mathbf{A}$ which is symmetric and positive definite,

$$\mathbf{A} = \begin{bmatrix} a_{11} & a_{12} & \cdots & a_{1n} \\ a_{21} & a_{22} & \cdots & a_{2n} \\ \vdots & & \ddots & \vdots \\ a_{n1} & a_{n2} & \cdots & a_{nn} \end{bmatrix}. \tag{12.4}$$

Finally, the square $\alpha(y - \beta)^2 = (y - \beta)\alpha(y - \beta)$ is replaced by the quadratic form

$$(\mathbf{x} - \mathbf{b})'\mathbf{A}(\mathbf{x} - \mathbf{b}) = \sum_{i=1}^{n} \sum_{j=1}^{n} a_{ij}(x_i - b_i)(x_j - b_j). \tag{12.5}$$

The density function for the multivariate normal distribution is therefore given by

$$f(x_1, \ldots, x_n) = Ke^{-(1/2)(\mathbf{x}-\mathbf{b})'\mathbf{A}(\mathbf{x}-\mathbf{b})}, \qquad -\infty < x < \infty, \tag{12.6}$$

where $K(>0)$ is chosen so that the integral of (12.6) over the entire n-dimensional Euclidean space of $x_1, \ldots, x_n$ is unity.

Observe first that, since $\mathbf{A}$ is by hypothesis positive definite,

$$(\mathbf{x} - \mathbf{b})'\mathbf{A}(\mathbf{x} - \mathbf{b}) > 0 \tag{12.7}$$

for all $\mathbf{x}$, so that $f(x_1, \ldots, x_n) \le K$, and is therefore bounded. Let us now determine the value of K by evaluating the n-fold integral,

$$K^* = \int_{-\infty}^{\infty} \cdots \int_{-\infty}^{\infty} e^{-(1/2)(\mathbf{x}-\mathbf{b})'\mathbf{A}(\mathbf{x}-\mathbf{b})}\, dx_1 \cdots dx_n. \tag{12.8}$$

Now, since $\mathbf{A}$ is positive definite, there exists a nonsingular matrix $\mathbf{C}$ such that

$$\mathbf{C}'\mathbf{A}\mathbf{C} = \mathbf{I}. \tag{12.9}$$

Therefore, let

$$\mathbf{C}\mathbf{y} = \mathbf{x} - \mathbf{b}, \tag{12.10}$$

where

$$\mathbf{y} = \begin{pmatrix} y_1 \\ \vdots \\ y_n \end{pmatrix}. \tag{12.11}$$

Then

$$(\mathbf{x}-\mathbf{b})'\mathbf{A}(\mathbf{x}-\mathbf{b}) = \mathbf{y}'\mathbf{C}'\mathbf{A}\mathbf{C}\mathbf{y} \qquad (12.12)$$
$$= \mathbf{y}'\mathbf{y}.$$

Since the transformation in (12.10) is linear, the absolute value of its Jacobian **J** is simply

$$|\mathbf{J}| = \text{mod}|\mathbf{C}'| \qquad (12.13)$$
$$= \text{mod}|\mathbf{C}|,$$

where "mod|**C**|" denotes the absolute value of the determinant of **C**. Thus, (12.8) becomes

$$K^* = \text{mod}|\mathbf{C}| \int_{-\infty}^{\infty} \cdots \int_{-\infty}^{\infty} e^{-(1/2)\mathbf{y}'\mathbf{y}}\, dy_1 \cdots dy_n. \qquad (12.14)$$

However, since

$$e^{-(1/2)\mathbf{y}'\mathbf{y}} = e^{-(1/2)\Sigma y_i^2} \qquad (12.15)$$
$$= \prod_{i=1}^{n} e^{-(1/2)y_i^2},$$

we can write (12.14) as

$$K^* = \text{mod}|\mathbf{C}| \int_{-\infty}^{\infty} \cdots \int_{-\infty}^{\infty} e^{-(1/2)y_1^2} \cdots e^{-(1/2)y_n^2}\, dy_1 \cdots dy_n$$
$$= \text{mod}|\mathbf{C}| \prod_{i=1}^{n} \int_{-\infty}^{\infty} e^{-(1/2)y_i^2}\, dy_i$$
$$= \text{mod}|\mathbf{C}| \prod_{i=1}^{n} \sqrt{2\pi} \qquad (12.16)$$

since

$$\frac{1}{\sqrt{2\pi}} \int_{-\infty}^{\infty} e^{-(1/2)y^2}\, dy = 1. \qquad (12.17)$$

Therefore,

$$K^* = \text{mod}|\mathbf{C}|(2\pi)^{n/2}. \qquad (12.18)$$

However,

$$|\mathbf{C}'\mathbf{A}\mathbf{C}| = |\mathbf{C}'| \cdot |\mathbf{A}| \cdot |\mathbf{C}|$$
$$= |\mathbf{A}| \cdot |\mathbf{C}'| \cdot |\mathbf{C}|$$
$$= |\mathbf{A}| \cdot |\mathbf{C}|^2$$
$$= \mathbf{I} \qquad \text{[from 12.9)]}. \qquad (12.19)$$

Hence,

$$|\mathbf{C}|^2 = |\mathbf{A}|^{-1} \tag{12.20}$$

and

$$\text{mod}|\mathbf{C}| = |\mathbf{A}|^{-1/2}. \tag{12.21}$$

Consequently,

$$K^* = |\mathbf{A}|^{-1/2}(2\pi)^{n/2}, \tag{12.22}$$

so that

$$K = |\mathbf{A}|^{1/2}(2\pi)^{-n/2}. \tag{12.23}$$

The density function of the multivariate normal thus becomes

$$f(x_1, \ldots, x_n) = |\mathbf{A}|^{1/2}(2\pi)^{-n/2}e^{-(1/2)(\mathbf{x}-\mathbf{b})'\mathbf{A}(\mathbf{x}-\mathbf{b})}. \tag{12.24}$$

12.2 MOMENTS OF THE DISTRIBUTION

Let us now obtain the moments of the multivariate normal distribution. However, as a preliminary to this, we shall derive a result that is applicable not only to the multivariate normal distribution, but to any multivariate distribution.

THEOREM 12.1
Let $\mathbf{X}' = (X_1, \ldots, X_n)$ be an n-dimensional random variable with density function $f(x_1, \ldots, x_n)$. Let

$$Y = \mathbf{B}X + \gamma, \tag{12.25}$$

where $\mathbf{B}$ is an $m \times n$, matrix and γ an $m \times 1$ vector, $\mathbf{B}$ and γ both being nonrandom. Then

$$E(Y) = \mathbf{B}E(X) + \gamma. \tag{12.26}$$

PROOF. Consider the ith element of Y,

$$Y_i = b_{i1}X_1 + b_{i2}X_2 + \cdots + b_{in}X_n + \gamma_i. \tag{12.27}$$

Then, from the definition of expected value,

$$\begin{aligned} E(Y_i) &= \int_{x_1} \cdots \int_{x_n} (b_{i1}X_1 + \cdots + b_{in}X_n + \gamma_i)f(x_1, \ldots, x_n)\, dx_1 \cdots dx_n \\ &= \int_x b_{i1}X_1 f(x)\, dx + \cdots + \int_x b_{in}X_n f(x)\, dx + \gamma_i \int_x f(x)\, dx \\ &= b_{i1}E(X_1) + \cdots + b_{in}E(X_n) + \gamma_i. \end{aligned} \tag{12.28}$$

where, to save space in the second line, a single integral is written in place of an n-fold integral. Since, by definition,

$$E(Y) = \begin{bmatrix} E(Y_1) \\ \vdots \\ E(Y_m) \end{bmatrix}, \tag{12.29}$$

this proves the theorem.

Return now to the transformation in (12.10), $\mathbf{C}Y = X - \mathbf{b}$. From the theorem just proved, we will have

$$E(X) = \mathbf{C}E(Y) + \mathbf{b}, \tag{12.30}$$

which means that we can obtain $E(X)$ by finding $E(Y)$. Since the absolute value of the Jacobian in the transformation, $\mathbf{C}y = x - \mathbf{b}$, is $|\mathbf{A}|^{-1/2}$, we have, then, for the density function of $Y' = (Y_1, \ldots, Y_n)$,

$$\begin{aligned} g(y_1, \ldots, y_n) &= \frac{1}{(2\pi)^{n/2}} e^{-(1/2)\mathbf{y}'\mathbf{y}} \\ &= g_1(y_1) \cdots g_n(y_n), \end{aligned} \tag{12.31}$$

where

$$g_i(y_i) = \frac{1}{\sqrt{2\pi}} e^{-(1/2)y_i^2}. \tag{12.32}$$

However, (12.32) will be recognized as the density function for a single normal variable with mean equal to zero and variance equal to one. Therefore, $E(Y_i) = 0$ for each Y_i, so that $E(Y) = 0$, and

$$E(X) = \mathbf{b}. \tag{12.33}$$

The next task is to derive the covariance matrix of X, which, in view of the foregoing, will be given by

$$\begin{aligned} E[(X - \boldsymbol{\mu})(X - \boldsymbol{\mu})'] &= E(\mathbf{C}YY'\mathbf{C}') \\ &= \mathbf{C}E(YY')\mathbf{C}'. \end{aligned} \tag{12.34}$$

The problem is thus reduced to finding $E(YY')$, a matrix whose typical element is

$$E(Y_i Y_j) = \int_{-\infty}^{\infty} \cdots \int_{-\infty}^{\infty} y_i y_j \left[\prod_{k=1}^{n} \frac{1}{\sqrt{2\pi}} e^{-(1/2)y_k^2} \right] dy_1 \cdots dy_n. \tag{12.35}$$

However, we have just established that each Y_i is distributed normally with mean zero and variance one. Hence, since

$$\frac{1}{\sqrt{2\pi}} \int_{-\infty}^{\infty} e^{-(1/2)y_k^2} \, dy_k = 1, \tag{12.36}$$

for $k \neq i, j$, (12.35) reduces to

$$E(Y_i^2) = \frac{1}{\sqrt{2\pi}} \int_{-\infty}^{\infty} y_i^2 e^{-(1/2)y_i^2}\, dy_i \qquad (12.37)$$
$$= 1$$

for $i = j$, and to

$$E(Y_i Y_j) = \frac{1}{\sqrt{2\pi}} \int_{-\infty}^{\infty} y_i e^{-(1/2)y_i^2}\, dy_i \cdot \frac{1}{\sqrt{2\pi}} \int_{-\infty}^{\infty} y_j e^{-(1/2)y_j^2}\, dy_j$$
$$= 0 \qquad (12.38)$$

for $i \neq j$. Combining (12.37) and (12.38), we therefore have

$$E(\boldsymbol{YY}') = \mathbf{I}, \qquad (12.39)$$

whence

$$E[(\boldsymbol{X} - \boldsymbol{\mu})(\boldsymbol{X} - \boldsymbol{\mu})'] = \mathbf{CC}'. \qquad (12.40)$$

However, from (12.9),

$$\mathbf{C}^{-1}\mathbf{A}^{-1}(\mathbf{C}')^{-1} = \mathbf{I}, \qquad (12.41)$$

so that

$$\mathbf{A}^{-1} = \mathbf{CC}', \qquad (12.42)$$

and consequently,

$$E[(\boldsymbol{X} - \boldsymbol{\mu})(\boldsymbol{X} - \boldsymbol{\mu})'] = \mathbf{A}^{-1}. \qquad (12.43)$$

Henceforth, we shall denote the covariance matrix of $\boldsymbol{X}$ by the $(n \times n)$ matrix $\boldsymbol{\Omega}$, that is, let

$$\boldsymbol{\Omega} = \mathbf{A}^{-1}. \qquad (12.44)$$

Since $\boldsymbol{\Omega}$ is a covariance matrix, it is necessarily symmetric and positive definite. Finally, combining (12.24), (12.33), (12.43), and (12.44), we have for the joint density function of $\boldsymbol{X}$,

$$f(x_1, \ldots, x_n) = |\boldsymbol{\Omega}|^{-1/2}(2\pi)^{-n/2} e^{-(1/2)(\mathbf{x}-\boldsymbol{\mu})'\boldsymbol{\Omega}^{-1}(\mathbf{x}-\boldsymbol{\mu})}. \qquad (12.45)$$

We shall now show that for $n = 2$ the density in (12.45) reduces to expression (9.46) of Chapter 9. For $\boldsymbol{\Omega}$, we have by definition,

$$\boldsymbol{\Omega} = \begin{bmatrix} E(X_1 - \mu_1)^2 & E[(X_1 - \mu_1)(X_2 - \mu_2)] \\ E[(X_1 - \mu_1)(X_2 - \mu_2)] & E(X_2 - \mu_2)^2 \end{bmatrix}$$
$$= \begin{bmatrix} \sigma_1^2 & \rho\sigma_1\sigma_2 \\ \rho\sigma_1\sigma_2 & \sigma_2^2 \end{bmatrix}, \qquad (12.46)$$

where $\rho(=\text{cov}(X_1, X_2)/\sigma_1\sigma_2)$ is the correlation between X_1 and X_2. Consequently,

$$\begin{aligned} |\mathbf{\Omega}| &= \sigma_1{}^2\sigma_2{}^2 - \rho^2\sigma_1{}^2\sigma_2{}^2 \\ &= \sigma_1{}^2\sigma_2{}^2(1-\rho^2) \end{aligned} \tag{12.47}$$

and

$$\mathbf{\Omega}^{-1} = \frac{1}{\sigma_1{}^2\sigma_2{}^2(1-\rho^2)}\begin{bmatrix} \sigma_2{}^2 & -\rho\sigma_1\sigma_2 \\ -\rho\sigma_1\sigma_2 & \sigma_1{}^2 \end{bmatrix}, \tag{12.48}$$

so that

$$(\mathbf{x}-\boldsymbol{\mu})'\mathbf{\Omega}^{-1}(\mathbf{x}-\boldsymbol{\mu}) = \frac{1}{1-\rho^2}\left[\frac{1}{\sigma_1{}^2}(x_1-\mu_1)^2 - \frac{2\rho}{\sigma_1\sigma_2}(x_1-\mu_1)(x_2-\mu_2) + \frac{1}{\sigma_2{}^2}(x_2-\mu_2)^2\right]. \tag{12.49}$$

Inserting (12.47) and (12.49) into (12.45) then yields the density function for the bivariate normal distribution of expression (9.46).

It will be noticed that the density function (12.45) is constant on ellipsoids defined by

$$(\mathbf{x}-\boldsymbol{\mu})'\mathbf{\Omega}^{-1}(\mathbf{x}-\boldsymbol{\mu}) = c \tag{12.50}$$

for $c > 0$ in an n-dimensional Euclidian space of $x_1, \ldots, x_n$. The shape and orientation of the ellipsoids are determined by $\mathbf{\Omega}$, and the size is determined by c. Let us consider the bivariate case just discussed. Let $y_i = (x_i - \mu_i)/\sigma_i$ $(i = 1, 2)$, so that the centers of the ellipses of constant density are located at the origin. These ellipses are defined by

$$\frac{1}{1-\rho^2}(y_1{}^2 - 2\rho y_1 y_2 + y_2{}^2) = c. \tag{12.51}$$

The y_1 and y_2 intercepts are equal. For $\rho > 0$, the major axis of the ellipse is along the line $y_1 = y_2$ with length equal to $2\sqrt{c(1+\rho)}$, and the minor axis has length $2\sqrt{c(1-\rho)}$. For $\rho < 0$, the major axis is along the line $y_1 = -y_2$ and has length $2\sqrt{c(1-\rho)}$; the minor axis has length $2\sqrt{c(1+\rho)}$. With the bivariate case, the density function describes a surface above the y_1y_2-plane. The contours of equal density are contours of equal altitude on a topographical map; for this case, the contours are all ellipses. They indicate the shape of the probability surface or hill. If $\rho > 0$, most of the hill will be in the first and third quadrants; if $\rho < 0$, most will be in the second and fourth quadrants. Upon transformation back to $x_i = y_i/\sigma_i + \mu_i$, each contour is expanded by the factor σ_i in the direction of the ith-axis, while the center of the contour is shifted to (μ_1, μ_2).

12.3 THE DISTRIBUTION OF LINEAR COMBINATIONS OF NORMALLY DISTRIBUTED VARIABLES

In this section, we shall discuss the basic properties of the multivariate normal distribution which make its study so worthwhile. In particular, we shall show (1) that the multivariate normal distribution is invariant to nonsingular linear transformations of its variates, (2) that, as we have already demonstrated for the bivariate normal in Chapter 9, the marginal and conditional distributions associated with it are also normal, and (3) that linear combinations of normal variable are again normal.

In the discussion to follow, we shall frequently use $N(\boldsymbol{\mu}, \boldsymbol{\Omega})$ to denote the multivariate normal function with mean vector $\boldsymbol{\mu}$ and covariance matrix $\boldsymbol{\Omega}$. In all cases, $\boldsymbol{\Omega}$ will be assumed to be nonsingular.

THEOREM 12.2
Let $\boldsymbol{X}$ (with n components) be distributed $N(\boldsymbol{\mu}, \boldsymbol{\Omega})$, and let $\mathbf{C}$ be an $n \times n$ nonsingular matrix. Then

$$\boldsymbol{Y} = \mathbf{C}\boldsymbol{X} \tag{12.52}$$

is distributed $N(\mathbf{C}\boldsymbol{\mu}, \mathbf{C}\boldsymbol{\Omega}\mathbf{C}')$.

PROOF. Employing a change of variables, the density of $\boldsymbol{Y}$ is obtained from the density of $\boldsymbol{X}$ by replacing $\mathbf{x}$ by

$$\mathbf{x} = \mathbf{C}^{-1}\mathbf{y} \tag{12.53}$$

and multiplying by the absolute value of the Jacobian of the transformation in (12.53), which is equal to $\operatorname{mod}|\mathbf{C}^{-1}|$. However,

$$\begin{aligned}
\operatorname{mod}|\mathbf{C}^{-1}| &= \frac{1}{\operatorname{mod}|\mathbf{C}|} \\
&= \left[\frac{1}{|\mathbf{C}|^2}\right]^{1/2} \\
&= \left[\frac{|\boldsymbol{\Omega}|}{|\mathbf{C}| \cdot |\boldsymbol{\Omega}| \cdot |\mathbf{C}|}\right]^{1/2} \\
&= \frac{|\boldsymbol{\Omega}|^{1/2}}{|\mathbf{C}\boldsymbol{\Omega}\mathbf{C}'|^{1/2}}.
\end{aligned} \tag{12.54}$$

The quadratic form in the exponent of $N(\boldsymbol{\mu}, \boldsymbol{\Omega})$ is

$$Q = (\mathbf{x} - \boldsymbol{\mu})'\boldsymbol{\Omega}^{-1}(\mathbf{x} - \boldsymbol{\mu}), \tag{12.55}$$

which the transformation in (12.53) carries into

$$\begin{aligned} Q &= (\mathbf{C}^{-1}\mathbf{y} - \boldsymbol{\mu})'\boldsymbol{\Omega}^{-1}(\mathbf{C}^{-1}\mathbf{y} - \boldsymbol{\mu}) \\ &= (\mathbf{C}^{-1}\mathbf{y} - \mathbf{C}^{-1}\mathbf{C}\boldsymbol{\mu})'\boldsymbol{\Omega}^{-1}(\mathbf{C}^{-1}\mathbf{y} - \mathbf{C}^{-1}\mathbf{C}\boldsymbol{\mu}) \\ &= [\mathbf{C}^{-1}(\mathbf{y} - \mathbf{C}\boldsymbol{\mu})]'\boldsymbol{\Omega}^{-1}[\mathbf{C}^{-1}(\mathbf{y} - \mathbf{C}\boldsymbol{\mu})] \\ &= (\mathbf{y} - \mathbf{C}\boldsymbol{\mu})'(\mathbf{C}^{-1})'\boldsymbol{\Omega}^{-1}\mathbf{C}^{-1}(\mathbf{y} - \mathbf{C}\boldsymbol{\mu}) \\ &= (\mathbf{y} - \mathbf{C}\boldsymbol{\mu})'(\mathbf{C}\boldsymbol{\Omega}\mathbf{C}')^{-1}(\mathbf{y} - \mathbf{C}\boldsymbol{\mu}). \end{aligned} \tag{12.56}$$

The density function for Y is therefore

$$\begin{aligned} &\operatorname{mod}|\mathbf{C}^{-1}|\ |\boldsymbol{\Omega}|^{-1}(2\pi)^{-n/2}\exp[-\tfrac{1}{2}(\mathbf{y} - \mathbf{C}\boldsymbol{\mu})'(\mathbf{C}\boldsymbol{\Omega}\mathbf{C}')^{-1}(\mathbf{y} - \mathbf{C}\boldsymbol{\mu})] \\ &\quad = |\mathbf{C}\boldsymbol{\Omega}\mathbf{C}'|^{-1/2}(2\pi)^{-n/2}\exp[-\tfrac{1}{2}(\mathbf{y} - \mathbf{C}\boldsymbol{\mu})'(\mathbf{C}\boldsymbol{\Omega}\mathbf{C}')^{-1}(\mathbf{y} - \mathbf{C}\boldsymbol{\mu})] \\ &\quad = N(\mathbf{C}\boldsymbol{\mu}, \mathbf{C}\boldsymbol{\Omega}\mathbf{C}'), \end{aligned} \tag{12.57}$$

as was to be shown.

Let us now consider two sets of random variables, $X_1, \ldots, X_m$ and $X_{m+1}, \ldots, X_n$, forming the vectors

$$\boldsymbol{X}^{(1)} = \begin{pmatrix} X_1 \\ \vdots \\ X_m \end{pmatrix}, \qquad \boldsymbol{X}^{(2)} = \begin{pmatrix} X_{m+1} \\ \vdots \\ X_n \end{pmatrix}, \tag{12.58}$$

and

$$\boldsymbol{X} = \begin{pmatrix} \boldsymbol{X}^{(1)} \\ \boldsymbol{X}^{(2)} \end{pmatrix}. \tag{12.59}$$

Assume that $\boldsymbol{X}^{(1)}$ and $\boldsymbol{X}^{(2)}$ have a joint normal distribution, with means

$$E(\boldsymbol{X}^{(1)}) = \boldsymbol{\mu}^{(1)}, \qquad E(\boldsymbol{X}^{(2)}) = \boldsymbol{\mu}^{(2)} \tag{12.60}$$

and covariance matrices

$$E[(\boldsymbol{X}^{(1)} - \boldsymbol{\mu}^{(1)})(\boldsymbol{X}^{(1)} - \boldsymbol{\mu}^{(1)})'] = \boldsymbol{\Omega}_{11}, \tag{12.61}$$

$$E[(\boldsymbol{X}^{(2)} - \boldsymbol{\mu}^{(2)})(\boldsymbol{X}^{(2)} - \boldsymbol{\mu}^{(2)})'] = \boldsymbol{\Omega}_{22}, \tag{12.62}$$

$$E[(\boldsymbol{X}^{(1)} - \boldsymbol{\mu}^{(1)})(\boldsymbol{X}^{(2)} - \boldsymbol{\mu}^{(2)})'] = \boldsymbol{\Omega}_{12}, \tag{12.63}$$

$$E[(\boldsymbol{X}^{(2)} - \boldsymbol{\mu}^{(2)})(\boldsymbol{X}^{(1)} - \boldsymbol{\mu}^{(1)})'] = \boldsymbol{\Omega}_{21}. \tag{12.64}$$

Let

$$\boldsymbol{\mu} = \begin{pmatrix} \boldsymbol{\mu}^{(1)} \\ \boldsymbol{\mu}^{(2)} \end{pmatrix}, \qquad \boldsymbol{\Omega} = \begin{bmatrix} \boldsymbol{\Omega}_{11} & \boldsymbol{\Omega}_{12} \\ \boldsymbol{\Omega}_{21} & \boldsymbol{\Omega}_{22} \end{bmatrix}. \tag{12.65}$$

THEOREM 12.3

Let $\boldsymbol{X}$ be distributed $N(\boldsymbol{\mu}, \boldsymbol{\Omega})$ with $\boldsymbol{X}$ defined as in (12.59) and $\boldsymbol{\mu}$ and $\boldsymbol{\Omega}$ as in (12.65). Then a necessary and sufficient condition for $\boldsymbol{X}^{(1)}$ and $\boldsymbol{X}^{(2)}$ to be distributed independently is to have $\boldsymbol{\Omega}_{12}(= \boldsymbol{\Omega}_{21}') = \mathbf{0}$.

PROOF. We shall first show that $\mathbf{\Omega}_{12} = \mathbf{0}$ is sufficient for $X^{(1)}$ and $X^{(2)}$ to be distributed independently. With $\mathbf{\Omega}_{12} = \mathbf{0}(= \mathbf{\Omega}_{21}')$, we then have for $\mathbf{\Omega}^{-1}$

$$\mathbf{\Omega}^{-1} = \begin{bmatrix} \mathbf{\Omega}_{11}{}^{-1} & \mathbf{0} \\ \mathbf{0} & \mathbf{\Omega}_{22}{}^{-1} \end{bmatrix}, \tag{12.66}$$

so that the quadratic form in the exponent of the density function for X becomes

$$\phi = (\mathbf{x}^{(1)} - \boldsymbol{\mu}^{(1)})'\mathbf{\Omega}_{11}{}^{-1}(\mathbf{x}^{(1)} - \boldsymbol{\mu}^{(1)}) + (\mathbf{x}^{(2)} - \boldsymbol{\mu}^{(2)})'\mathbf{\Omega}_{22}{}^{-1}(\mathbf{x}^{(2)} - \boldsymbol{\mu}^{(2)}). \tag{12.67}$$

Also, for $|\mathbf{\Omega}|$,

$$|\mathbf{\Omega}| = |\mathbf{\Omega}_{11}| \cdot |\mathbf{\Omega}_{22}|. \tag{12.68}$$

Consequently the density function for X can be written as

$$\begin{aligned} &|\mathbf{\Omega}|^{-1}(2\pi)^{-n/2} \exp[-\tfrac{1}{2}(\mathbf{x} - \boldsymbol{\mu})'\mathbf{\Omega}^{-1}(\mathbf{x} - \boldsymbol{\mu})] \\ &= |\mathbf{\Omega}_{11}|^{-1} \cdot |\mathbf{\Omega}_{22}|^{-1}(2\pi)^{-n/2} \exp[-\tfrac{1}{2}(\mathbf{x}^{(1)} - \boldsymbol{\mu}^{(1)})'\mathbf{\Omega}_{11}{}^{-1} \\ &\quad \times (\mathbf{x}^{(1)} - \boldsymbol{\mu}^{(1)}) - \tfrac{1}{2}(\mathbf{x}^{(2)} - \boldsymbol{\mu}^{(2)})'\,\mathbf{\Omega}_{22}{}^{-1}(\mathbf{x}^{(2)} - \boldsymbol{\mu}^{(2)})] \\ &= |\mathbf{\Omega}_{11}|^{-1}(2\pi)^{-m/2} \exp[-\tfrac{1}{2}(\mathbf{x}^{(1)} - \boldsymbol{\mu}^{(1)})'\mathbf{\Omega}_{11}{}^{-1}(\mathbf{x}^{(1)} - \boldsymbol{\mu}^{(1)})] \\ &\quad \times |\mathbf{\Omega}_{22}|^{-1} \cdot (2\pi)^{-(n-m)/2} \exp[-\tfrac{1}{2}(\mathbf{x}^{(2)} - \boldsymbol{\mu}^{(2)})'\mathbf{\Omega}_{11}{}^{-1} \\ &\quad \times (\mathbf{x}^{(2)} - \boldsymbol{\mu}^{(2)})] \\ &= f_1(x_1, \ldots, x_m) \cdot f_2(x_{m+1}, \ldots, x_n), \end{aligned} \tag{12.69}$$

thus establishing that $X^{(1)}$ and $X^{(2)}$ are distributed independently.

Turning now to necessity, we employ the formula for the inversion of a matrix by partitioning[2]

$$\mathbf{\Omega}^{-1} = \begin{bmatrix} \mathbf{\Omega}_{11}{}^{-1} + \mathbf{\Omega}_{11}{}^{-1}\mathbf{\Omega}_{12}\mathbf{H}^{-1}\mathbf{\Omega}_{21}\mathbf{\Omega}_{11}{}^{-1} & -\mathbf{\Omega}_{11}{}^{-1}\mathbf{\Omega}_{12}\mathbf{H}^{-1} \\ -\mathbf{H}^{-1}\mathbf{\Omega}_{21}\mathbf{\Omega}_{11}{}^{-1} & \mathbf{H}^{-1} \end{bmatrix}. \tag{12.70}$$

where

$$\mathbf{H} = \mathbf{\Omega}_{22} - \mathbf{\Omega}_{21}\mathbf{\Omega}_{11}{}^{-1}\mathbf{\Omega}_{12}. \tag{12.71}$$

The quadratic form in the exponent of the density function for X can thus be written

$$\begin{aligned} Q &= (\mathbf{x} - \boldsymbol{\mu})'\mathbf{\Omega}^{-1}(\mathbf{x} - \boldsymbol{\mu}) \\ &= (\mathbf{x}^{(1)} - \boldsymbol{\mu}^{(1)})'\mathbf{\Omega}_{11}{}^{-1}(\mathbf{x}^{(1)} - \boldsymbol{\mu}^{(1)}) + (\mathbf{x}^{(1)} - \boldsymbol{\mu}^{(1)})' \\ &\quad \times \mathbf{\Omega}_{11}{}^{-1}\mathbf{\Omega}_{12}\mathbf{H}^{-1}\mathbf{\Omega}_{21}\mathbf{\Omega}_{11}{}^{-1}(\mathbf{x}^{(1)} - \boldsymbol{\mu}^{(1)}) \\ &\quad - 2(\mathbf{x}^{(1)} - \boldsymbol{\mu}^{(1)})'\mathbf{\Omega}_{11}{}^{-1}\mathbf{\Omega}_{12}\mathbf{H}^{-1}(\mathbf{x}^{(2)} - \boldsymbol{\mu}^{(2)}) + (\mathbf{x}^{(2)} - \boldsymbol{\mu}^{(2)})' \\ &\quad \times \mathbf{H}^{-1}(\mathbf{x}^{(2)} - \boldsymbol{\mu}^{(2)}). \end{aligned} \tag{12.72}$$

[2] See Theorem A.5.12 of Appendix 5.

Inspection of this expression and (12.71) shows that it will reduce to (12.67), and therefore to a form which allows the density function for X to be written as in (12.69), only if

$$\mathbf{\Omega}_{12}(=\mathbf{\Omega}_{21}') = \mathbf{0}.$$

The reader will observe that expression (12.69) also shows that $X^{(1)}$ and $X^{(2)}$ are distributed $N(\boldsymbol{\mu}^{(1)}, \mathbf{\Omega}_{11})$ and $N(\boldsymbol{\mu}^{(2)}, \mathbf{\Omega}_{22})$, respectively; that is, the marginal distributions of $X^{(1)}$ and $X^{(2)}$ are themselves multivariate normal. The next theorem establishes that this result holds independently of whether $X^{(1)}$ and $X^{(2)}$ are independent.

THEOREM 12.4
Let X be distributed $N(\boldsymbol{\mu}, \mathbf{\Omega})$, with X, $\boldsymbol{\mu}$, and $\mathbf{\Omega}$ defined according to (12.59) and (12.65). Then, the marginal distribution of $X^{(2)}$ is $N(\boldsymbol{\mu}^{(2)}, \mathbf{\Omega}_{22})$; similarly, the marginal distribution of $X^{(1)}$ is $N(\boldsymbol{\mu}^{(1)}, \mathbf{\Omega}_{11})$.

PROOF. Our procedure in proving this theorem will be to transform $X^{(1)}$ and $X^{(2)}$ to two new sets of variables, $Y^{(1)}$ and $Y^{(2)}$, which are uncorrelated with one another. Accordingly, let

$$\mathbf{y}^{(1)} = \mathbf{x}^{(1)} + \mathbf{M}\mathbf{x}^{(2)}, \tag{12.73}$$

$$\mathbf{y}^{(2)} = \mathbf{x}^{(2)}, \tag{12.74}$$

where $\mathbf{M}$ is chosen in such a way that the components of $Y^{(1)}$ have zero covariance with those of $Y^{(2)}$. The matrix $\mathbf{M}$ must accordingly satisfy the equation

$$\begin{aligned}
\mathbf{0} &= E[Y^{(1)} - E(Y^{(1)})][Y^{(2)} - E(Y^{(2)})]' \\
&= E[(X^{(1)} + \mathbf{M}X^{(2)} - \boldsymbol{\mu}^{(1)} - \mathbf{M}\boldsymbol{\mu}^{(2)})(X^{(2)} - \boldsymbol{\mu}^{(2)})'] \\
&= E\{[(X^{(1)} - \boldsymbol{\mu}^{(1)}) + \mathbf{M}(X^{(2)} - \boldsymbol{\mu}^{(2)})](X^{(2)} - \boldsymbol{\mu}^{(2)})'\} \\
&= \mathbf{\Omega}_{12} + \mathbf{M}\mathbf{\Omega}_{22}.
\end{aligned} \tag{12.75}$$

Hence,

$$\mathbf{M} = -\mathbf{\Omega}_{12}\mathbf{\Omega}_{22}^{-1} \tag{12.76}$$

and

$$\mathbf{y}^{(1)} = \mathbf{x}^{(1)} - \mathbf{\Omega}_{12}\mathbf{\Omega}_{22}^{-1}\mathbf{x}^{(2)}. \tag{12.77}$$

The vector

$$\begin{aligned}
Y &= \begin{pmatrix} Y^{(1)} \\ Y^{(2)} \end{pmatrix} \\
&= \begin{bmatrix} \mathbf{I} & -\mathbf{\Omega}_{12}\mathbf{\Omega}_{22}^{-1} \\ \mathbf{0} & \mathbf{I} \end{bmatrix} \begin{pmatrix} X^{(1)} \\ X^{(2)} \end{pmatrix}
\end{aligned} \tag{12.78}$$

is a nonsingular linear transformation of $\boldsymbol{X}$, and therefore has a normal distribution with mean

$$E(\boldsymbol{Y}) = E\begin{pmatrix}\boldsymbol{Y}^{(1)}\\ \boldsymbol{Y}^{(2)}\end{pmatrix}$$

$$= \begin{bmatrix}\boldsymbol{\mu}^{(1)} - \boldsymbol{\Omega}_{12}\boldsymbol{\Omega}_{22}^{-1}\boldsymbol{\mu}^{(2)}\\ \boldsymbol{\mu}^{(2)}\end{bmatrix}$$

$$= \begin{pmatrix}\boldsymbol{\nu}^{(1)}\\ \boldsymbol{\nu}^{(2)}\end{pmatrix} \quad \text{(say)}, \tag{12.79}$$

and covariance matrix

$$E[\boldsymbol{Y} - \boldsymbol{\nu})(\boldsymbol{Y} - \boldsymbol{\nu})']$$

$$= E\begin{bmatrix}(\boldsymbol{Y}^{(1)} - \boldsymbol{\nu}^{(1)})(\boldsymbol{Y}^{(1)} - \boldsymbol{\nu}^{(1)})' & (\boldsymbol{Y}^{(2)} - \boldsymbol{\nu}^{(2)})(\boldsymbol{Y}^{(1)} - \boldsymbol{\nu}^{(1)})'\\ (\boldsymbol{Y}^{(1)} - \boldsymbol{\nu}^{(1)})(\boldsymbol{Y}^{(2)} - \boldsymbol{\nu}^{(2)})' & (\boldsymbol{Y}^{(2)} - \boldsymbol{\nu}^{(2)})(\boldsymbol{Y}^{(2)} - \boldsymbol{\nu}^{(2)})'\end{bmatrix}$$

$$= \begin{bmatrix}\boldsymbol{\Omega}_{11} - \boldsymbol{\Omega}_{21}\boldsymbol{\Omega}_{22}^{-1}\boldsymbol{\Omega}_{12} & \mathbf{0}\\ \mathbf{0} & \boldsymbol{\Omega}_{22}\end{bmatrix}, \tag{12.80}$$

since

$$\begin{aligned}
&E[(\boldsymbol{Y}^{(1)} - \boldsymbol{\nu}^{(1)})(\boldsymbol{Y}^{(1)} - \boldsymbol{\nu}^{(1)})']\\
&\quad = E[(\boldsymbol{X}^{(1)} - \boldsymbol{\Omega}_{12}\boldsymbol{\Omega}_{22}^{-1}\boldsymbol{X}^{(2)} - \boldsymbol{\mu}^{(1)} + \boldsymbol{\Omega}_{12}\boldsymbol{\Omega}_{22}^{-1}\boldsymbol{\mu}^{(2)})\\
&\qquad \cdot (\boldsymbol{X}^{(1)} - \boldsymbol{\Omega}_{12}\boldsymbol{\Omega}_{22}^{-1}\boldsymbol{X}^{(2)} - \boldsymbol{\mu}^{(1)} + \boldsymbol{\Omega}_{12}\boldsymbol{\Omega}_{22}^{-1}\boldsymbol{\mu}^{(2)})']\\
&\quad = E\{[\boldsymbol{X}^{(1)} - \boldsymbol{\mu}^{(1)} - \boldsymbol{\Omega}_{12}\boldsymbol{\Omega}_{22}^{-1}(\boldsymbol{X}^{(2)} - \boldsymbol{\mu}^{(2)})][\boldsymbol{X}^{(1)} - \boldsymbol{\mu}^{(1)}\\
&\qquad - \boldsymbol{\Omega}_{12}\boldsymbol{\Omega}_{22}^{-1}(\boldsymbol{X}^{(2)} - \boldsymbol{\mu}^{(2)})]'\}\\
&\quad = \boldsymbol{\Omega}_{11} - \boldsymbol{\Omega}_{12}\boldsymbol{\Omega}_{22}^{-1}\boldsymbol{\Omega}_{21} - \boldsymbol{\Omega}_{12}\boldsymbol{\Omega}_{22}^{-1}\boldsymbol{\Omega}_{21}\\
&\qquad + \boldsymbol{\Omega}_{12}\boldsymbol{\Omega}_{22}^{-1}\boldsymbol{\Omega}_{22}\boldsymbol{\Omega}_{22}^{-1}\boldsymbol{\Omega}_{21}\\
&\quad = \boldsymbol{\Omega}_{11} - \boldsymbol{\Omega}_{12}\boldsymbol{\Omega}_{22}^{-1}\boldsymbol{\Omega}_{21}.
\end{aligned} \tag{12.81}$$

Thus, $\boldsymbol{Y}^{(1)}$ and $\boldsymbol{Y}^{(2)}$ are independent, and by Theorem 12.2 $\boldsymbol{Y}^{(2)} = \boldsymbol{X}^{(2)}$ has the marginal distribution $N(\boldsymbol{\mu}^{(2)}, \boldsymbol{\Omega}_{22})$. Moreover, it is clear that the above argument with $\boldsymbol{X}^{(1)}$ and $\boldsymbol{X}^{(2)}$ interchanged will show that $\boldsymbol{X}^{(1)}$ has marginal distribution $N(\boldsymbol{\mu}^{(1)}, \boldsymbol{\Omega}_{11})$. This proves the theorem.

Let us now consider the transformation

$$\boldsymbol{Y} = \mathbf{D}\boldsymbol{X}, \tag{12.82}$$

where $\boldsymbol{Y}$ has m components and $\mathbf{D}$ is an $m \times n$ matrix, with $m \leq n$. From Theorem 12.1, the expected value of $\boldsymbol{Y}$ is

$$E(\boldsymbol{Y}) = \mathbf{D}[(E(\boldsymbol{X})], \tag{12.83}$$

while its covariance matrix is

$$E[(Y - \mathbf{D}\boldsymbol{\mu})(Y - \mathbf{D}\boldsymbol{\mu})'] = E[\mathbf{D}(X - \boldsymbol{\mu})(X - \boldsymbol{\mu})'\mathbf{D}']$$
$$= \mathbf{D}\boldsymbol{\Omega}\mathbf{D}'. \tag{12.84}$$

The case $m = n$ and $\mathbf{D}$ nonsingular has already been treated in Theorem 12.2. Accordingly, suppose that $m < n$; assume also that $\mathbf{D}$ has rank m. In this case, we can find an $(n - m) \times n$ matrix $\mathbf{E}$, such that

$$\begin{pmatrix} Y \\ Z \end{pmatrix} = \begin{pmatrix} \mathbf{D} \\ \mathbf{E} \end{pmatrix} X \tag{12.85}$$

is a nonsingular transformation. Then, by Theorem 12.2, Y and Z have a joint normal distribution, and by the theorem just proved Y has a marginal normal distribution. Thus, we have proven the following theorem:

THEOREM 12.5
Let X be distributed $N(\boldsymbol{\mu}, \boldsymbol{\Omega})$, and let $Y = \mathbf{D}X$, where $\mathbf{D}$ is an $m \times n$ matrix of rank $m \leq n$. Then Y is distributed $N(\mathbf{D}\boldsymbol{\mu}, \mathbf{D}\boldsymbol{\Omega}\mathbf{D}')$.

A special case of this theorem which is of particular interest is when Y represents the sum of normal variables

$$Y = \sum_{i=1}^{n} X_i$$
$$= \boldsymbol{\iota}'X, \tag{12.86}$$

where $\boldsymbol{\iota}' = (1, \ldots, 1)$ denotes the summing vector of dimension n. In this case, we have:

COROLLARY 12.5.1
Let $Y = \boldsymbol{\iota}'X$, where X in $N(\boldsymbol{\mu}, \boldsymbol{\Omega})$. Then Y is distributed $N(\boldsymbol{\iota}'\boldsymbol{\mu}, \boldsymbol{\iota}'\boldsymbol{\Omega}\boldsymbol{\iota})$.

Our final effort in this chapter will be to show that the conditional distributions derived from the multivariate normal distribution are again multivariate normal.

THEOREM 12.6
Let X be distributed $N(\boldsymbol{\mu}, \boldsymbol{\Omega})$, and let X, $\boldsymbol{\mu}$, and $\boldsymbol{\Omega}$ be partitioned as in (12.59) and (12.65). Then the distribution of $X^{(1)}$ given $X^{(2)} = \mathbf{x}^{(2)}$ is normal with mean $\boldsymbol{\mu}^{(1)} + \boldsymbol{\Omega}_{12}\boldsymbol{\Omega}_{22}^{-1}(\mathbf{x}^{(2)} - \boldsymbol{\mu}^{(2)})$ and covariance matrix $\boldsymbol{\Omega}_{11} - \boldsymbol{\Omega}_{12}\boldsymbol{\Omega}_{22}^{-1}\boldsymbol{\Omega}_{21}$.

PROOF. In proving this theorem, we shall return to the algebra employed in proving Theorem 12.3. That is, let

$$Y^{(1)} = X^{(1)} - \Omega_{12}\Omega_{22}^{-1}X^{(2)}, \tag{12.87}$$

$$Y^{(2)} = X^{(2)}. \tag{12.88}$$

From (12.79) and (12.80), the joint density of $Y^{(1)}$ and $Y^{(2)}$ is given by

$$g(\mathbf{y}^{(1)}, \mathbf{y}^{(2)}) = N(Y^{(1)} | \boldsymbol{\mu}^{(1)} - \Omega_{12}\Omega_{22}^{-1}\boldsymbol{\mu}^{(2)}, \Omega_{11} - \Omega_{12}\Omega_{22}^{-1}\Omega_{21})N(y^{(2)} | \boldsymbol{\mu}^{(2)}, \Omega_{22}), \tag{12.89}$$

where $N(\mathbf{z} | \mathbf{a}, \mathbf{B})$ in this expression represents a multivariate normal density function with mean $\mathbf{a}$ and covariance matrix $\mathbf{B}$. The density function for $X^{(1)}$ and $X^{(2)}$ can be obtained from (12.89) by substituting $\mathbf{x}^{(1)} - \Omega_{12}\Omega_{22}^{-1}\mathbf{x}^{(2)}$ for $\mathbf{y}^{(1)}$ and $\mathbf{x}^{(2)}$ for $\mathbf{y}^{(2)}$, the Jacobian of this transformation being one. The resulting density for $X^{(1)}$ and $X^{(2)}$ is then

$$f(\mathbf{x}^{(1)}, \mathbf{x}^{(2)}) = |\Omega_{11\cdot 2}|^{-1/2}(2\pi)^{-m/2} \exp[-\tfrac{1}{2}\mathbf{z}'\Omega_{11\cdot 2}^{-1}\mathbf{z}] \cdot |\Omega_{22}|^{-1/2}(2\pi)^{-(n-m)/2} \times \exp[-\tfrac{1}{2}(\mathbf{x}^{(2)} - \boldsymbol{\mu}^{(2)})'\Omega_{22}^{-1}(\mathbf{x}^{(2)} - \boldsymbol{\mu}^{(2)})], \tag{12.90}$$

where

$$\mathbf{z} = \mathbf{x}^{(1)} - \boldsymbol{\mu}^{(1)} - \Omega_{12}\Omega_{22}^{-1}(\mathbf{x}^{(2)} - \boldsymbol{\mu}^{(2)}) \tag{12.91}$$

and

$$\Omega_{11\cdot 2} = \Omega_{11} - \Omega_{12}\Omega_{22}^{-1}\Omega_{21}. \tag{12.92}$$

The conditional density of $X^{(1)}$ given that $X^{(2)} = \mathbf{x}^{(2)}$ is then given by the ratio of (12.90) to the marginal density of $X^{(2)}$, which from Theorem 12.3 is equal to $N(\mathbf{x}^{(2)} | \boldsymbol{\mu}^{(2)}, \Omega_{22})$. Consequently,

$$f(\mathbf{x}^{(1)} | \mathbf{x}^{(2)}) = |\Omega_{11\cdot 2}|^{-1}(2\pi)^{-m/2} \cdot \exp[-\tfrac{1}{2}\mathbf{z}'\Omega_{11\cdot 2}^{-1}\mathbf{z}] = N(\mathbf{z} | \boldsymbol{\mu}^{(1)} + \Omega_{12}\Omega_{22}^{-1}(\mathbf{x}^{(2)} - \boldsymbol{\mu}^{(2)}), \Omega_{11} - \Omega_{12}\Omega_{22}^{-1}\Omega_{21}), \tag{12.93}$$

which proves the theorem.

REFERENCES

Anderson, T. W., *Introduction to Multivariate Statistical Analysis*, Wiley, 1958, chap. 2.

Mood, A. M. and Graybill, F. A., *Introduction to the Theory of Statistics*, 2nd ed., McGraw-Hill, 1963, chap. 9.

CHAPTER 13

DISTRIBUTIONS ASSOCIATED WITH THE NORMAL

In this chapter, we shall discuss three distributions which are closely associated with the normal distribution, namely, the chi-square, t, and F-distributions. All of these arise in connection with sampling from a normal population, which will be taken up in Chapter 15; their examination is therefore an important preliminary to that chapter.

13.1 CHI-SQUARE DISTRIBUTION

Let

$$Y_i = \frac{X_i - \mu_i}{\sigma_i}, \tag{13.1}$$

where X_i is normal with mean μ_i and variance σ_i^2, and suppose that we are interested in the distribution of the quantity

$$U = \sum_{i=1}^{k} Y_i^2. \tag{13.2}$$

We shall obtain the density function for U from its moment-generating function.

The moment-generating for U is, by definition, given by

$$M_U(t) = \int_{y_1} \cdots \int_{y_k} e^{t\Sigma y_i^2} f(y_1, \ldots, y_k)\, dy \cdots dy_k\,, \tag{13.3}$$

where $f(y_1, \ldots, y_k)$ is the joint density function of $y_1, \ldots, y_k$. In this case, $y_1, \ldots, y_k$ is $N(\mathbf{0}, \mathbf{I})$, so that

$$M_U(t) = (2\pi)^{-k/2} \int_{y_1} \cdots \int_{y_k} e^{t\Sigma y_i^2} e^{-(1/2)\Sigma y_i^2}\, dy_1 \cdots dy_k. \qquad (13.4)$$

Through rearrangement, the multiple integral on the right can be rewritten as the product of k integrals of the form

$$\frac{1}{\sqrt{2\pi}} \int_{y_i} e^{-(1/2)(1-2t)y_i^2}\, dy_i.$$

If this integral is divided and multiplied by $(1 - 2t)^{1/2}$,

$$\frac{1}{(1-2t)^{1/2}} \left[\left(\frac{1-2t}{2\pi} \right)^{1/2} \int_{y_i} e^{-(1/2)(1-2t)y_i^2}\, dy_i \right],$$

it is seen to be equal to $1/(1 - 2t)^{1/2}$, since the term within brackets represents the area beneath a normal density with mean zero and variance $1/(1 - 2t)$. Consequently, there being k such terms, it follows that

$$M_U(t) = \left(\frac{1}{1-2t} \right)^{k/2}. \qquad (13.5)$$

Returning to formula (8.49), we will recognize (13.5) as being the moment-generating function of a gamma distribution with $\alpha = (k/2) - 1$ and $\beta = 2$. In view of this, we therefore conclude that the density function of U is

$$f(u; k) = \frac{1}{[(k/2) - 1]!\,2^{k/2}}\, u^{(k/2)-1} e^{-(1/2)u}, \qquad u > 0. \qquad (13.6)$$

This particular form of the gamma distribution is called the *chi-square distribution with k degrees of freedom*, and refers to the distribution of the sum of squares of k standardized normal variates. Its name follows from the fact that the variable U is usually designated by the square of the Greek chi,

$$\chi^2 = \sum_{i=1}^{k} \left(\frac{X_i - \mu_i}{\sigma_i} \right)^2. \qquad (13.7)$$

The phrase "degrees of freedom" refers to the number of independent squares in the sum in (13.7); however, we may think of it as simply a name for the parameter k in (13.6). Frequently, the chi-square distribution with k degrees of freedom is denoted by $\chi_{(k)}^2$.

From formulas (8.50) and (8.51), the mean and variance of the chi-square distribution are

$$\mu = k, \tag{13.8}$$

$$\sigma^2 = 2k. \tag{13.9}$$

13.2 *t*-DISTRIBUTION

Consider the quantity

$$T = \frac{(X - \mu)/\sigma}{\sqrt{U/k}}, \tag{13.10}$$

where X is distributed normally with mean μ and variance σ^2 and U is distributed as chi-square, *independently* of X, with k degrees of freedom. Thus, T is the quotient of a standardized normal variate and the square root of the ratio of a chi-square to its number of degrees of freedom. We seek the distribution of T.

Let $Y = (X - \mu)/\sigma$; then, since X and U are independent, the joint density function of Y and U is

$$f(y, u) = \frac{1}{\sqrt{2\pi}} e^{-(1/2)y^2} \cdot \frac{1}{[(k-2)/2]!2^{k/2}} u^{(k-2)/2} e^{-(1/2)u}. \tag{13.11}$$

To find the distribution of T, we substitute for y in (13.11) and then integrate out u. Accordingly, let Y and U be transformed to new variables T and V by the transformations

$$y = t\sqrt{\frac{v}{k}}, \tag{13.12}$$

$$u = v. \tag{13.13}$$

Then

$$\frac{\partial y}{\partial t} = \sqrt{\frac{v}{k}}, \qquad \frac{\partial y}{\partial v} = \frac{t}{2k}\left(\frac{v}{k}\right)^{-1/2},$$

$$\frac{\partial u}{\partial t} = 0, \qquad \frac{\partial u}{\partial v} = 1, \tag{13.14}$$

so that the Jacobian of the inverse transformation from t and v to y and u is $\sqrt{v/k}$. Hence, the joint density function of T and V is

$$g(t, v) = \frac{1}{\sqrt{2\pi}} e^{-t^2 v/2k} \cdot \frac{1}{[(k-2)/2]!2^{k/2}} v^{(k-2)/2} e^{-(1/2)v} \cdot \sqrt{\frac{v}{k}}. \tag{13.15}$$

We now proceed to integrate out v:

$$h(t) = \frac{1}{\sqrt{k\pi}} \cdot \frac{1}{[(k-2)/2]!2} \int_0^\infty e^{-(1/2)(t^2/k+1)v} \left(\frac{v}{2}\right)^{(k-1)/2} dv. \tag{13.16}$$

Let

$$A = \frac{1}{2} \int_0^\infty e^{-(1/2)(t^2/k+1)v} \left(\frac{v}{2}\right)^{(k-1)/2} dv. \tag{13.17}$$

Next, let

$$\alpha - \frac{(k-1)}{2} \tag{13.18}$$

and

$$z = \frac{1}{2}\left(1 + \frac{t^2}{k}\right)v. \tag{13.19}$$

Then

$$\frac{v}{2} = \left(1 + \frac{t^2}{k}\right)^{-1} z, \tag{13.20}$$

so that

$$\left(\frac{v}{2}\right)^\alpha = \left(1 + \frac{t^2}{k}\right)^{-\alpha} z^\alpha. \tag{13.21}$$

Also,

$$d\left(\frac{v}{2}\right) = \left(1 + \frac{t^2}{k}\right)^{-1} dz. \tag{13.22}$$

With these substitutions, (13.17) becomes

$$\begin{aligned} A &= \left(1 + \frac{t^2}{k}\right)^{-(k+1)/2} \int_0^\infty z^\alpha e^{-z}\, dz \\ &= \left(1 + \frac{t^2}{k}\right)^{-(k+1)/2} \left[\frac{(k-1)}{2}\right]!, \end{aligned} \tag{13.23}$$

since the integral on the right is equal to $\Gamma(\alpha + 1)$.

Returning now to (13.16), we consequently finally find for the density function for T,

$$h(t) = \frac{1}{\sqrt{k\pi}} \cdot \frac{[(k-1)/2]!}{[(k-2)/2]!} \left(1 + \frac{t^2}{k}\right)^{-(k+1)/2}, \qquad -\infty < t < \infty. \tag{13.24}$$

This is the density function for the *t-distribtuion*, and was first derived by W. S. Gossett (under the name of "Student") early in this century. As shall be seen in Chapter 15 the t-distribution comes into play in sampling from a normal population with unknown variance. The t-distribution depends on a single parameter k and has been extensively tabulated.

From an examination of (13.24), it is evident that the density function for the t-distribution is symmetrical about the point $t = 0$, which means that all odd moments, if they exist, vanish. Thus, in particular, the mean of the distribution (for $k \geq 2$) is equal to zero. The even moments are given by

$$\mu_{2r} = k^r \frac{\Gamma(r + \frac{1}{2})\Gamma(k/2 - r)}{\Gamma(\frac{1}{2})\Gamma(k/2)}, \qquad r = 0, 1, \ldots, m < \frac{k}{2}, \tag{13.25}$$

Moments of order $\geq k$ do not exist for this distribution.

13.3 *F*-DISTRIBUTION

The third and final distribution to be discussed in this chapter arises as the distribution of a quantity which is the quotient of two independently distributed chi-square variates, each of which is divided by its degrees of freedom. That is, it is the distribution of the variable

$$Y = \frac{U/m}{V/n}, \tag{13.26}$$

where U is distributed as chi-square with m degrees of freedom and V as chi-square with n degrees of freedom, the two distributions being independent of one another.

Since U and V are independent, their joint density is given by

$$f(u, v) = \frac{1}{\Gamma(m/2)\Gamma(n/2)2^{(m+n)/2}} u^{(m-2)/2} v^{(n-2)/2} e^{-(u+v)/2}, \tag{13.27}$$

for $u, v > 0$. As in the derivation of the t-distribution, we employ a change of variables from U and V to Y and W, where

$$u = \frac{mwy}{\mathrm{n}}, \tag{13.28}$$

$$v = w, \tag{13.29}$$

and then integrate out the unwanted quantity w. Since the Jacobian of the transformation in (13.28)–(13.29) is equal to $(m/n)w$, the density function for Y and W is therefore found to be

$$g(y, w) = \frac{1}{\Gamma(m/2)\Gamma(n/2)2^{(m+n)/2}}\left(\frac{mwy}{n}\right)^{(m-2)/2} w^{(n-2)/2} e^{-(mwy/n+w)/2}\left(\frac{m}{n}\right)w$$

$$= \frac{1}{\Gamma(m/2)\Gamma(n/2)}\left(\frac{m}{n}\right)^{m/2} y^{(m-2)/2}\frac{1}{2}\left(\frac{w}{2}\right)^{(m+n-2)/2} e^{-(w+wmy/n)/2}. \tag{13.30}$$

We now wish to integrate out w:

$$h(y) = \frac{1}{\Gamma(m/2)\Gamma(n/2)}\left(\frac{m}{n}\right)^{m/2} y^{(m-2)/2}\frac{1}{2}\int_0^\infty\left(\frac{w}{2}\right)^{(m+n-2)/2} e^{-w(1+my/n)/2}\,dw. \tag{13.31}$$

Proceeding as in the evaluation of the integral in (13.16), let

$$A = \frac{1}{2}\int_0^\infty\left(\frac{w}{2}\right)^{(m+n-2)/2} e^{-w(1+my/n)/2}\,dw. \tag{13.32}$$

Next, let

$$z = \frac{w}{2}\left(1 + \frac{m}{n}y\right), \tag{13.33}$$

$$\alpha = \frac{m+n}{2} - 1. \tag{13.34}$$

Then

$$\frac{w}{2} = \left(1 + \frac{m}{n}y\right)^{-1} z \tag{13.35}$$

and

$$\left(\frac{w}{2}\right)^{\alpha} = \left(1 + \frac{m}{n}y\right)^{-(m+n-2)/2} z^{\alpha}. \tag{13.36}$$

Also,

$$d\left(\frac{w}{2}\right) = \left(1 + \frac{m}{n}y\right)^{-1} dz, \tag{13.37}$$

whence

$$A = \left(1 + \frac{m}{n}y\right)^{-(m+n)/2}\int_0^\infty z^{\alpha} e^{-z}\,dz$$

$$= \left(1 + \frac{m}{n}y\right)^{-(m+n)/2}\Gamma\left[\frac{(m+n)}{2}\right]. \tag{13.38}$$

Consequently, returning to (13.31), we find for the density function for Y,

$$h(y) = \frac{\Gamma[(m+n)/2]}{\Gamma(m/2)\Gamma(n/2)}\left(\frac{m}{n}\right)^{m/2}\left(1+\frac{m}{n}y\right)^{(m+n)/2} y^{(m-2)/2}. \qquad (13.39)$$

A random variable having the function in (13.39) as its density function is said to follow the F-distribution, and, as we shall find in Chapter 18, it finds major use in testing hypotheses regarding the equality of means and homogeneity of variances in different normal populations. The distribution depends on two parameters, m and n, which are, respectively, the numbers of degrees of freedom of the chi-square variates in the numerator and denominator.

The F-distribution is skewed to the right, and its range is always zero to infinity. It is J-shaped for $m \leq 2$, but becomes unimodal for $m > 2$. The mean and variance of the distribution are

$$\mu = \frac{n}{n-2}, \qquad n > 2, \qquad (13.40)$$

$$\sigma^2 = \frac{2n^2(m+n-2)}{m(n-2)^2(n-4)}, \qquad n > 4. \qquad (13.41)$$

The variance does not exist for $n \leq 4$, while the mean does not exist for $n \leq 2$.

Finally, it is useful to remark that as m and n become large, the F-distribution has the normal distribution as its limit. This is also true, incidentally, of the chi-square distribution and the t-distribution; that is, as the number of degrees of freedom in these distributions tends to infinity, their limit form is the normal.

REFERENCES

Freeman, H., *An Introduction to Statistical Inference*, Addison-Wesley, 1963, chap. 23.

Mood, A. M. and Graybill, F. A., *Introduction to the Theory of Statistics*, 2nd ed., McGraw-Hill, 1963, chap. 10.

CHAPTER 14

SAMPLING FROM GENERAL POPULATIONS

In this chapter we begin a study of sampling. Until now we have been dealing with the general theory of probability, general probability distributions, and several specific distributions that are frequently encountered in practice. Now we undertake a discussion that is motivated by an argument that implies a physical operation: We wish to draw a sample of data from a population and then study the behavior of the sample data in light of *known* or *hypothesized* characteristics of the population. Later, we will turn the analysis around and use the information given in a sample to hazard conclusions about *unknown* characteristics of the population from which the sample was drawn. This is usually the real world problem and is the central problem of statistical inference. Some examples should make this clear.

1. In preparing an index of consumer prices it is necessary to have a set of numbers with which to weight the importance of the several commodities whose prices comprise the index. A natural set of weights to use are the proportions of total expenditure on all commodities that consumers spend on the commodities in question. However, it would be prohibitively expensive to obtain this information from the entire population of consuming units, and so a representative sample is called for.
2. At the Ministerio de Hacienda in Bogota, Colombia, the processing of tax returns is rapidly becoming mechanized. For a large

number of tax payers, information on gross income, net income, taxable income, various classes of deductions, and so on, are now stored on magnetic tape. However, there is still a significant number of tax payers whose returns are still processed manually and whose data therefore do not appear on tape. Such tax information is potentially extremely valuable to policymakers because a proper analysis of the data for a given year will provide information on average and marginal tax rates, importance of the various types of deductions, and so on. The availability of high-speed computers makes the analysis of all the tax returns stored on tape feasible, but account should also be taken of the returns processed manually. One way, of course, would be to code all of the manually processed returns and transmit them to tape also, while another way would be to analyze each of them by hand. But the number of returns involved renders each of these solutions impossible. The feasible solution, therefore, is to draw a sample from the population of manual returns, blow up the sample figures to population proportions, and then combine those figures with the ones obtained from the machine processed returns.

To implement these procedures, however, it is necessary to have a foundation based on the behavior of samples drawn from a population with all relevant characteristics known. Our purpose in this chapter is to develop nomenclature and to derive a number of basic results that are obtained when the population being sampled is described by a general probability distribution. *Inter alia*, this chapter will derive the weak law of large numbers, which is of fundamental importance to the study of sampling, and in addition will provide a statement and proof of the classical central limit theorem. However, we begin by introducing the concept of random sampling.

14.1 RANDOM SAMPLING

Suppose that we have a population that is defined as the totality of all possible values of a random variable X. Let the density function of X be $f(x)$. A sample of size n drawn from this population will yield a set of n numbers

$$x_1, x_2, \ldots, x_n.$$

We require that the sampling process by which these numbers are obtained be such that they give estimates of characteristics of $f(x)$ with *determinable errors of estimates*.

It is the latter requirement that severely restricts the ways and means of selecting the n numbers. For example, if the population refers to persons and our desire is to estimate the mean weight of the population, choosing the apparent heaviest and apparent lightest persons in the population and then averaging their weights for the estimate of the mean may have much to recommend it. But we are unable to provide an estimate of the error of such an estimate, and so we generally reject this method of sampling (although this does not mean that in the absence of a measure of sampling error we always reject a sampling method that is otherwise believed to be good). However, if certain procedures are observed, there does exist a method of sampling which permits probabilistic estimates of sampling errors to be evaluated, although somewhat unfortunately the method differs between sampling from an infinite population and sampling from a finite population without replacement. Both are generally described as *random sampling*.

Consider a single random variable X, either discrete or continuous, with density function $f(x)$, and assume that we repeatedly draw samples of size n from the population described by $f(x)$, that is, we replicate (make repeatedly) n trials on the random variable X. (If the population is finite, the n sample values are assumed to be replaced after each replication.) The sample data of each replication will be of the form of an n-tuple:

$$\begin{array}{ll} \text{replication 1:} & x_1^{(1)}, x_2^{(1)}, \ldots, x_n^{(1)}, \\ \text{replication 2:} & x_1^{(2)}, x_2^{(2)}, \ldots, x_n^{(2)}, \\ \text{replication 3:} & x_1^{(3)}, x_2^{(3)}, \ldots, x_n^{(3)}, \\ \vdots & \vdots \quad \vdots \qquad \vdots \end{array} \tag{14.1}$$

We now assume that the outcome of any replication,

$$x_1^{(i)}, x_2^{(i)}, \ldots, x_n^{(i)},$$

can be regarded as a value of the n-dimensional random variable

$$X_1, X_2, \ldots, X_n,$$

with joint density function $g(x_1, x_2, \ldots, x_n)$. This is a critical assumption, for it connects the outcomes of the physical process of drawing a sample with the abstract theory of probability.

We can now define random sampling: If the sampling operation which produces (14.1) meets the two following conditions, then the sampling is said to be *random*:

1. The outcomes of the n trials within any replication must be statistically independent of each other, that is, we must have

$$g(x_1, x_2, \ldots, x_n) = g_1(x_1)g_2(x_2) \cdots g_n(x_n), \tag{14.2}$$

2. The marginal density functions are identical and equal to f, the density function of X in the population,

$$g_1 = g_2 = \cdots = g_n = f. \tag{14.3}$$

If these two conditions are met, then the joint density function of the sample variables, $g(x_1, x_2, \ldots, x_n)$, is expressible in terms of the density function of the random variable of the population $f(x)$. Thus the sample, which is an n-dimensional random variable, is connected in probability to the population from which it comes, and its properties depend on the properties of f.

As example of random sampling, we can refer again to two experiments that we have discussed before, namely, the flipping of a coin and the rolling of a die. With each of these, we can *in principle* perform the experiment an indefinitely large number of times (assuming that the coin or die never wears out), so that we can view the populations as being infinite, and we can, without difficulty, ensure that the trials are independent. For the coin, therefore, if we denote the outcome by X with probabilities of p and $1 - p$ for heads and tails, respectively, the joint density function for a sample of n flips composed of r heads and $n - r$ tails is

$$\begin{aligned} g(x_1, x_2, \ldots, x_n) &= g_1(x_1)g_2(x_2) \cdots g_n(x_n) \\ &= p^r(1 - p)^{n-r}. \end{aligned} \tag{14.4}$$

Similarly, with the rolling of a die, if we assume that the die is fair so that $f(x) = \frac{1}{6}$ $(x = 1, 2, \ldots, 6)$, the joint density function for *any* sample of n rolls is

$$g(x_1, x_2, \ldots, x_n) = (\tfrac{1}{6})^n. \tag{14.5}$$

Unfortunately, there does not exist a set of precautions that guarantees that any physical operation of sampling will satisfy the mathematical conditions of (14.2) and (14.3). We can only make every effort to draw the sample in a way that each unit drawn does not affect any other, thus hoping to achieve independence in a probability sense, and to take care that all units are in fact drawn from the same population. The former is usually the more difficult to ensure.

Random Sampling from a Finite Population

When sampling from a finite population, we must distinguish two cases, namely, sampling with replacement and sampling without replacement. If the samples are drawn with replacement, the population is in essence infinite from the point of view of the sampling procedure in that its density function is unaffected by the sampling. Expressions (14.2) and (14.3) remain the conditions for random sampling, and, assuming that the population contains N units, these are fulfilled if the n units in the sample are drawn in such a way that each of the N units has probability $1/N$ of being drawn. For example, if the population is comprised of 4 different objects a, b, c, d, a random sample of size 2 can be drawn by ensuring that the first of the two places in the sample can be any of the 4 letters, with probability $\frac{1}{4}$ for each, and that the second place can also be any of the 4 letters with equal probabilities. Note that the number of ordered samples is N^n, which in the present case is equal to $4^2 = 16$:

$$\begin{matrix} aa & ab & ac & ad \\ ba & bb & bc & bd \\ ca & cb & cc & cd \\ da & db & dc & dd \end{matrix}$$

with each ordered sample having probability $(\frac{1}{4})^2$ of being drawn.

On the other hand, if a sample of size n were to be drawn without replacement from a finite population, we would have an example of non-random sampling, since the probabilities would change from one trial to the next. For example, if we have a population of N objects, each object has probability $1/N$ of being drawn on the first draw. On the second draw, however, the probability of being drawn for each of the remaining $N - 1$ objects is increased to $1/(N - 1)$, and so on.

In this situation, a modification of the above definition of random sampling is necessary, and the one that is usually made is to define randomness as follows:

> Sampling without replacement from a finite population is random if all possible sequences of n trials, that is, all possible samples of size n, have the same probability of being drawn. (14.6)

By this definition the population to be sampled thus becomes the synthesis of all possible samples of size n from the original population.

As an example, we can take the population considered above which consists of the 4 objects a, b, c, d. A sample of size 2 drawn without replacement from this population would then be deemed to be done by a random procedure if the 12 possible sequences

$$\begin{matrix} ab & ac & ad \\ ba & bc & bd \\ ca & cb & cd \\ da & db & dc \end{matrix}$$

have equal probability of being drawn. One sampling procedure that would ensure that this condition would be met would be the following. Select the first object in such a way that the probability of each object being selected is $1/N$. Then select the second object from the remaining $N - 1$ objects in a way that each has probability $1/(N - 1)$ of being drawn. Continuing, it is easy to see that each sample of size n will have probability

$$\frac{1}{N} \cdot \frac{1}{N-1} \cdots \frac{1}{N-n+1}$$

of being drawn. Samples so drawn, without replacement from a finite population, meet the above condition of randomness, and, what is especially important, errors of inference based on the data of such samples can be determined.

However, the remark that "all samples drawn at random of size n are equally likely" must be interpreted with care. We have seen that it is meaningful when sampling without replacement, but what it might mean when sampling from a population described by a normal distributon is obscure. And it does not mean that in sampling, for instance, from a binomial population with $p \neq \frac{1}{2}$ that the 8 possible outcomes in a sample of size 3,

$$111 \quad 110 \quad 101 \quad 100 \quad 011 \quad 010 \quad 001 \quad 000,$$

are equally likely to be drawn. The sampling process that produces these 3-tuples is random, but the 3-tuples are not equally likely.

On the other hand, it is true that each sample from a binomial population that contains a fixed number of successes has the same probability. In the above example, outcomes 2, 3, and 5 each yield 2 successes in a sample of 3 and each has the probability ppq, the latter following from the independence of the drawings. Similarly, even though the probabilities do not remain constant from one drawing to the next, n drawings from a hypergeometric population also lead to equally likely samples in this sense. For example, samples of size 3 yielding 2 successes,

$$110 \quad 101 \quad 011,$$

are drawn without replacement from a hypergeometric population containing A successes and B failures, $A + B - N$, with the respective probabilities

$$\frac{A}{N} \cdot \frac{A-1}{N-1} \cdot \frac{B}{N-2}, \qquad \frac{A}{N} \cdot \frac{B}{N-1} \cdot \frac{A-1}{N-2}, \qquad \frac{B}{N} \cdot \frac{A}{N-1} \cdot \frac{A-1}{N-2},$$

which are all equal.

Finally, we should emphasize that in random sampling, it is the sampling method and not the sample that is random. A sample obtained by a method that meets the conditions that have been set out in (14.2), (14.3), or (14.6), and thus satisfies the requirements of random sampling, may be a strange sample indeed. Randomness is not synonymous with representativeness.

Use of Random Number Tables in Sampling

Frequently we may have a population from which we wish to sample, such as the manually processed tax returns discussed at the beginning of this chapter, but there is some problem as to how to go about actually doing it. One procedure would be to number each unit in the population, also write the numbers on small slips of paper (one to each slip of paper), put the numbers in a hat, shake them up, and then blindfolded, draw from the hat slips equal in number to the size of sample desired. Such an operation has much to recommend it when the population is small and the circumstances informal, but in scientific work better control over the conditions of randomness is needed. The use of a random number table provides such control, and has the added advantage of being a simpler operation physically than "drawing from a hat."

A number of random number tables now exist, probably the most widely known being those published by the RAND Corporation. The RAND tables were generated from a normal random variable, but others have been generated from uniform, binomial, and Poisson variates. Also, a variety of random number generators for use on high-speed computers have been constructed, so that "pseudo" random numbers in nearly any shape and form are now readily available.

Using a random number table to draw a sample is simple. As with "drawing from the hat," we first number the elements in the population that is to be sampled. Next, the size of the sample is set. Then, from a table of random numbers generated by a uniform distribution and whose entries contain at least as many digits as there are elements in the population being sampled, we proceed to draw a set of random numbers equal in size to the postulated sample size. Any convenient rule can be used in selecting the random numbers; we might, for example, take every tenth number in the table beginning with the one that corresponds to the current day of the month. If the population contains N elements, but the random numbers contain $R > N$ digits, then $R - N$ of the digits must be ignored. It does not matter which $R - N$ are ignored, so long as they are always the same for each number. Once the random numbers are drawn, the final task is to select the elements of the population whose numbers correspond to the random numbers. These then comprise the sample.

14.2 THE JOINT DISTRIBUTION OF A SAMPLE DRAWN AT RANDOM

Suppose that a population is described by a random variable X with density function $f(x)$. Let a sample of two values of X, x_1, and x_2 be drawn at random. The pair of numbers (x_1, x_2) determines a point in the real plane, and the population of all such pairs of numbers that might have been drawn forms a bivariate population. We are interested in finding the distribution of this bivariate population in terms of the original distribution $f(x)$.

The joint distribution of X_1 and X_2 must be some function, say $g(x_1, x_2)$, such that for all a_1, a_2, b_1, b_2,

$$P(a_1 < X_1 < b_1, < a_2 < X_2 < b_2) = \int_{a_1}^{b_1} \int_{a_2}^{b_2} g(x_1, x_2)\, dx_2\, dx_1. \tag{14.7}$$

By the definition of random sampling, however, X_1 and X_2 are independent in the statistical sense, so that $g(x_1, x_2)$ factors into the product of the marginal distribution of X_1 and the marginal distribution of X_2. In the present case, therefore,

$$g(x_1, x_2) = f(x_1)f(x_2). \tag{14.8}$$

As an example, suppose that x is a binomial variable that takes the values 0 and 1 with probabilities q and p, respectively. Then

$$f(x) = \binom{1}{x} p^x q^{1-x}, \qquad x = 0, 1, \tag{14.9}$$

which, since

$$\binom{1}{0} = \binom{1}{1} = 1,$$

can be written as

$$f(x) = p^x q^{1-x}. \tag{14.10}$$

Therefore, the joint density function for a sample of 2 randomly drawn (with replacement) from $f(x)$ is

$$g(x_1, x_2) = p^{x_1+x_2} q^{2-x_1-x_2}, \qquad x_1, x_2 = 0, 1, \tag{14.11}$$

which is defined at the 4 points (0, 0), (0, 1), (1, 0), (1, 1) in the x_1x_2-plane. It should be noted that this density is not what we would obtain by drawing 2 elements from a binomial population and counting the successes, say Y. That density function is

$$h(y) = \binom{2}{y} p^y q^{2-y}, \qquad = y\,0, 1, 2, \tag{14.12}$$

which differs from (14.5) in that it is the distribution of the single random variable $Y = X_1 + X_2$; Equation (14.11) gives the joint distribution of the two random variables X_1 and X_2.

Through the same reasoning as with a sample of size 2, we find that the joint density function of samples of size n, $x_1, x_2, \ldots, x_n$, drawn (with replacement if from a finite population) at random from a population with density function $f(x)$ is

$$g(x_1, x_2, \ldots, x_n) = f(x_1) \cdot f(x_2) \cdots f(x_n). \tag{14.13}$$

This again gives the density function for the sample in the order drawn.

14.3 SAMPLE MOMENTS

One of the central problems in statistics is the following: It is desired to study a population described by the density function $f(x, \theta)$, where θ is a parameter. The form of the function f is known, but the parameter θ is not. Therefore, a sample of size n, $x_1, x_2, \ldots, x_n$, is drawn at random from the population with the idea of using some function of the sample, say $h(x_1, x_2, \ldots, x_n)$, to "estimate" the unknown parameter θ. The problem is to determine which function will be (in a certain sense) "best" to use, and this will be explored in detail in a later chapter. In this section, we shall examine certain functions of random samples, namely, the sample moments. To begin the discussion, however, we shall define a *statistic*:

> *Statistic:* A statistic is a function of observable random variables that do not contain any unknown parameters.

For example, if a random variable x is normally distributed with unknown parameters mean μ and variance σ^2, x, $x + 5$, and $x^2 + \ln x$ are statistics, while $x - \mu$ and x/σ are not, since they are functions of the unknown parameters. Since a statistic is a function of random variables, it also is a random variable, and therefore has a distribution. For many statistics, and fortunately for the most important in practice, their distributions can be derived from the distributions that describe the parent populations. But, more about this later.

Let $x_1, x_2, \ldots, x_n$ be a sample drawn at random from the population described by the density function $f(x)$. Then the rth *sample moment* about the point $a = 0$ is defined as

$$m_r' = \frac{1}{n} \sum_{i=1}^{n} x_i^r, \qquad r = 0, 1, 2, \ldots. \tag{14.14}$$

In particular, when $r = 1$, we have the sample mean, which is usually denoted by $\bar{x}$:

$$\bar{x} = \frac{1}{n} \sum_{i=1}^{n} x_i. \tag{14.15}$$

If we define the moments about the point $a = \bar{x}$, then we have the *adjusted* sample moments

$$m_r = \frac{1}{n} \sum_{i=1}^{n} (x_i - \bar{x})^r, \qquad r = 0, 1, 2, \ldots. \tag{14.16}$$

For $r = 2$, we have the sample variance, which is usually denoted by s^2:

$$s^2 = \frac{1}{n} \sum_{i=1}^{n} (x_i - \bar{x})^2. \tag{14.17}$$

Expanding the right-hand side of (14.17), we find that

$$\begin{aligned} \frac{1}{n} \sum_{i=1}^{n} (x_i - \bar{x})^2 &= \frac{1}{n} \sum_{i=1}^{n} (x_i^2 - 2x_i\bar{x} + \bar{x}^2) \\ &= \frac{1}{n} \left[\sum_{i=1}^{n} x_i^2 - 2\bar{x} \sum_{i=1}^{n} x_i + n\bar{x}^2 \right] \\ &= \frac{1}{n} \left[\sum_{i=1}^{n} x_i^2 - 2n\bar{x}^2 + n\bar{x}^2 \right] \\ &= \frac{1}{n} \left[\sum_{i=1}^{n} x_i^2 - n\bar{x}^2 \right]. \end{aligned} \tag{14.18}$$

Therefore,

$$s^2 = \frac{1}{n} [\sum x_i^2 - n\bar{x}^2], \tag{14.19}$$

an expression which is analogous to formula (6.12) for the population variance σ^2.

The sample mean and variance are the two most widely used of the sample moments, and are additional examples of statistics, since they depend only on sample values and not on population parameters.

An important property of the sample mean is that on the average it is equal to the population mean—that is, if μ is the population mean, then $E(\overline{X}) = \mu$. This property is called *unbiasedness*, and is very easy to show, for we have

$$\begin{aligned}
E(\overline{X}) &= \frac{1}{n} E\left[\sum_{i=1}^{n} X_i\right] \\
&= \frac{1}{n} \sum_{i=1}^{n} E(X_i) \\
&= \frac{1}{n} \sum_{i=1}^{n} \mu \qquad \text{[since } E(X_i) = \mu \text{ for all } i\text{]} \\
&= \frac{1}{n} \cdot n\mu \\
&= \mu. \qquad (14.20)
\end{aligned}$$

Next, we shall derive the variance of the sample mean. Let var $X = \sigma^2$. Then

$$\begin{aligned}
\text{var } \overline{X} &= \text{var } \frac{1}{n} \sum_{i=1}^{n} X_i \\
&= \frac{1}{n^2} \text{var } \sum_{i=1}^{n} X_i \\
&= \frac{1}{n^2} \sum_{i=1}^{n} \text{var } X_i \\
&= \frac{1}{n^2} \sum_{i=1}^{n} \sigma^2 \qquad (\text{since var } X_i = \sigma^2 \text{ for all } i) \\
&= \frac{\sigma^2}{n}. \qquad (14.21)
\end{aligned}$$

Thus, we see that the distribution of the sample mean has mean equal to the population mean and variance equal to the population variance divided by n, the size of the sample. This is true, moreover, no matter what the distribution of the parent population, so long as the latter has finite variance and the sampling is random.

The fact that the variance of the sample mean, usually written as $\sigma_{\bar{x}}^2$, is equal to σ^2/n is an extremely important result, for it implies that whatever the parent distribution (again so long as σ^2 is finite), the distribution of the sample mean becomes more and more concentrated on the population mean as the sample size increases. As a result, it follows that we can say with certainty that the sample mean becomes a better and better estimate of the population mean as the sample size increases. This is essentially a statement of the weak law of large numbers, which we shall now discuss and prove.

14.4 WEAK LAW OF LARGE NUMBERS

Let $f(x)$ be the density function of the continuous random variable X with mean μ and finite variance σ^2. The mean is unknown (the variance can also be unknown; for the present discussion, all that matters is that it be finite), and the problem is to estimate it. Speaking loosely, $\mu[= E(X)]$ is the average of all possible values of the random variable X. On the other hand, in any real world problem we can only observe a finite number of values of X, and a crucial question then is: Can a reliable inference about $E(X)$, the "average" of an infinity of values of X, be made on the basis of a finite number of values (say a sample of size n)? The answer is yes, and it follows from the weak law of large numbers.

THEOREM 14.1 (Weak Law of Large Numbers)
Let a parent population be described by the density function $f(x)$ with mean μ and variance $\sigma^2 < \infty$. Then for any two arbitrary small numbers ε and δ, $\varepsilon > 0$ and $0 < \delta < 1$, there exists an integer n such that if a sample of size n or larger is drawn at random from $f(x)$, then the probability is greater than $1 - \delta$ (that is, is as close to 1 as desired) that $\bar{x}_n$, the mean of the sample, deviates from μ by less than ε (that is, is arbitrarily close to μ)—that is,

$$P(|\bar{x}_n - \mu| < \varepsilon) > 1 - \delta. \tag{14.22}$$

PROOF. The proof follows from an application of Chebyshev's inequality (Theorem 6.1), which we recall, is

$$P\left(|\bar{x}_n - \mu| < \frac{h\sigma}{\sqrt{n}}\right) > 1 - \frac{1}{h^2}. \tag{14.23}$$

Choose h such that $1/h^2 = \delta$, or so that $h = 1/\sqrt{\delta}$. Then choose n such that $h\sigma/\sqrt{n} = \varepsilon$. Then $n = \sigma^2/\delta\varepsilon^2$. When these values are substituted into (14.23), Equation (14.22) follows, and the theorem is proved.

As an example, consider a distribution which has an unknown mean and variance equal to 1. Then, how large a sample must be drawn (at random) in order that the probability be at least 0.95 that the sample mean $\bar{x}$ will not deviate by more than 0.4 from the population mean? We have $\sigma^2 = 1$, $\varepsilon = 0.4$, and $\delta = 0.05$, so that

$$n = \frac{1}{0.05(0.16)} = 125. \tag{14.24}$$

In other words, the sample size must be at least 125.

The importance of the weak law of large numbers cannot be overstressed for it provides a rigorous link via probability from random sampling, which is a physical operation, to a density function, which is a mathematical concept. And most important of all, it shows how inductive inferences can be made to a population and the reliability of the inference measured in terms of probability.

14.5 MEAN AND VARIANCE OF SAMPLE MEAN WHEN SAMPLING WITHOUT REPLACEMENT FROM A FINITE POPULATION

The results of the two preceding sections have been with reference to samples randomly drawn from an infinite population. In this section, we deal with a finite population where the samples are drawn without replacement. Let the size of the sample be n and the number of elements in the population be N. If the sample were drawn with replacement, there would be N^n possible samples in the sample space. Without replacement, however, the first element of the sample, say x_1, can be drawn in N ways, the second element, say x_2, in only $N - 1$ ways, x_3 in only $N - 2$ ways, and so on. Therefore, there are only $N!(N - n)!$ possible samples in the sample space.

Assume that each possible sample has the same probability p. Then

$$p = \frac{(N - n)!}{N!}. \tag{14.25}$$

By definition,

$$\mu = \sum_{i=1}^{N} \frac{x_i}{N}, \qquad \sigma^2 = \sum_{i=1}^{N} \frac{(x_i - \mu)^2}{N},$$

and the mean of a sample of size n drawn at random from this population is

$$\bar{x} = \sum_{i=1}^{n} \frac{x_i}{n}.$$

Thus,

$$E(\bar{X}) = E \sum_{i=1}^{n} \frac{X_i}{n}$$

$$= \frac{1}{n} \sum_{i=1}^{n} E(X_i). \qquad (14.26)$$

We shall now show that $E(X_i) = \mu$ for all i, even though the sampling is without replacement. From the definition of expected value, we have for X_2, say,

$$E(X_2) = x_1 \cdot P(X_2 = x_1) + x_2 \cdot P(X_2 = x_2) + \cdots + x_N \cdot P(X_2 = x_N). \qquad (14.27)$$

Consider the first term on the right. Altogether there are $N!/(N - n)!$ possible samples of size n, and of these $(N - 1) \cdot (N - 2) \cdots (N - n + 1)$ can have $X_2 = x_1$. For if X_2 is to be x_1, X_2 can assume only one population value (x_1 itself), X_1 can then take only one of the remaining $N - 1$ population values, X_3 one of the remaining $N - 2$, and so on. Therefore,

$$P(X_2 = x_1) = \frac{(N - 1)(N - 2) \cdots (N - n + 1)(N - n)!}{N!}$$

$$= \frac{(N - 1)!}{N!}$$

$$= \frac{1}{N}. \qquad (14.28)$$

Similar considerations will lead to the fact that for all i and j,

$$P(X_i = x_j) = \frac{1}{N}, \qquad (14.29)$$

from which it follows that

$$E(X_i) = \mu \qquad (14.30)$$

for all i. Therefore,

$$E(\bar{X}) = \frac{1}{n} \sum_{i=1}^{n} E(X_i)$$

$$= \mu. \qquad (14.31)$$

Thus, as with sampling with replacement, the expected value of the sample mean when sampling without replacement is equal to the population mean. The sample mean in both cases is therefore an unbiased estimate of the population mean.

For the sample variance, we have from the definition,

$$\begin{aligned}\sigma_{\bar{x}}{}^2 &= E(\bar{X} - \mu)^2 \\ &= E\left[\sum_{i=1}^{n} \frac{X_i}{n} - \mu\right]^2 \\ &= \frac{1}{n^2} E\left[\sum_{i=1}^{n} (X_i - \mu)\right]^2 . \end{aligned} \tag{14.32}$$

Squaring the sum within the brackets and noting that the cross products do *not* vanish, we can write the result in the form

$$\sigma_{\bar{x}}{}^2 = \left(\frac{1}{n^2}\right) \sum_{i=1} \sum_{j=1} \rho_{ij} \sigma_i \sigma_j , \tag{14.33}$$

where $\sigma_i{}^2$ is the variance of X_i, $\sigma_j{}^2$ the variance of X_j, and ρ_{ij} the correlation between X_i and $X_j[\equiv E\{(X_i - \mu)(X_j - \mu)/\sigma_i \sigma_j\}]$.

For $i = j$, $\rho_{ij}\sigma_i\sigma_j = \sigma_i{}^2(=\sigma^2)$, so that the sum on the right can be written as

$$\sigma_{\bar{x}}{}^2 = \frac{1}{n^2}\left[n\sigma^2 + \sum_{\substack{i=1 \\ i \neq j}} \sum_{j=1} \rho_{ij}\sigma_i\sigma_j\right]. \tag{14.34}$$

For $i \neq j$, the typical element in the sum is

$$\rho_{ij}\sigma_i\sigma_j = E[X_i - \mu)(X_j - \mu)]. \tag{14.35}$$

The expectation is obtained by multiplying each possible value taken on by $X_i - \mu$ and $X_j - \mu$ by the joint probability of that pair of values, and then summing these products over all possible pairs of values of $X_i - \mu$ and $X_j - \mu$, namely,

$$\begin{aligned} E[(X_i - \mu)(X_j - \mu)] &= \sum_{\substack{i=1 \\ i \neq j}}^{N} \sum_{j=1}^{N} (x_i - \mu)(x_j - \mu) \cdot \frac{1}{N} \cdot \frac{1}{(N-1)} \\ &= \frac{1}{N(N-1)} \left[\sum_{i=1}^{N} \sum_{j=1}^{N} (x_i - \mu)(x_j - \mu) - \sum (x_i - \mu)^2\right] \\ &= -\frac{1}{N(N-1)} \left[\sum_{i=1}^{N} (x_i - \mu)^2\right], \end{aligned} \tag{14.36}$$

since $\sum\sum(x_i-\mu)(x_j-\mu)=0$. Therefore,

$$\rho_{ij}\sigma_i\sigma_j = -\frac{1}{N-1}\sum_{i=1}^{N}\frac{(x_i-\mu)^2}{N},$$
$$= \frac{\sigma^2}{N-1}. \tag{14.37}$$

However, since $\sigma_i=\sigma_j$, it follows that

$$\rho_{ij} = -\frac{1}{N-1}. \tag{14.38}$$

Hence, going back to (14.34), we have

$$\sigma_{\bar{x}}{}^2 = \frac{1}{n^2}\left[n\sigma^2 - \sum_{\substack{i=1\\ i\neq j}}^{n}\sum_{j=1}^{n}\frac{\sigma^2}{(N-1)}\right]$$
$$= \frac{1}{n^2}\left[n\sigma^2 - \frac{n(n-1)\sigma^2}{(N-1)}\right]$$
$$= \frac{\sigma^2}{n}\left(\frac{N-n}{N-1}\right). \tag{14.39}$$

Consequently, we find that the variance of the sample mean when the sampling is from a finite population without replacement differs by the factor $(N-n)/(N-1)$ from when the sampling is with replacement. Since for $n>1$, this factor is always less than 1, it follows that for a finite population the sample mean when sampling without replacement always has a smaller variance than when sampling with replacement. The reader will remember that a similar relationship was obtained in comparing the variances of the binomial and hypergeometric distributions in Section 7.4.

14.6 CENTRAL LIMIT THEOREM FOR THE NORMAL DISTRIBUTION

Assume that we have an infinite population described by the density function $f(x)$ with mean μ and variance $\sigma^2<\infty$. Let $\bar{x}_n$ denote the mean of a sample of size n drawn at random from $f(x)$. From Section 14.3, we know that the expected value of $\bar{X}_n$ is μ and that its variance is σ^2/n. This is true, no matter what the form of the density function f. Let $g(\bar{x}_n)$ be the density function for $\bar{X}_n$. What can we say about the form of g?

First, let us observe that if $f(x)$ is the density function for the normal distribution, then it follows from Corollary 12.5.1 that $\bar{X}_n$ will also have the normal distribution. In particular, $\bar{X}_n$ will be distributed $N(\mu, \sigma^2/n)$. However, one of the more remarkable results in mathematical statistics is the fact that as $n \to \infty$ this result holds irrespective of the form of f, so long that is, as the variance of f is finite. This is the celebrated central limit theorem for the normal distribution, which we shall now prove. The proof here in the text will be restricted to the case of density functions which possess moment-generating functions. A proof for the more general case where the density function simply has a finite variance is given in Appendix 6.

THEOREM 14.2 (Central Limit Theorem for Normal Distribution) Let $X_i, \ldots, X_n$ denote the items of a sample of size n drawn at random from a distribution which has mean μ, finite variance σ^2, and moment-generating function $M(t)$ that exists for $|t| < h$. Then the random variable,

$$\bar{Y}_n = \frac{\bar{X}_n - \mu}{\sigma/\sqrt{n}}, \tag{14.40}$$

has a limiting distribution which is normal with mean zero and variance one.

PROOF. Before giving the proof, we should clarify what is meant by a limiting distribution. Let $Z_1, Z_2, \ldots, Z_n, \ldots$ be a sequence of random variables, and let $F_1, F_2, \ldots, F_n, \ldots$ be the corresponding sequence of distribution functions. The sequence $Z_1, Z_2, \ldots, Z_n, \ldots$ is said to converge in distribution to a random variable, say Z, if the sequence $F_1, F_2, \ldots, F_n, \ldots$ converges to the distribution function of Z, $F(z)$, at the continuity points of the latter.

Turning now to the task at hand, our proof of the theorem will consist of showing that the moment-generating function for $\bar{Y}_n$ has as its limit as $n \to \infty$ the moment-generating function for the standard normal distribution, namely,

$$m(t) = e^{(1/2)t^2}. \tag{14.41}$$

Since, by hypothesis, $M(t) = E(e^{tX})$ exists for $|t| < h$, then so also does

$$\begin{aligned} m(t) &= E[e^{t(X-\mu)}] \\ &= e^{-\mu t}M(t). \end{aligned} \tag{14.42}$$

By Taylor's formula, there exists a number ξ between o and t such that

$$m(t) = m(o) + m'(o)t + \tfrac{1}{2}m''(\xi)t^2. \tag{14.43}$$

However, in view of the definition of m, $m(o) = 1$ and $m'(o) = 0$, so that (14.43) can be written

$$m(t) = 1 + \tfrac{1}{2}m''(\xi)t^2. \qquad (14.44)$$

We now add and subtract $\sigma^2 t^2/2$, so that

$$m(t) = 1 + \frac{\sigma^2 t^2}{2} + \frac{1}{2}[m''(\xi) - \sigma^2]t^2. \qquad (14.45)$$

Next, consider $M(t; n)$, where

$$\begin{aligned} M(t; n) &= E\left\{\exp\left[t\left(\frac{\bar{X}_n - \mu}{\sigma/\sqrt{n}}\right)\right]\right\} \\ &= E\left\{\exp\left[t\left(\frac{\sum X_i - n\mu}{\sigma\sqrt{n}}\right)\right]\right\} \\ &= E\prod_{i=1}^{n}\left\{\exp\left[t\left(\frac{X_i - \mu}{\sigma\sqrt{n}}\right)\right]\right\} \\ &= \left[E\left\{\exp\left[t\left(\frac{X - \mu}{\sigma\sqrt{n}}\right)\right]\right\}\right]^n \\ &= \left[m\left(\frac{t}{\sigma\sqrt{n}}\right)\right]^n, \qquad -h < \frac{t}{\sigma\sqrt{n}} < h, \end{aligned} \qquad (14.46)$$

since by hypothesis the X_i are are identically and independently distributed. Now, in (14.45), replace t with $t/\sigma\sqrt{n}$ to obtain

$$m\left(\frac{t}{\sigma\sqrt{n}}\right) = 1 + \frac{t^2}{2n} + \frac{[m''(\xi) - \sigma^2]t^2}{2n\sigma^2}, \qquad (14.47)$$

where ξ is now between 0 and $t/\sigma\sqrt{n}$, with $|t| < h\sigma\sqrt{n}$. Accordingly,

$$M(t; n) = \left[1 + \frac{t^2}{2n} + \frac{[m''(\xi) - \sigma^2]t^2}{2n\sigma^2}\right]^n. \qquad (14.48)$$

We now want the limit of this expression as $n \to \infty$. To obtain this, note that

$$\lim_{n\to\infty} \frac{[m''(\xi) - \sigma^2]t^2}{2n\sigma^2} = 0. \qquad (14.49)$$

This being the case, it then follows from Theorem A.4.4 of Appendix 4 that

$$\begin{aligned} \lim_{n\to\infty} M(t; n) &= \lim_{n\to\infty}\left(1 + \frac{t^2}{2n} + \frac{[m''(\xi) - \sigma^2]t^2}{2n\sigma^2}\right)^n \\ &= e^{(1/2)t^2}. \end{aligned} \qquad (14.50)$$

Since this is the moment-generating function for the normal distribution with mean zero and variance one, the theorem is proven.

While the above proof is restricted to distributions possessing moment-generating functions, the proof given in Appendix 6 removes this restriction. However, even the theorem proven there is not the most general form that the central limit theorem can take, for the restriction can also be eliminated that the X_i in the standardized sum,

$$Z_n = \sum_{i=1}^{n} \frac{X_i - \mu_i}{\sigma_i}, \tag{14.51}$$

be identically distributed. So long as the X_i are independent, $\sigma_i < \infty$, and no individual $(X_i - \mu_i)/\sigma_i$ have a nonzero probability of being "large" relative to the rest, the limiting distribution of Z_n will be $N(0, 1)$. For the details, the interested reader is referred to Dhrymes (1970, pp. 103–109) and the references cited there.

The practical importance of the central limit theorem is that much of the time researchers are interested in quantities that can reasonably be viewed as sums of random variables that are independently distributed. This being the case, the theorem assures us that, no matter what the form of the underlying distributions so long as they have finite variances, these quantities (with suitable standardization) will be approximately normally distributed, even for moderate values of n. For example, for a reference population that is uniform, the sample mean is for all practical purposes distributed normally for samples of size 12. The central limit theorem thus provides deep justification for the extensive attention that the normal distribution receives.

REFERENCES

Dhrymes, P. J., *Econometrics*, Harper & Row, 1970, chap. 3.

Freeman, H., *An Introduction to Statistical Inference*, Addison-Wesley, 1963, chaps. 19, 20.

Hogg, R. V. and Craig, A. T., *Introduction to Mathematical Statistics*, 3rd ed., Macmillan, 1970, chap. 5.

Mood, A. M. and Graybill, F. A., *Introduction to the Theory of Statistics*, 2nd ed., McGraw-Hill, chap. 7.

CHAPTER 15

SAMPLING FROM A NORMAL POPULATION

This chapter continues the discussion of the preceding chapter, but assumes that the population being sampled is described by the normal distribution. We have already noted that the sample mean in this case will be exactly normally distributed no matter what the sample size, so that there is really nothing further to be said with regard to the sample mean. Let us begin, therefore, with the distribution of the difference of the means of two samples drawn (at random) from normal distributions.

15.1 DISTRIBUTION OF THE DIFFERENCE OF TWO INDEPENDENT SAMPLE MEANS

It is not infrequent that a research worker finds himself in the situation where he has two populations and he wishes to know whether the populations have a common mean. A fundamental ingredient in solving this problem is knowledge of the distribution of the difference between two independent sample means, and we will derive this distribution in this section on the assumption that both populations are normal. (A complete analysis of this problem will be given in Chapter 18.) In doing so, we shall once again make use of the moment-generating function.

By definition, the moment-generating function for the difference $\bar{x} - \bar{y}$ between the means of two samples drawn from two populations is

$$M(t) = \int_x \int_y e^{(\bar{x}-\bar{y})} f(x, y)\, dy\, dx, \tag{15.1}$$

which, for $\bar{x}$, $\bar{y}$ independent, is equal to

$$M(t) = \int e^{t\bar{x}} f_x(x, y)\,dx \cdot \int e^{-t\bar{y}} f_y(x, y)\,dy$$
$$= M_{\bar{x}}(t) M_{\bar{y}}(-t). \tag{15.2}$$

For two normal populations, this expression is equal to

$$M(t) = \exp\,[t(\mu_x - \mu_y)] \cdot \exp\left[\frac{t^2}{2}\left(\frac{\sigma_x{}^2}{n_x} + \frac{\sigma_y{}^2}{n_y}\right)\right], \tag{15.3}$$

where μ_x, μ_y, $\sigma_x{}^2$, $\sigma_y{}^2$, n_x, and n_y are the means, variances, and sample sizes, respectively. Thus, it follows from (15.3) that the distribution of $\bar{X} - \bar{Y}$ is normal with mean $\mu_x - \mu_y$ and variance $\sigma_x{}^2/n_x + \sigma_y{}^2/n_y$.

A similar argument applied to the standardized difference of the two means,

$$\frac{\bar{X} - \mu_x - (\bar{Y} - \mu_y)}{(\sigma_x{}^2/n_x - \sigma^2/n_y)^{1/2}},$$

shows that it, too, is normal, with zero mean and unit variance.

15.2 DISTRIBUTION OF THE SAMPLE VARIANCE

Let us next focus on the distribution of the sample variance, which, to recall from Chapter 13, is defined as

$$S^2 = \frac{\sum(X_i - \bar{x})^2}{n}, \tag{15.4}$$

We now view S^2 as a random variable, and seek its distribution (again on the assumption that the X_i are independent and normal).

To begin with, we express S^2 in terms of deviations about the population mean μ:

$$\begin{aligned} nS^2 &= \sum [X_i - \mu - (\bar{X} - \mu)]^2 \\ &= \sum [(X_i - \mu)^2 - 2(X_i - \mu)(\bar{X} - \mu) + (\bar{X} - \mu)^2] \\ &= \sum (X_i - \mu)^2 - n(\bar{X} - \mu)^2, \end{aligned} \tag{15.5}$$

since $2\sum(X_i - \mu)(\bar{X} - \mu) = 2n(\bar{X} - \mu)^2$. Next, we divide both sides of (15.5) by σ^2:

$$\frac{nS^2}{\sigma^2} = \sum\left(\frac{X_i - \mu}{\sigma}\right)^2 - \left(\frac{\bar{X} - \mu}{\sigma/\sqrt{n}}\right)^2, \tag{15.6}$$

which transforms the problem into finding the distribution of nS^2/σ^2. It is this distribution that we shall now derive.

Let

$$Y_i = \frac{X_i - \mu}{\sigma}, \tag{15.7}$$

$$U = \left(\frac{\bar{X} - \mu}{\sigma/\sqrt{n}}\right)^2 = \frac{1}{n}(\sum Y_i)^2 = n\bar{Y}^2, \tag{15.8}$$

$$V = \sum(Y_i - \bar{Y})^2 = \sum\left(\frac{X_i - \bar{X}}{\sigma}\right)^2 = \frac{nS^2}{\sigma^2}. \tag{15.9}$$

The joint moment-generating function[1] for U and V is then

$$M(t_1, t_2) = E(e^{t_1 U + t_2 V})$$

$$= (2\pi)^{-n/2} \int_{y_1} \cdots \int_{y_n} \exp\left[\left(\frac{t_1}{n}\right)(\sum y_i)^2 + t_2 \sum(y_i - \bar{y})^2 - \frac{1}{2}\sum y^2\right] dy_1 \ldots dy_n. \tag{15.10}$$

The exponent in the integral may be written as

$$\frac{t_1}{n}(\sum y_i)^2 + t_2\sum(y_i - \bar{y})^2 - \frac{1}{2}\sum y_i^2$$

$$= -\frac{1}{2}[\sum y_i^2 - \frac{2t_1}{n}(\sum y_i)^2 - 2t_2\sum(y_i - \bar{y})^2]$$

$$= -\frac{1}{2}[\sum y_i^2 - \frac{2t_1}{n}(\sum y_i)^2 - 2t_2\sum y_i^2 + 2nt_2\bar{y}^2]$$

$$= -\frac{1}{2}[(1 - 2t_2)\sum y_i^2 - \frac{2(t_1 - t_2)}{n}(\sum y_i)^2]$$

$$= -\frac{1}{2}\sum_{i=1}^{n}\sum_{j=1}^{n} r_{ij}\, y_i\, y_{ij}; \tag{15.11}$$

[1] See Section 9.6.

where

$$r_{ii} = 1 - 2t_2 - \frac{2(t_1 - t_2)}{n}$$
$$= a \qquad \text{(say)} \qquad i = j, \tag{15.12}$$
$$r_{ij} = -\frac{2(t_1 - t_2)}{n}$$
$$= b \qquad \text{(say)} \qquad i \neq j. \tag{15.13}$$

Next, let **R** be the $n \times n$ matrix

$$\mathbf{R} = (r_{ij}). \tag{15.14}$$

The moment-generating function in (15.11) can now be written

$$M(t_1, t_2) = (2\pi)^{-n/2} \int_{\mathbf{y}} e^{-(1/2)\mathbf{y}'\mathbf{R}\mathbf{y}}\, d\mathbf{y}, \tag{15.15}$$

$\mathbf{y}'$ now being interpreted as the $1 \times n$ vector $(y_1, \ldots, y_n)$.

Let us now divide and multiply the right-hand side of (15.15) by $|\mathbf{R}|^{1/2}$:

$$M(t_1, t_2) = |\mathbf{R}|^{-1/2}\left\{|\mathbf{R}|^{1/2}(2\pi)^{-n/2} \int_{\mathbf{y}} e^{-(1/2)\mathbf{y}'\mathbf{R}\mathbf{y}}\, d\mathbf{y}\right\}. \tag{15.16}$$

Provided that **R** is positive definite, the term within braces represents the area beneath the density function for a multivariate normal distribution and is therefore equal to one. However, it must be established that **R** is positive definite. Now, **R** is a matrix with a's on the diagonal and b's elsewhere, and accordingly it can be shown (by a tedious inductive argument) that its latent roots consist of $a - b$ repeated $n - 1$ times and $a + (n - 1)b$. Hence its determinant is equal to

$$|\mathbf{R}| = (a - b)^{n-1}[a + (n - 1)b]. \tag{15.17}$$

For **R** to be positive definite, a necessary and sufficient condition is for its latent roots to be positive. Now, from (15.12) and (15.13), $a - b = 1 - 2t_2$ can be assumed to be positive, since we can require t_2 to be less than $\frac{1}{2}$. Hence, $(a - b)$ is positive. What about $a + (n - 1)b$? Returning again to (15.12) and (15.13), $a + (n - 1)b = 1 - 2t_2 - 2(t_1 - t_2) = 1 - 2t_1$. As with t_2, we can require t_1 to be less than $\frac{1}{2}$. Hence, **R** is indeed positive definite. Consequently, we can write (15.16) as

$$M(t_1, t_2) = |\mathbf{R}|^{-1/2}, \qquad t_1, t_2 < \tfrac{1}{2}, \tag{15.18}$$

which, in view of (15.17), can ultimately be reduced to

$$M(t_1, t_2) = \left(\frac{1}{1 - 2t_1}\right)^{1/2} \left(\frac{1}{1 - 2t_2}\right)^{(n-1)/2}. \tag{15.19}$$

If we return to formula (13.5), we will recognize the first factor on the right-hand side in (15.19) as being the moment-generating function for the chi-square distribution with 1 degree of freedom, and the second factor as the moment-generating function for the chi-square distribution with $n - 1$ degrees of freedom. Thus, we find that the joint moment-generating function for U and V reduces to the product of the moment-generating functions for U and V taken singly. This being the case, it follows from Theorem 9.1 that U and V are distributed independently, U as ${\chi_{(1)}}^2$ and V as $\chi^2_{(n-1)}$. Since V is equal to nS^2/σ^2, we have finally found, therefore, that

$$\frac{nS^2}{\sigma^2} = \sum_{i=1}^{n} \left(\frac{X_i - \bar{X}}{\sigma}\right)^2 \tag{15.20}$$

is distributed as chi-square with $n - 1$ degrees of freedom. We shall find in Chapter 18 that this result is fundamental to the testing of hypotheses about population variances.

It will be noted that both $\sum[(X_i - \mu)/\sigma]^2$ and $\sum[(X_i - \bar{X})/\sigma]^2$ are distributed as chi-square, but that the latter has $n - 1$, rather than n, degrees of freedom. It is sometimes said that since the sample mean instead of the population mean is used in $\sum[(X_i - \bar{X})/\sigma]^2$, 1 degree of freedom is used up in estimating μ by $\bar{X}$. Although $\sum(x_i - \bar{x})^2$ is a sum of squares, the squares are not all functionally independent. The relation $\sum x_i = n\bar{x}$ enables us to compute any one of the deviations $x_i - \bar{x}$, given the other $n - 1$.

15.3 DISTRIBUTION OF THE SAMPLE MEAN WHEN THE POPULATION VARIANCE IS UNKNOWN

Thus far in this chapter, we have assumed that the population variance is known. In many practical situations, however, the population variance as well as the mean are unknown, although the assumption that the population is normal may still be taken as valid. When the variance is unknown, the standardized normal variate, $Y = (X - \mu)/\sigma$, can no longer be used to test hypotheses (in a way yet to be developed) about μ. We turn instead to the quantity

$$Z = \frac{\bar{X} - \mu}{[S^2/(n-1)]^{1/2}}, \tag{15.21}$$

where S^2 is the sample variance as defined in (15.4). We shall show that Z has the t-distribution with $n - 1$ degrees of freedom.

In establishing this, we begin by dividing both the numerator and denominator of Z by $\sigma/\sqrt{n}$, obtaining

$$Z = \frac{\dfrac{\bar{X} - \mu}{\sigma/\sqrt{n}}}{\left[\dfrac{nS^2}{\sigma^2}\Big/(n-1)\right]^{1/2}}. \tag{15.22}$$

Since X is normal by hypothesis, the numerator of this expression has the standard normal distribution, while, in view of the preceding section, the term nS^2/σ^2 in the denominator is distributed as chi-square with $n - 1$ degrees of freedom. Z is therefore the quotient of a standard normal variate and the square root of a chi-square variate divided by its degrees of freedom, and will accordingly have the t-distribution provided that the numerator and denominator are distributed independently. However, this turns out in fact to be the case since the square of $(\bar{X} - \mu)/(\sigma/\sqrt{n})$ is equal to U in expression (15.8), and we have just shown that U and nS^2/σ^2 [$=V$ in expression (15.9)] are distributed independently. Consequently, Z has the t-distribution with $n - 1$ degrees of freedom.

An unknown parameter which is not the parameter of interest, as is the case with σ^2 in this context, is frequently referred to as a "nuisance" parameter. In forming the quantity Z as a ratio as in (15.22), the nuisance parameter is cancelled out since it appears in both the numerator and denominator. Such a procedure is sometimes referred to as "studentization."

15.4 ADDITIONAL THEOREMS RELATED TO SAMPLING FROM THE NORMAL DISTRIBUTION[2]

In this section, we shall prove several additional theorems that are related to the normal distribution, and which are of frequent use. In particular, these theorems are fundamental to the study of statistical inference in the general linear model, a model which is employed extensively in the social sciences, especially economics.

THEOREM 15.1
Let the $n \times 1$ random vector $\boldsymbol{X}$ be distributed $N(\mathbf{0}, \mathbf{I})$ and let $\boldsymbol{Y} = \mathbf{C}'\boldsymbol{X}$, where $\mathbf{C}$ is an $n \times n$ orthogonal matrix. Then, $\boldsymbol{Y}$ is also distributed $N(\mathbf{0}, \mathbf{I})$.

[2] This section requires the following concepts from linear algebra: reduction of a real matrix to diagonal form, positive definite matrices, and idempotent matrices. Readers whose backgrounds are weak in these areas of linear algebra should consult Appendix 5.

PROOF. From Theorem 12.5, Y is distributed $N(\mathbf{0}, \mathbf{C}'\mathbf{C})$. However, since $\mathbf{C}$ is orthogonal, $\mathbf{C}' = \mathbf{C}^{-1}$, so that $\mathbf{C}'\mathbf{C} = \mathbf{I}$.

THEOREM 15.2
Let the $n \times 1$ random vector X be distributed $N(\mathbf{0}, \mathbf{I})$, and let the $n \times n$ matrix $\mathbf{A}$ be idempotent of rank r. Then the quadratic form $X'\mathbf{A}X$ is distributed as chi-square with r degrees of freedom.

PROOF. Let $\mathbf{C}$ be the orthogonal matrix that diagonalizes $\mathbf{A}$ into

$$[\hat{\lambda}] = \begin{bmatrix} \mathbf{I} & \mathbf{0} \\ \mathbf{0} & \mathbf{0} \end{bmatrix}, \tag{15.23}$$

where $\mathbf{I}$ is an $r \times r$ identity matrix and the $\mathbf{0}$'s are appropriately dimensioned matrices of zeros. (That the latent roots of $\mathbf{A}$ are either 1 or 0 follows from the idempotency of $\mathbf{A}$.) Let $Y = \mathbf{C}'X$, and partition Y so that

$$Y = \begin{bmatrix} Y_1 \\ Y_2 \end{bmatrix} = \begin{bmatrix} Y_1 \\ \vdots \\ Y_r \\ Y_{r+1} \\ \vdots \\ Y_n \end{bmatrix}. \tag{15.24}$$

Since X is standard normal, by the theorem just proved, Y is standard normal also, and, by Theorem 12.3, so also is Y_1. Hence, $Y_1' Y_1$ is distributed as $\chi^2_{(r)}$. However, since $\mathbf{CC}' = \mathbf{I}$,

$$\begin{aligned} X'\mathbf{A}X &= X'(\mathbf{CC}')\mathbf{A}(\mathbf{CC}')X \\ &= (\mathbf{C}'X)'\mathbf{C}'\mathbf{AC}(\mathbf{C}'X) \\ &= Y'[\hat{\lambda}]\,Y \qquad\qquad (15.25) \\ &= (Y_1', Y_2')\begin{bmatrix} \mathbf{I} & \mathbf{0} \\ \mathbf{0} & \mathbf{0} \end{bmatrix}\begin{pmatrix} Y_1 \\ Y_2 \end{pmatrix} \\ &= Y_1' Y_1, \end{aligned}$$

thus proving the theorem.

THEOREM 15.3
Let the $n \times 1$ vector X be $N(\mathbf{0}, \mathbf{I})$, let $\mathbf{A}$ be an $n \times n$ idempotent matrix of rank r, let $\mathbf{B}$ be an $m \times n$ matrix, and suppose that $\mathbf{BA} = \mathbf{0}$. Then the linear form $\mathbf{B}X$ is distributed independently of the quadratic form $X'\mathbf{A}X$.

PROOF. Let $\mathbf{C}$ be the orthogonal matrix that diagonalizes $\mathbf{A}$ into $[\hat{\lambda}]$ as defined in (15.23) and let $Y = \mathbf{C}'X$, where Y is partitioned as in (15.24). Then, from (15.25), $X'\mathbf{A}X = Y_1'Y_1$, so that the quadratic form is expressed in terms of the first r elements of the standard normal vector Y.

Next, define $\mathbf{F} = \mathbf{BC}$; then

$$\begin{aligned} \mathbf{F}[\hat{\lambda}] &= \mathbf{BC}\mathbf{C}'\mathbf{AC} \\ &= \mathbf{BAC} \\ &= \mathbf{0}, \end{aligned} \tag{15.26}$$

since $\mathbf{BA} = \mathbf{0}$. Partitioning $\mathbf{F}$ as $(\mathbf{F}_1, \mathbf{F}_2)$, where $\mathbf{F}_1$ is $m \times r$ and $\mathbf{F}_2$ is $m \times (n - r)$, we have from (15.26),

$$\begin{aligned} (\mathbf{F}_1, \mathbf{F}_2)\begin{bmatrix} \mathbf{I} & \mathbf{0} \\ \mathbf{0} & \mathbf{0} \end{bmatrix} &= (\mathbf{F}_1, \mathbf{0}) \\ &= (\mathbf{0}, \mathbf{0}), \end{aligned} \tag{15.27}$$

whence $\mathbf{F}_1 = \mathbf{0}$. From this, it then follows that

$$\begin{aligned} \mathbf{B}X &= \mathbf{BC}\mathbf{C}'X \\ &= \mathbf{F}Y \\ &= (\mathbf{0}, \mathbf{F}_2)\begin{bmatrix} Y_1 \\ Y_2 \end{bmatrix} \\ &= \mathbf{F}_2 Y_2 \end{aligned} \tag{15.28}$$

is expressed in terms of the last $n - r$ elements of the standard normal vector Y. Since the elements of Y are independent, two functions of Y that have no two of these elements in common are independent. This being the case for $X'\mathbf{A}X$ and $\mathbf{B}X$, they are therefore distributed independently.

THEOREM 15.4

Let the $n \times 1$ vector X be distributed $N(\mathbf{0}, \mathbf{I})$, let $\mathbf{A}$ be an $n \times n$ idempotent matrix of rank r, let $\mathbf{B}$ be an $n \times n$ matrix of rank s, and suppose that $\mathbf{BA} = \mathbf{0}$. Then $X'\mathbf{A}X$ is distributed independently of $X'\mathbf{B}X$.

PROOF. Let $\mathbf{C}$ be an orthogonal matrix which diagonalizes $\mathbf{A}$ as in the proof of Theorem 15.2, and let Y be defined and partitioned as there. Then, $X'\mathbf{A}X = Y_1'Y_1$, as before.

Next, define $\mathbf{G} = \mathbf{C}'\mathbf{B}\mathbf{C}$; $\mathbf{G}$ is symmetrical, and

$$\begin{aligned} \mathbf{G}[\hat{\lambda}] &= \mathbf{G}\mathbf{C}'\mathbf{A}\mathbf{C} \\ &= \mathbf{C}'\mathbf{B}\mathbf{C}\mathbf{C}'\mathbf{A}\mathbf{C} \\ &= \mathbf{C}'\mathbf{B}\mathbf{A}\mathbf{C} \\ &= \mathbf{0}, \end{aligned} \tag{15.29}$$

since $\mathbf{BA} = \mathbf{0}$. Consequently, partitioning $\mathbf{G}$ as

$$\mathbf{G} = \begin{bmatrix} \mathbf{G}_1 & \mathbf{G}_2 \\ \mathbf{G}_2' & \mathbf{G}_3 \end{bmatrix}, \tag{15.30}$$

where $\mathbf{G}_1$ is $r \times r$, $\mathbf{G}_2$ $r \times (n - r)$, and $\mathbf{G}_3 (n - r) \times (n - r)$, we have

$$\begin{aligned} \begin{bmatrix} \mathbf{G}_1 & \mathbf{G}_2 \\ \mathbf{G}_2' & \mathbf{G}_3 \end{bmatrix} \begin{bmatrix} \mathbf{I} & \mathbf{0} \\ \mathbf{0} & \mathbf{0} \end{bmatrix} &= \begin{bmatrix} \mathbf{G}_1 & \mathbf{0} \\ \mathbf{G}_2' & \mathbf{0} \end{bmatrix} \\ &= \begin{bmatrix} \mathbf{0} & \mathbf{0} \\ \mathbf{0} & \mathbf{0} \end{bmatrix} \end{aligned} \tag{15.31}$$

from (15.29). Whence $\mathbf{G}_1 = \mathbf{0}$, $\mathbf{G}_2' = \mathbf{0}$, and $\mathbf{G}_2 = \mathbf{0}$. Then

$$\begin{aligned} X'\mathbf{B}X &= X'\mathbf{C}\mathbf{C}'\mathbf{B}\mathbf{C}\mathbf{C}'X \\ &= Y'\mathbf{G}\,Y \\ &= (Y_1', \; Y_2') \begin{bmatrix} \mathbf{0} & \mathbf{0} \\ \mathbf{0} & \mathbf{G}_3 \end{bmatrix} \begin{bmatrix} Y_1 \\ Y_2 \end{bmatrix} \\ &= Y_2'\mathbf{G}_3 Y_2 \end{aligned}$$

is expressed in terms of the last $n - r$ elements of the standard normal vector Y. Since $X'\mathbf{A}X$ is expressed in terms of the first r elements of Y, it follows that $X'\mathbf{A}X$ and $X'\mathbf{B}X$ are distributed independently.

THEOREM 15.5
Let X, $\mathbf{A}$, and $\mathbf{B}$ be defined as in Theorem 15.4. The ratio of quadratic forms divided by their respective ranks,

$$F = \frac{X'\mathbf{A}X}{r} \Big/ \frac{X'\mathbf{B}X}{s}, \tag{15.32}$$

has the F-distribution with r and s degrees of freedom.

Proof. Theorem 15.4 establishes that $X'\mathbf{A}X$ and $X'\mathbf{B}X$ are distributed independently, while Theorem 15.2 establishes that $X'\mathbf{A}X$ and

$X'\mathbf{B}X$ are distributed as chi-square with r and s degrees of freedom, respectively.

REFERENCES

Goldberger, A. S., *Econometric Theory*, Wiley, 1964, pp. 109–113.
Hogg, R. V. and Craig, A. T., *Introduction to Mathematical Statistics*, 3rd ed., Macmillan, 1970, chap. 12.

CHAPTER 16

POINT ESTIMATION

We mentioned earlier that one of the central problems in statistics is to devise ways to use the information given in a sample to make inferences about unknown characteristics of the parent population. There are various approaches to this, and in this chapter we shall study one particular approach, namely, point estimation.

16.1 POINT ESTIMATION DEFINED

By *point estimation*, we mean the following: Suppose that we have a population described by a single parameter density function $f(x; \theta)$, θ being the parameter. Assume that θ is unknown and that our task is to estimate it. We draw a sample of size n at random from $f(x; \theta)$—the form of f is assumed to be known—and then proceed to develop a statistic (i.e., a quantity that depends only on the sample),

$$\hat{\theta} = \hat{\theta}(x_1, x_2, \ldots, x_n), \tag{16.1}$$

to provide an estimate of θ. If $\hat{\theta}$ is a single value, we say that $\hat{\theta}$ is a *point* estimate of θ. Instead of a single value, we could construct two estimates (or, indeed, an entire distribution of estimates) of θ, say, $\hat{\theta}_1$, and $\hat{\theta}_2$, with

$\hat{\theta}_1 < \hat{\theta}_2$. In this case, we could use $\hat{\theta}_1$ and $\hat{\theta}_2$ to form an *interval* estimate of θ. Interval estimation will be discussed in Chapter 19.

As an example of point estimation, consider a population that is normal with unknown mean μ and known variance σ^2. We draw a sample of size n and use the sample mean $\bar{x}$ to estimate μ. Is this the only estimate of μ that can be constructed from the sample information? A little thought shows that it is not, for we could also use only the first observation, the last, or the median. Indeed, there are an indefinitely large number of estimates that can be constructed from the sample information, but some are better than others and part of our task in this chapter is to develop criteria that define a good estimator.

A second example of point estimation is one that is of particular relevance in economics. Suppose that we know that the beer consumption of an individual is determined by the level of his income and the price of beer relative to the prices of other commodities he buys through the relationship

$$c = \alpha + \beta x + \gamma p, \tag{16.2}$$

where c = amount consumed of beer, x = level of income, p = relative price of beer, and α, β, and γ are parameters. Although income and price are the primary determinants of beer consumption, they are not the only ones. However, these other factors are individually small, usually unobservable, and vary from one person to another and from one time period to the next, As a consequence, we lump these factors together in a single term which we shall denote by u and treat as a random variable. Accordingly, we write the demand function for beer as

$$c = \alpha + \beta x + \gamma p + u, \tag{16.3}$$

where u is assumed to be a random variable with mean zero and variance σ^2. The problem, however, is that α, β, γ, and σ^2 are usually unknown and have to be estimated on the basis of observations on c, x, and p. Thus, in order to use Equation (16.3) in forecasting the amount of beer consumed in a year, say, we need point estimates of α, β, and γ.

At this point, we should make clear the difference between an estimate and an estimator. An *estimator* is a function of sample information that is used to estimate a population parameter. The sample mean, for example, can be used as an estimator of the population mean. Since an estimator is a function of random variables, it, too, is a random variable and therefore has a distribution. An *estimate*, on the other hand, is simply a particular value of the estimator based on a particular sample.

16.2 SOME DESIRABLE PROPERTIES OF ESTIMATORS

In this section, we shall discuss several criteria that are often used in assessing the "goodness" of an estimator. The criteria discussed are unbiasedness, consistency, sufficiency, minimum variance, and minimum mean-square error.

Unbiasedness

If $\hat{\theta}$ is an estimator of θ, it is said to be *unbiased* if

$$E(\hat{\theta}) = \theta \tag{16.4}$$

for all sample sizes and for all values of θ. As we have already seen, the sample mean is an unbiased estimator of the population mean no matter what distribution describes the population. Other examples of unbiased estimators of the population mean are a single observation and the median and mode for symmetrical distributions.

As an example of an estimator that is biased, consider the sample variance,

$$S^2 = \frac{\sum (X_i - \overline{X})^2}{n}, \tag{16.5}$$

as an estimator of the population variance σ^2. Subtracting and adding μ within the parentheses, we have

$$\begin{aligned} S^2 &= \frac{1}{n} \sum [(X_i - \mu) - (\overline{X} - \mu)]^2 \\ &= \frac{1}{n} \sum [(X_i - \mu)^2 - 2(X_i - \mu)(\overline{X} - \mu) + (\overline{X} - \mu)^2] \\ &= \frac{1}{n} \sum (X_i - \mu)^2 - (\overline{X} - \mu)^2. \end{aligned} \tag{16.6}$$

Therefore,

$$\begin{aligned} E(S^2) &= E\left[\frac{1}{n} \sum (X_i - \mu)^2 - (\overline{X} - \mu)^2\right] \\ &= \sigma^2 - \frac{\sigma^2}{n} \\ &= \frac{n-1}{n} \sigma^2. \end{aligned} \tag{16.7}$$

Hence, we see that the sample variance is a biased estimator of the population variance. However, we can easily obtain an unbiased estimator of σ^2 by multiplying S^2 by $n/(n-1)$—that is, by dividing $\sum (X_i - \overline{X})^2$ by $n - 1$ instead of by n.

Being unbiased means that an estimator yields a correct estimate on the average, although it implies nothing about how close to the true value any particular estimate will be. Unbiased estimators are particularly important in situations where the same methodology is used repeatedly in making decisions, as, for example, the case of where economic policy is geared to the estimated unemployment rate. Biased estimators in such circumstances lead to an accumulation of errors.

Consistency

A property that in the abstract it seems particularly desirable for an estimator to have is that it becomes more likely to be close to the true value as the sample size increases. This intuitive attribute is formalized in the notion of consistency.

Consistency. An estimator $\hat{\theta}_n$ of θ is said to be *consistent* if, for arbitrarily small ε, $\delta > 0$, there exists a value N such that for $n > N$,

$$P(|\hat{\theta} - \theta| < \varepsilon) > 1 - \delta. \tag{16.8}$$

Since the condition just stated implies that

$$\lim_{n \to \infty} P(|\hat{\theta}_n - \theta| < \varepsilon) = 1, \tag{16.9}$$

consistency is alternatively referred to as *convergence in probability*.[1] Accordingly, we say that $\hat{\theta}$ has θ as its *probability limit*, which we write as

$$p \lim \hat{\theta} = \theta. \tag{16.10}$$

Unlike unbiasedness, which is a property that can hold for any sample size, consistency is strictly a large sample, or *asymptotic*, property. However, it is sometimes mistakenly thought that an estimator which is consistent is necessarily asymptotically unbiased, but as the following example

[1] Convergence in probability should not be confused with convergence with probability 1. $\hat{\theta}$ is said to *converge with probability* 1 to θ if

$$P\left\{\lim_{n \to \infty} |\hat{\theta}_n - \theta| < \varepsilon\right\} = 1$$

for any small $\varepsilon > 0$. Convergence with probability 1 is a much stronger condition than convergence in probability, for it refers to the behavior of the entire limit, not just an individual term. For a discussion, see Dhrymes (1970, pp. 86–87).

shows, this is not the case. Let $\hat{\theta}$ be an estimator whose distribution is given by

$$P(\hat{\theta} = \theta) = 1 - \frac{1}{n^\alpha},$$

$$P(\hat{\theta} = n) = \frac{1}{n^\alpha}, \qquad \alpha > 0. \tag{16.11}$$

It is clear that $p \lim \hat{\theta} = \theta$, and hence that $\hat{\theta}$ is consistent. However,

$$\begin{aligned} E(\hat{\theta}) &= \theta\left(1 - \frac{1}{n^\alpha}\right) + n\left(\frac{1}{n^\alpha}\right) \\ &= \theta\left(1 - \frac{1}{n^\alpha}\right) + n^{1-\alpha}, \end{aligned} \tag{16.12}$$

so that

$$\lim_{n\to\infty} E(\hat{\theta}) = \theta + \lim_{n\to\infty} n^{1-\alpha}. \tag{16.13}$$

Thus, despite the fact that $\hat{\theta}$ is a consistent estimator of θ for any α, it is asymptotically unbiased only for $\alpha < 1$. Indeed, for $\alpha > 1$, its asymptotic expectation does not even exist.

The following theorem provides a sufficient condition for an estimator to be consistent.

THEOREM 16.1

If, as $n \to \infty$,

$$E(\hat{\theta}) = \theta \qquad \text{and} \qquad {\sigma_{\hat{\theta}}}^2 \to 0, \tag{16.14}$$

then $\hat{\theta}$ is a consistent estimator of θ.

PROOF. From Chebyshev's inequality, we have

$$P[|\hat{\theta} - E(\hat{\theta})| < k\sigma_{\hat{\theta}}] > 1 - \frac{1}{k^2} \tag{16.15}$$

for any $k > 0$. Let $k\sigma_{\hat{\theta}} = c$. Then

$$P[|\hat{\theta} - E(\hat{\theta})| < c] > 1 - \frac{{\sigma_{\hat{\theta}}}^2}{c^2} \tag{16.16}$$

for any $c > 0$. Since for any value of θ,

$$\begin{aligned} |\hat{\theta} - E(\hat{\theta})| &= |\hat{\theta} - \theta + \theta - E(\hat{\theta})| \\ &\geq |\hat{\theta} - \theta| - |\theta - E(\hat{\theta})|, \end{aligned} \tag{16.17}$$

we have for $|\hat{\theta} - E(\hat{\theta})| < c$,

$$|\hat{\theta} - \theta| < c + |\theta - E(\hat{\theta})|. \tag{16.18}$$

Consequently,

$$P[|\hat{\theta} - \theta| < c + |\theta - E(\hat{\theta})|] > 1 - \frac{\sigma_{\hat{\theta}}^2}{c^2}. \tag{16.19}$$

However, (16.14) implies that for arbitrary d, $e > 0$ there exists an N such that for $n > N$,

$$|\theta - E(\hat{\theta})| < d \qquad \text{and} \qquad \sigma_{\hat{\theta}}^2 < e. \tag{16.20}$$

Hence (16.19) becomes, for $n > N$,

$$\begin{aligned} P[|\hat{\theta} - \theta| < c + d] &> P[|\hat{\theta} - \theta| < c + |\theta - E(\hat{\theta})|] \\ &> 1 - \frac{\sigma_{\hat{\theta}}^2}{c^2} \\ &> 1 - \frac{e}{c^2}, \end{aligned} \tag{16.21}$$

for c, d, and e arbitrary. Therefore, we can satisfy (16.8) for $n > N$ by taking $\varepsilon = c + d$ and $\delta = e/c^2$.

As an example of an estimator which is consistent, consider the mean $\overline{X}$ of a sample of size n drawn at random from a distribution with mean μ and variance $\sigma^2 < \infty$. Since $E(\overline{X}) = \mu$ for all n and $\lim_{n\to\infty} \text{var}\,(\overline{X}) = \lim_{n\to\infty} \sigma^2/n = 0$, Theorem 16.1 obviously applies, and $\overline{X}$ is a consistent estimator of μ.

Sufficiency

Our assumption in this chapter is that the only information about an unknown population parameter is that provided in a sample. Consequently, in forming a point estimate of the parameter, we are reducing the information in $X_1, \ldots, X_n$ to a single quantity. Intuitively, it seems reasonable that an estimator which makes "full use" of the information in the sample is to be preferred over one that does not. The definition of a sufficient statistic makes this notion precise.

Sufficiency. Let $X_1, \ldots, X_n$ denote a sample drawn at random from the distribution with density function $f(x; \theta)$, and let $Y = u(X_1, \ldots, X_n)$ be a statistic. Let $Y^* = u^*(X_1, \ldots, X_n)$ be any other statistic not a function of Y. If, for every Y^*, the conditional distribution of Y^* given Y does not involve θ, then Y is said to be a *sufficient statistic* for θ.

Much of the time, ascertaining from the definition of sufficiency whether or not a statistic is sufficient turns out to be very difficult. In many instances, however, the question can be settled by the following theorem.

THEOREM 16.2 (Fisher-Neyman Factorization Theorem) Let $X_1, \ldots, X_n$ be a sample drawn at random from the distribution with density function $f(x;\theta)$ and let $Y_1 = u_1(X_1, \ldots, X_n)$ be a statistic whose density function is given by $g_1(y_1;\theta)$. Then Y_1 is a sufficient statistic for θ if and only if the joint density function of $X_1, \ldots, X_n$, $f(x_1;\theta)\ldots f(x_n;\theta)$ can be factored,

$$f(x_1;\theta)\ldots f(x_n;\theta) = g_1(y_1;\theta)H(x_1, \ldots, x_n), \tag{16.22}$$

where, for every fixed value of y_1, $H(x_1, \ldots, x_n)$ does not depend on θ.

PROOF. Suppose, first, that $f(x_1;\theta)\ldots f(x_n;\theta)$ can be factored as in (16.22) and consider the change of variables, $y_1 = u_1(x_1, \ldots, x_n)$, $\ldots$, $y_n = u_n(x_1, \ldots, x_n)$, with inverse transformation, $x_1 = w_1(y_1, \ldots, y_n)$, $\ldots$, $x_n = w_n(y_1, \ldots, y_n)$, and Jacobian $|J|$. Let $g(y_1, \ldots, y_n)$ be the joint density function of $Y_1, \ldots, Y_n$. Then

$$g(y_1, \ldots, y_n) = g_1(y_1;\theta)H[w_1(y_1, \ldots, y_n), \ldots, w_n(y_1, \ldots, y_n)]|J|. \tag{16.23}$$

Since $g_1(y_1;\theta)$ is, by hypothesis, the marginal density function of Y_1, $H(w_1, \ldots, w_n)|J|$ must therefore be the conditional density of $Y_2, \ldots, Y_n$ given $Y_1 = y_1$, say $h(y_2, \ldots, y_n|y_1;\theta)$. However by hypothesis, $H(x_1, \ldots, x_n) = H(w_1, \ldots, w_n)$ does not depend on θ and, since θ does not appear in the transformations, $u_1, \ldots, u_n$, neither does θ appear in $|J|$. Consequently, $h(y_2, \ldots, y_n|y_1;\theta)$ is independent of θ, thereby establishing that Y_1 is a sufficient statistic for θ.

Turning now to the proof of the converse, suppose that

$$g(y_1, \ldots, y_n) = g_1(y_1;\theta)h(y_2, \ldots, y_n|y_1), \tag{16.24}$$

where $h(y_2, \ldots, y_n|y_1)$ is assumed not to depend on θ. Now let $y_1, \ldots, y_n$ be the "old" variables and $x_1, \ldots, x_n$ the "new" and let $J^*(=J^{-1})$ be the associated Jacobian. The joint density function of $X_1, \ldots, X_n$ is then given by

$$\begin{aligned} f(x_1;\theta)\ldots f(x_n;\theta) &= g[u_1(x_1, \ldots, x_n), \ldots, u_n(x_1, \ldots, x_n)]|J^*| \\ &= g_1(y_1;\theta)h[u_1(x_1, \ldots, x_n), \ldots, u_n(x_1, \ldots, x_n)]|J^*|. \end{aligned} \tag{16.25}$$

Since neither $h(u_1, \ldots, u_n)$ nor $|J^*|$ involve θ, we can write

$$f(x; \theta) \ldots f(x_n; \theta) = g_1(y_1; \theta)H(x_1, \ldots, x_n), \tag{16.26}$$

where $H(x_1, \ldots, x_n) = h(u_1, \ldots, u_n)|J^*|$ is independent of θ. This completes the proof of the theorem.[2]

Although the theorem just proved greatly simplifies the search for a sufficient statistic, its application can nevertheless be very tedious since it requires knowledge of the density function of the statistics in question. However, as the following extension of the Fisher-Neyman theorem shows, actually having to derive this density function is unnecessary.

THEOREM 16.3
Let $X_1, \ldots, X_n$ be a sample drawn at random from the density $f(x; \theta)$ and let $Y_1 = u_1(X_1, \ldots, X_n)$ be a statistic. Then Y_1 is a sufficient statistic for θ if and only if there exist two nonnegative functions, k_1 and k_2, such that the joint density function of $X_1, \ldots, X_n$, $f(x_1; \theta) \ldots f(x_n; \theta)$ can be factored

$$f(x_1; \theta) \ldots f(x_n; \theta) = k_1(y_1; \theta)k_2(x_1, \ldots, x_n), \tag{16.27}$$

where for every fixed value of y_1, $k_2(x_1, \ldots, x_n)$ does not depend on θ.

PROOF. To begin with, suppose that $f(x_1; \theta) \cdots f(x_n; \theta)$ can be factored as in (16.27) and consider the same change of variables as in the proof of Theorem 16.2. As before, let $g(y_1, \ldots, y_n)$ be the joint density function of $Y_1, \ldots, Y_n$. Hence,

$$g(y_1, \ldots, y_n) = k_1(y_1; \theta)k_2[w_1(y_1, \ldots, y_n), \ldots, w_n(y_1, \ldots, y_n)]|J|. \tag{16.28}$$

The marginal density of Y_1 is then given by

$$\begin{aligned} g_1(y_1; \theta) &= \int_{y_2} \cdots \int_{y_n} g(y_1, \ldots, y_n)\, dy_2 \ldots dy_n \\ &= k_1(y_1; \theta) \int_{y_2} \cdots \int_{y_n} |J|\, k_2(w_1, \ldots, w_n)\, dy_2 \ldots dy_n. \end{aligned} \tag{16.29}$$

By hypothesis, the function k_2 does not depend on θ; neither is θ involved in the Jacobian J nor in the limits of integration. Conse-

[2] For discrete X, the proof is identical except that there is no Jacobian.

quently, the $(n-1)$-fold integral in the right-hand side of (16.29) is a function of y_1 alone, say $m(y_1)$. Hence we have

$$g_1(y_1;\theta) = k_1(y_1;\theta)m(y_1). \tag{16.30}$$

If $m(y_1) = 0$, then $g_1(y_1;\theta) = 0$, also. If $m(y_1) > 0$, we can write

$$k_1[u_1(x_1, \ldots, x_n;\theta)] = \frac{g_1[u_1(x_1, \ldots, x_n;\theta)]}{m[u_1(x_1, \ldots, x_n)]}, \tag{16.31}$$

and the assumed factorization becomes

$$f(x_1;\theta)\ldots f(x_n;\theta) = g_1[u_1(x_1, \ldots, x_n;\theta)]\,\frac{k_2(x_1, \ldots, x_n)}{m[u_1(x_1, \ldots, x_n)]}. \tag{16.32}$$

Since neither k_2 nor m depends on θ, then, in accordance with Theorem 16.2, Y_1 is a sufficient statistic for θ.

The converse is proven simply by noting that if Y_1 is sufficient for θ, the factorization can be realized by taking k_1 to be the density function of Y_1, namely, the function g_1.

Although an estimator is, by definition, a statistic, sufficiency has been defined with reference to statistics rather than to estimators in order to emphasize the fact that the concept is motivated by the desire to roll up, as it were, all of a sample's information regarding a particular parameter into a single quantity. Obviously, an estimator can itself be a sufficient statistic, but, in general, it is more likely to be based on a function of a sufficient statistic. However, the following theorem establishes that such an estimator, too, is a sufficient statistic.

THEOREM 16.4

Let $X_1, \ldots, X_n$ denote a sample drawn from the distribution with density function $f(x;\theta)$. Let $Y = u(X_1, \ldots, X_n)$ be a sufficient statistic for θ and consider the transformation

$$Z = \phi(Y), \tag{16.33}$$

where ϕ has a single-valued inverse and does not depend on θ. Then Z, too, is sufficient statistic for θ.

PROOF. Since Y is sufficient for θ, we can, in view of Theorem 16.3, factor the joint density function of $X_1, \ldots, X_n$ into

$$f(x_1;\theta)\ldots f(x_n;\theta) = k_1(y;\theta)k_2(x_1, \ldots, x_n), \tag{16.34}$$

where k_2 does not depend on θ. Let Φ be the (single-valued) inverse transformation of ϕ, so that $y = \Phi(z)$. Then

$$k_1(y;\theta) = k_1[\Phi(z);\theta] = k_1{}^*(z;\theta). \tag{16.35}$$

Consequently,

$$f(x_1;\theta)\ldots f(x_n;\theta) = k_1{}^*(z;\theta)k_2(x_1,\ldots,x_n). \tag{16.36}$$

Since k_2 is not affected, it follows that Z is a sufficient statistic for θ, as was to be shown.

The next theorem establishes that, in cases where one exists, a sufficient statistic is unique in the sense that any other sufficient statistic is functionally related to it.

THEOREM 16.5
Let Y_1 and Y_2 be sufficient statistics for a parameter θ. Then Y_1 and Y_2 are functionally related.

PROOF. Since Y_1 and Y_2 are both sufficient for θ, we will have from the definition of sufficiency,

$$g_1(y_1;\theta)h_1(y_2 \mid y_1) = f(y_1, y_2;\theta) = g_2(y_2;\theta)h_2(y_1 \mid y_2). \tag{16.37}$$

Consequently, we can write

$$y_2 = k(y_1, \theta), \tag{16.38}$$

for some function k. However, Y_1 and Y_2 are functions of the observations only, and do not depend on θ. Hence, from (16.38), Y_2 is functionally related to Y_1, and the theorem is proven.

Let us now consider some examples of a sufficient statistic. As a first example, let $X_1, \ldots, X_n$ be a sample of size n drawn at random from the normal distribution with mean μ and variance 1. The joint density function for $X_1, \ldots, X_n$ is then

$$g(x_1,\ldots,x_n;\theta) = (2\pi)^{-n/2}\exp[-\tfrac{1}{2}\textstyle\sum (x_i - \mu)^2]. \tag{16.39}$$

However,

$$\textstyle\sum (x_i - \mu)^2 = \sum (x_i - \bar{x})^2 + n(\bar{x} - \mu)^2, \tag{16.40}$$

so that

$$g(x_1, \ldots, x_n; \theta) = (2\pi)^{-n/2} \exp\left[-\frac{1}{2}\sum (x_i - \bar{x})^2 - \frac{n(\bar{x} - \mu)^2}{2}\right]$$

$$= (2\pi)^{-n/2} \exp\left[-\frac{n}{2}(\bar{x} - \mu)^2\right] \exp\left[-\frac{1}{2}\sum (x_i - \bar{x})^2\right]. \tag{16.41}$$

Thus, $g(x_1, \ldots, x_n; \theta)$ factors into two terms, one involving $\bar{x}$ and μ and the other involving only sample information. Whence, in accordance with Theorem 16.3, $\bar{X}$ is a sufficient statistic for μ.[3]

As a second example, let $X_1, \ldots, X_n$ represent a sample drawn at random from the distribution with density function

$$f(x; \theta) = \theta x^{\theta - 1}, \qquad 0 < x < 1, \tag{16.42}$$

where $\theta > 0$. We shall use Theorem 16.3 to show that the product $Y = X_1 X_2 \ldots X_n$ is a sufficient statistic for θ. The joint density function of $X_1, \ldots, X_n$ is

$$\theta^n (x_1 x_2 \ldots x_n)^{\theta - 1} = [\theta^n (x_1 x_2 \ldots x_n)^\theta] \frac{1}{x_1 x_2 \ldots x_n}, \tag{16.43}$$

where $0 < x_i < 1$, $i = 1, n$. In Theorem 16.3, let $k_1[u_1(x_1, \ldots, x_n); \theta] = \theta^n (x_1 x_2 \ldots x_n)^\theta$ and $k_2(x_1, \ldots x_n) = (x_1 x_2 \ldots x_n)^{-1}$. Since, for any fixed value of u_1, k_2 does not depend on θ, $Y = X_1 X_2, \ldots X_n$ is a sufficient statistic for θ.

Minimum Variance

It was noted in our discussion of unbiasedness that an estimator being unbiased tells us nothing about how close the estimate yielded by the estimator may be to the true value of the parameter of interest. Accordingly, it is clear that we must look at the variance of an estimator as well as its mean. *Ceteris paribus*, an estimator with a small variance is to be preferred to one with a large variance. We shall develop this principle in detail in Section 16.3.

[3] Since $(2\pi)^{-n/2} \exp\left[\frac{-n(\bar{x} - \mu)^2}{2}\right]$ is actually the density function for $\bar{X}$, Theorem 16.2 is applicable also.

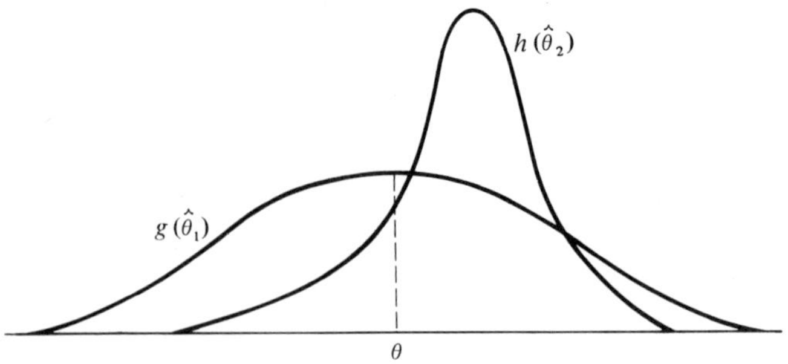

FIGURE 16.1 θ_1 unbiased, with large variance; θ_2 biased, but with much smaller variance.

Minimum Mean-Square Error

It is easy to conceive of situations where a biased estimator with a small variance is to be preferred to one that is unbiased, but with a large variance. Figure 16.1 illustrates a case in point; $\hat{\theta}_1$ is unbiased, but has a very large variance, while $\hat{\theta}_2$ is biased, but has a much smaller variance. In such situations, it is relevant to consider estimators which have *minimum mean-square error*, where mean-square error is defined as

$$MSE = E(\hat{\theta} - \theta)^2. \tag{16.44}$$

A little manipulation shows that

$$\begin{aligned} E(\hat{\theta} - \theta)^2 &= E[\hat{\theta} - E(\hat{\theta}) + E(\hat{\theta}) - \theta]^2 \\ &= E[\hat{\theta} - E(\hat{\theta})]^2 + [E(\hat{\theta}) - \theta]^2 \\ &= \text{var}(\hat{\theta}) + (\text{bias})^2. \end{aligned} \tag{16.45}$$

Thus, since the cross-product term drops out, the mean-square error is equal to the variance of the estimator plus the square of its bias. A minimum mean-square error estimator, therefore, is an estimator which minimizes expression (16.44). If $\hat{\theta}$ is unbiased, it is evident from (16.45) that the criterion of minimum MSE reduces to that of minimum variance unbiasedness.

16.3 MINIMUM VARIANCE UNBIASED ESTIMATION

In this section, we shall adopt a particular principle of estimation, namely, that of selecting out of the class of unbiased estimators the member which has minimum variance, and will develop the theory which underlies it. Although minimum variance unbiased is not the only principle of estimation that offers itself—maximum likelihood will be discussed in the next chapter, least squares in Chapter 20, and Bayesian estimation in Chapter 22—it obviously has much intuitive appeal and has received a great deal of attention.

The theory underlying minimum variance unbiased estimation is based primarily on two theorems. The first of these, due to Cramer and Rao, provides the lower bound to the variance of an unbiased estimator—and therefore establishes the ideal to which a minimum variance unbiased estimator can aspire—while the second theorem, due to Rao and Blackwell, is instrumental in showing that, if a sufficient statistic for the parameter in question exists, then the minimum variance unbiased estimator is a function of it. We begin with the theorem of Cramer and Rao.

The Cramer-Rao Lower Bound

If the density function of a population is $f(x; \theta)$, we know that the density function of the joint distribution of the elements of a sample size n drawn at random from this population is equal to $f(x_1; \theta) \dots f(x_n; \theta)$. Let

$$L(x_1, \dots, x_n; \theta) = f(x_1; \theta) \dots f(x_n; \theta). \tag{16.46}$$

We shall refer to $L(x_1, \dots, x_n; \theta)$ as the *likelihood function* of the sample,[4] and shall frequently write it simply as L.

Since L is a density, we must have

$$\int_{x_1} \cdots \int_{x_n} L(x_1, \dots, x_n; \theta)\, dx_1 \dots dx_n = 1. \tag{16.47}$$

Let us assume that the first two derivatives of L with respect to θ exist for all θ, and suppose that we differentiate both sides of (16.47) with respect to θ, obtaining,[5]

$$\int \cdots \int \frac{\partial L}{\partial \theta}\, dx_1 \dots dx_n = 0, \tag{16.48}$$

[4] R. A. Fisher, who invented this terminology, reserved the name likelihood function for the case where $x_1, \dots, x_n$ are known and θ is unknown, calling L the probability of the sample when θ is known. With a view to keeping a simple notation, we shall use likelihood function to describe both cases.

[5] Differentiation under the integral requires that the limits of integration be independent of θ and, in addition, that certain conditions be satisfied as to uniform convergence. We assume all of these conditions to be met.

which we may rewrite as

$$E\left(\frac{\partial \ln L}{\partial \theta}\right) = \int \cdots \int \left(\frac{1}{L}\frac{\partial L}{\partial \theta}\right) L\, dx_1 \ldots dx_n \tag{16.49}$$

$$= 0.$$

Differentiating again, we have

$$\int \cdots \int \left[\left(\frac{1}{L}\frac{\partial L}{\partial \theta}\right)\frac{\partial L}{\partial \theta} + L\frac{\partial}{\partial \theta}\left(\frac{1}{L}\frac{\partial L}{\partial \theta}\right)\right] dx_1 \ldots dx_n = 0 \tag{16.50}$$

or

$$\int \cdots \int \left[\left(\frac{1}{L}\frac{\partial L}{\partial \theta}\right)^2 + \frac{\partial^2 \ln L}{\partial \theta^2}\right] L\, dx_1 \ldots dx_n = 0, \tag{16.51}$$

from which it follows that

$$E\left[\left(\frac{1}{L}\frac{\partial L}{\partial \theta}\right)^2\right] = -E\left(\frac{\partial^2 \ln L}{\partial \theta^2}\right). \tag{16.52}$$

Consider, now, an unbiased estimator $\hat{\theta}$ of θ. Then

$$E(\hat{\theta}) = \int \cdots \int \hat{\theta} L\, dx_1 \ldots dx_n \tag{16.53}$$

$$= \theta.$$

Next, let us differentiate (16.53) with respect to θ, the result being

$$\int \cdots \int \hat{\theta}\frac{\partial L}{\partial \theta} dx_1 \ldots dx_n = \int \cdots \int \hat{\theta}\frac{\partial \ln L}{\partial \theta} L\, dx_1 \ldots dx_n$$

$$= 1, \tag{16.54}$$

which, in view of (16.48) we can rewrite as

$$1 = \int \cdots \int (\hat{\theta} - \theta)\frac{\partial \ln L}{\partial \theta} L\, dx_1 \ldots dx_n. \tag{16.55}$$

By the Schwarz inequality, we have from (16.55),

$$1 \leq \int \cdots \int (\hat{\theta} - \theta)^2 L\, dx_1 \ldots dx_n \int \cdots \int \left(\frac{\partial \ln L}{\partial \theta}\right)^2 L\, dx_1 \ldots dx_n, \tag{16.56}$$

which, upon rearrangement, becomes

$$E(\hat{\theta} - \theta)^2 \geq 1/E\left(\frac{\partial \ln L}{\partial \theta}\right)^2, \tag{16.57}$$

or, in view of (16.52), equivalently,

$$E(\hat{\theta} - \theta)^2 \geq -1/E\left(\frac{\partial^2 \ln L}{\partial \theta^2}\right). \tag{16.58}$$

Expression (16.57) [or, equivalently, (16.58)] is the fundamental inequality on the variance of an unbiased estimator usually referred to as the Cramer-Rao inequality. We thus have proven the following.

THEOREM 16.6 (Cramer-Rao)
Let $X_1, \ldots, X_n$ denote a sample drawn at random from the distribution with density function $f(x; \theta)$. Let $L(x_1, \ldots, x_n; \theta)$ be the likelihood function of the sample, and assume that it is twice differentiable and, in addition, satisfies the conditions set out in footnote 5. Finally, let $\hat{\theta}$ be an estimator of θ. Then the variance of $\hat{\theta}$ must satisfy

$$\text{var}(\hat{\theta}) \geq 1/E\left(\frac{\partial \ln L}{\partial \theta}\right)^2, \tag{16.59}$$

or, equivalently,

$$\text{var}(\hat{\theta}) \geq -1/E\left(\frac{\partial^2 \ln L}{\partial \theta^2}\right). \tag{16.60}$$

An estimator whose variance actually attains the Cramer-Rao bound is said to be an *efficient* estimator. However, as the following theorem shows, an efficient estimator in this sense may not always exist.

THEOREM 16.7
With reference to Theorem 16.6, the variance of $\hat{\theta}$ actually attains the Cramer-Rao bound if and only if $\partial \ln L/\partial\theta$ can be written in the form

$$\frac{\partial \ln L}{\partial \theta} = A(\theta) \cdot (\hat{\theta} - \theta), \tag{16.61}$$

where A is independent of $x_1, \ldots, x_n$, but may depend on θ.

PROOF. The inequality in (16.59) arose purely from the use of the Schwarz inequality, and the necessary and sufficient condition that the Schwarz inequality be an equality is for $\hat{\theta} - \theta$ to be proportional to $\partial \ln L/\partial\theta$ for all possible $x_1, \ldots, x_n$—that is, if $\partial \ln L/\partial\theta$ can be written as

$$\frac{\partial \ln L}{\partial \theta} = A(\theta) \cdot (\hat{\theta} - \theta),$$

for A independent of $x_1, \ldots, x_n$, but possibly depending on θ. This proves the theorem.

In view of (16.49), it is to be noted that from (16.61) we have

$$E\left(\frac{\partial \ln L}{\partial \theta}\right)^2 = [A(\theta)]^2 E(\hat{\theta} - \theta)^2. \tag{16.62}$$

Since, in this case, (16.57) holds as an equality, it consequently follows that

$$\operatorname{var}(\hat{\theta}) = \frac{1}{A(\theta)}. \tag{16.63}$$

Thus, in situations where the Cramer-Rao lower bound can be reached, the minimum variance unbiased estimator can be found directly from Theorem 16.7.

Let us now consider some examples. Suppose, first, that $f(x;\ \theta)$ is $N(\theta, \sigma^2)$, where σ^2 is known, and that we wish to estimate θ. In this case,

$$L = (2\pi\sigma^2)^{-n/2} \exp\left\{-\frac{1}{2}\sum\left[\frac{(x_i - \theta)}{\sigma}\right]^2\right\}, \tag{16.64}$$

so that

$$\begin{aligned} \ln L &= -\frac{n}{2}\ln(2\pi\sigma^2) - \frac{1}{2}\sum\left[\frac{(x_i - \theta)}{\sigma}\right]^2 \\ &= -\frac{n}{2}\ln(2\pi\sigma^2) - \frac{1}{2}\sum\left(\frac{x_i - \bar{x}}{\sigma}\right)^2 - \frac{n}{2}\left(\frac{\bar{x} - \theta}{\sigma}\right)^2. \end{aligned} \tag{16.65}$$

Differentiating with respect to θ, we then have

$$\frac{\partial \ln L}{\partial \theta} = \frac{n}{\sigma^2}(\bar{x} - \theta), \tag{16.66}$$

which has the form in (16.61) with $A(\theta) = n/\sigma^2$ and $\hat{\theta} = \bar{x}$. Consequently, we conclude that $\bar{X}$ provides an unbiased estimator of θ whose variance actually attains the Cramer-Rao bound.

For a second example, let

$$f(x;\theta) = \frac{1}{\pi}\frac{1}{[1 + (x - \theta)^2]}, \qquad -\infty < x < \infty. \tag{16.67}$$

Taking logarithms and then differentiating with respect to θ, we have

$$\frac{\partial \ln L}{\partial \theta} = 2 \sum \frac{(x_i - \theta)}{1 + (x_i - \theta)^2}. \tag{16.68}$$

Since this expression cannot be made commensurate with (16.61), we accordingly conclude that, whatever it is, the minimum variance unbiased estimator of the parameter of this distribution does not attain the Cramer-Rao bound.

As a third and final example, let us consider the parameter λ in the Poisson density

$$f(x; \theta) = e^{-\lambda} \frac{\lambda^x}{x!}. \tag{16.69}$$

In this case,

$$\begin{aligned} \frac{\partial \ln L}{\partial \lambda} &= -n + \frac{\sum x_i}{\lambda} \\ &= \frac{n}{\lambda} (\bar{x} - \lambda) \\ &= A(\lambda)(\hat{\lambda} - \lambda), \end{aligned} \tag{16.70}$$

with $A(\lambda) = n/\lambda$ and $\hat{\lambda} = \bar{x}$. Since $\bar{X}$ is unbiased, it therefore follows that $\bar{X}$ provides an estimator whose variance attains the Cramer-Rao lower bound.

The Cramer-Rao Lower Bound and Sufficiency

Theorem 16.2, which provides the necessary and sufficient condition for sufficiency, has a consequence of immediate interest to the attainment of the Cramer-Rao lower bound. On taking logarithms of both sides of (16.22), we have

$$\frac{\partial \ln L}{\partial \theta} = \frac{\partial \ln g(y; \theta)}{\partial \theta}. \tag{16.71}$$

On comparing this expression with (16.59), a moment's thought should convince the reader the Cramer-Rao bound can be reached only if there exists a sufficient statistic. In fact, (16.61) is simply a special case of (16.71) when

$$\frac{\partial \ln g(y; \theta)}{\partial \theta} = A(\theta)(y - \theta). \tag{16.72}$$

Thus, it follows that in cases where the Cramer-Rao lower bound can be attained, the minimum variance unbiased estimator is a sufficient statistic. Moreover, in view of Theorem 16.5, it is also unique.

The Rao-Blackwell Theorem

As we have already seen, when the Cramer-Rao bound is attainable, the minimum variance estimator can actually be found from Theorem 16.7. However, when the Cramer-Rao bound cannot be reached, the search for the minimum variance estimator must proceed along other lines. Material assistance in narrowing the scope of this search is provided by the following theorem.

THEOREM 16.8 (Rao-Blackwell)
Let X and Y denote random variables where Y has mean μ and variance ${\sigma_y}^2$. Let $E(Y|x) = \phi(x)$. Then $E[\phi(X)] = \mu$ and $\sigma^2_{\phi(x)} \leq {\sigma_y}^2$.

PROOF. Let $f(x, y)$, $f_1(x)$, $f_2(y)$, and $h(y|x)$ denote the joint density function of X and Y, the marginal density of X, the marginal density of Y, and the conditional density of Y for $X = x$, respectively. Then

$$\begin{aligned} E(Y|x) &= \int_y yh(y|x)\,dy \\ &= \frac{\int_y yf(x, y)\,dy}{f_1(x)} \\ &= \phi(x), \end{aligned} \tag{16.73}$$

so that

$$\int_y yf(x, y)\,dy = \phi(x)f_1(x). \tag{16.74}$$

Consequently,

$$\begin{aligned} E[\phi(X)] &= \int_x \phi(x)f_1(x)\,dx \\ &= \int_x \left[\int_y yf(x, y)\,dy\right] dx \\ &= \int_y y\left[\int_x f(x, y)\,dx\right] dy \\ &= \int_y yf_2(y)\,dy \\ &= \mu, \end{aligned} \tag{16.75}$$

which proves the first part of the theorem.

For the second part, we have

$$\begin{aligned}\sigma_y^2 &= E(Y-\mu)^2 \\ &= E[Y-\phi(X)+\phi(X)-\mu]^2 \\ &= E[Y-\phi(X)]^2 + E[\phi(X)-\mu]^2 \\ &\quad + 2E\{[Y-\phi(X)][\phi(X)-\mu]\}. \end{aligned} \tag{16.76}$$

However,

$$\begin{aligned} &E\{[Y-\phi(X)][\phi(X)-\mu]\} \\ &\quad = \int_x \int_y [y-\phi(x)][\phi(x)-\mu] f(x, y)\, dy\, dx \\ &\quad = \int_x \int_y [y-\phi(x)][\phi(x)-\mu] h(y \mid x) f_1(x)\, dy\, dx \\ &\quad = \int_x [\phi(x)-\mu]\left[\int_y [y-\phi(x)] h(y \mid x)\, dy\right] f_1(x)\, dx \\ &\quad = 0, \end{aligned} \tag{16.77}$$

since $\phi(x)$ is the mean of $h(y \mid x)$, and the integral in the brackets is accordingly zero. Hence,

$$\begin{aligned}\sigma_y^2 &= \sigma_{\phi(x)}^2 + E[Y-\phi(X)]^2 \\ &\geq \sigma_{\phi(x)}^2, \end{aligned} \tag{16.78}$$

since $E[Y-\phi(X)]^2$ is necessarily nonnegative. This completes the proof of the theorem.[6]

As an example of the Rao-Blackwell theorem, let X and Y be bivariate normal with means μ_1 and μ_2, variances σ_1^2 and σ_2^2, and correlation coefficient ρ. All of these quantities are assumed to be known. In this case, $E(Y) = \mu = \mu_2$ and $\sigma_y^2 = \sigma_2^2$. From Section 9.4, we know that $E(Y \mid x)$ is linear in x and given by

$$\begin{aligned} E(Y \mid x) &= \theta(x) \\ &= \mu_2 + \rho \frac{\sigma_2}{\sigma_1}(x-\mu_1). \end{aligned} \tag{16.79}$$

[6] The proof for discrete X and Y is similar except that integrals are replaced by summations.

Thus, $\theta(X) = \mu_2 + \rho(\sigma_2/\sigma_1)(X - \mu_1)$ and $E[\theta(X)] = \mu_2$, as stated in the theorem. Moreover,

$$\begin{aligned}
\sigma^2_{\phi(x)} &= E[\theta(X) - \mu_2]^2 \\
&= E\left[\rho \frac{\sigma_2}{\sigma_1}(X - \mu_1)\right]^2 \\
&= \rho^2 \sigma_2{}^2. \qquad (16.80)
\end{aligned}$$

Since $|\rho| \le 1$, we accordingly have the inequality, $\sigma^2_{\phi(x)} \le \sigma_2{}^2$.

As before, let $X_1, \ldots, X_n$ denote a sample drawn at random from the distribution with density function $f(x; \theta)$, and let $Y_1 = u_1(X_1, \ldots, X_n)$ be a sufficient statistic for θ. Let $Y_2 = u_2(X_1, \ldots, X_n)$ be another statistic, not a function of Y_1 alone, which is an unbiased estimator of θ, that is, $E(Y_2) = \theta$. Consider $E(Y_2 | y_1)$. This expectation is a function of y_1, say $\phi(y_1)$. Since Y_1 is assumed to be a sufficient statistic for θ, the conditional distribution of Y_2, given $Y_1 = y_1$, will not involve θ, so that $E(Y_2 | y_1) = \phi(y_1)$ will be a function of y_1 alone, and hence is a statistic. In accordance with the Rao-Blackwell theorem, $\phi(Y_1)$ is unbiased,[7] and, since Y_2 is not a function of Y_1 alone, the variance of $\phi(Y_1)$ will be strictly less than the variance of Y_2. Consequently, we have proven the following.

THEOREM 16.9

Let $X_1, \ldots, X_n$ denote a sample of size n drawn at random from the distribution with density function $f(x; \theta)$. Let $Y_1 = u_1(X_1, \ldots, X_n)$ be a sufficient statistic for θ, and let $Y_2 = u_2(X_1, \ldots, X_n)$ be another statistic, not a function of Y_1 alone, which is an unbiased estimator of θ. Then $\phi(y_1) = E(Y_2 | y_1)$ defines a statistic, which:

1. Is a function of the sufficient statistic for θ (and is itself a sufficient statistic if ϕ possesses a single-valued inverse).
2. Is an unbiased estimator of θ.
3. Has a variance which is strictly smaller than the variance of Y_2.

This theorem tells us that in our search for a minimum variance estimator for a parameter, we can, if a sufficient statistic for the parameter exists, restrict our attention to functions of that sufficient statistic. For if we begin with a statistic Y_2 which is an unbiased estimator of the parameter and which is not a function of the sufficient statistic Y_1 alone, we can,

[7] If the function ϕ has a single-valued inverse, then $\phi(Y_1)$, being a function of a sufficient statistic, is also a sufficient statistic.

in terms of variance, always improve upon Y_2 by computing $\phi(y_1) = E(Y_2 \mid y_1)$ and using $\phi(Y_1)$ as the estimator of θ. $\phi(Y_1)$ will be unbiased and will have a variance smaller than the variance of Y_2.

However, Theorem 16.9 does not state that, of all unbiased estimators of a parameter, $\phi(Y_1)$ will be the one with minimum variance. For this to be true, the distribution of Y_1 must be a member of a family of distributions which possess the property of completeness. This concept is now defined.

> **Complete families of density functions.** Let $\{f(x;\theta);\theta \in \Omega\}$ denote a family of density functions of either the discrete or continuous type, depending on a parameter θ, $\theta \in \Omega$. Let $u(x)$ be any continuous function of x, independent of θ. If for all $\theta \in \Omega$,
>
> $$E[u(X)] = \int_x u(x) f(x;\theta)\, dx \qquad (16.81)$$
> $$= 0$$
>
> implies that
>
> $$u(x) = 0 \qquad (16.82)$$
>
> identically in x (except possibly over a set with probability zero), then the family $\{f(x;\theta);\theta \in \Omega\}$ is called a *complete family of density functions.*

As an illustration of a family of density functions which is complete, consider the family $\{f(x;\theta);\ 0 < \theta < \infty\}$, where

$$f(x;\theta) = \begin{cases} \dfrac{1}{\theta}, & 0 < x < \theta, \\ 0, & \text{otherwise.} \end{cases} \qquad (16.83)$$

Let $u(x)$ be a continuous function of x, but independent of θ, and suppose that $E[u(X)] = 0$ for all θ. We want to show that this condition requires that $u(x) = 0$ for all x, $0 < x < \infty$. [In this case, since for each x, there exists at least one member of the family of distributions $\{f(x;\theta);\ 0 < \theta < \infty\}$ which is positive, our task can be amended to showing that $u(x) = 0$ at every point x at which at least one member of the family of density functions is positive.]

Our assumption is that

$$E[u(X)] = \int_x u(x) f(x;\theta)\, dx$$
$$= \int_0^\theta u(x)\frac{1}{\theta}\, dx$$
$$= 0, \qquad (16.84)$$

for $0 < \theta < \infty$. Thus,

$$\int_0^\theta u(x)\,dx = 0 \tag{16.85}$$

for all θ. Differentiating both sides of (16.85) with respect to θ, we have from the fundmental theorem of calculus,

$$u(\theta) = 0, \tag{16.86}$$

for all θ, $0 < \theta < \infty$. Since $u(\theta) = 0$ for all θ, it thus follows that $u(x) = 0$ for all x, $0 < x < \infty$, as was to be shown. Consequently, the family $\{f(x; \theta);\ 0 < \theta < \infty\}$ forms a complete family of density functions.

In statistical applications of the concept of completeness, the family of distributions that we are interested in is often the sampling distribution of a statistic Y, say $g(y; \theta)$. We then call Y a *complete statistic* if, for all θ, $E[\phi(Y)] = 0$ implies $\phi(y) = 0$ identically in y (except possibly on a set with probability zero).

Let us assume that the parameter θ in the density function $f(x; \theta)$, for a member of some set Ω, has a sufficient statistic $Y_1 = u_1(X_1, \ldots, X_n)$, where $X_1, \ldots, X_n$ is a sample of size n drawn at random from this distribution. Let the density function of Y_1 be $g_1(y_1; \theta)$, $\theta \in \Omega$. From Theorem 16.9, we know that if there is any statistic Y_2 (not a function of Y_1 alone) which is an unbiased estimator of θ, then there is at least one function of Y_1 that is an unbiased estimator of θ and which has a variance strictly smaller than the variance of Y_2. Accordingly, our search for a minimum variance estimator can be restricted to functions of Y_1. Suppose that it has been verified that a certain continuous function of Y_1, $\phi(Y_1)$, not a function of θ, is such that $E[\phi(Y_1)] = \theta$ for all θ, $\theta \in \Omega$. Let $\psi(Y_1)$ be another continuous function of Y_1 such that $E[\psi(Y_1)] = \theta$, also, for all values of θ. Hence,

$$E[\phi(Y_1) - \psi(Y_1)] = 0, \tag{16.87}$$

for all θ, $\theta \in \Omega$. If the family of density functions $\{g_1(y_1; \theta);\ \theta \in \Omega\}$ is complete, the continuous function $\Phi(y_1) = \phi(y_1) - \psi(y_1)$ is equal to zero at each point y_1 where at least one member of the family is positive. That is, at all points of nonzero probability density, we have for every continuous unbiased estimator $\psi(Y_1)$,

$$\phi(y_1) = \psi(y_1). \tag{16.88}$$

In view of (16.88), $\phi(y_1)$ can be said do be the *unique* continuous function of Y_1 which is an unbiased estimator of θ. This being the case, it then follows from the Rao-Blackwell theorem that, of all unbiased estimators of θ, $\phi(Y_1)$ is the estimator with smallest variance.

The foregoing is summarized in the following theorem.

THEOREM 16.10

Let $X_1, \ldots, X_n$ denote a sample of size n drawn at random from a distribution that has a density function $f(x; \theta)$, $\theta \in \Omega$. Let $Y_1 = u_1(X_1, \ldots, X_n)$ be a sufficient statistic for θ, and let the family of density functions $\{g_1(y_1; \theta); \theta \in \Omega\}$, where $g_1(y_1; \theta)$ is the density function of Y_1, be complete. If there is a continuous function of Y_1 which is an unbiased estimator of θ, then this funtion of Y_1 is the unique minimum variance unbiased estimator of θ. Here, "unique" is used in the sense described in the preceding paragraph.

This theorem thus establishes a fundamental connection between minimum variance unbiased estimation and sufficient statistics in situations where the latter exist. It should be noted, though, that the version of the theorem proven here is unduly restrictive in that continuity is unnecessary. However, not assuming continuity involves problems of a measure-theoretic nature that, at the present level of exposition, are best avoided. The interested reader is referred to the paper of Lehmann and Scheffé (1950).

REFERENCES

Dhrymes, P. J., *Econometrics*, Harper & Row, 1970, chap. 3.

Freeman, H., *An Introduction to Statistical Inference*, Addison-Wesley, 1963, chap. 25.

Hogg, R. V. and Craig, A. T., *Introduction to Mathematical Statistics*, 3rd ed., Macmillan, 1970, chap. 7.

Kendall, M. G. and Stuart, A., *The Advanced Theory of Statistics*, vol. II, Charles Griffin and Co., 1961, chap. 17.

Lehmann, E. L. and Scheffé, H., "Completeness, Similar Regions, and Unbiased Estimation," *Sankhya*, vol. 10, 1950.

CHAPTER 17

MAXIMUM LIKELIHOOD ESTIMATION

In the preceding chapter, we defined a number of properties of estimators that are commonly agreed to be desirable, and discussed, as a principle of estimation, the search for the unbiased estimator that has minimum variance. In this chapter, we discuss another method, or principle, of estimation which, because of its essentially constructive character, has achieved widespread popularity and use. This is the method of maximum likelihood.

17.1 MAXIMUM LIKELIHOOD ESTIMATION DEFINED

As before, let $X_1, \ldots, X_n$ denote a sample of size n drawn at random from the distribution with density function $f(x; \theta)$. We assume the form of f to be known, but the value of θ to be unknown. Also, as before, let $L(x_1, \ldots, x_n; \theta)$ denote the likelihood function of the sample, namely,

$$L(x_1, \ldots, x_n; \theta) = f(x_1; \theta) \cdots f(x_n; \theta). \tag{17.1}$$

Now, for given values of $x_1, \ldots, x_n$, $L(x_1, \ldots, x_n; \theta)$ will vary with θ. This being the case, suppose, then, that we take as our estimator of θ the $\hat{\theta}$ (assuming that one exists) which renders $L(x_1, \ldots, x_n; \theta)$ a maximum. Since it maximizes the likelihood function, such an estimator is called the *maximum likelihood estimator*.

In cases where the likelihood function is differentiable, the maximum likelihood estimator can be found by solving the equation

$$\frac{dL(\theta)}{d\theta} = 0, \qquad (17.2)$$

provided, of course, that the resulting value of θ also satisfies

$$\frac{d^2L(\theta)}{d\theta^2} < 0, \qquad (17.3)$$

so that we have a maximum, rather than a minimum. In cases where the likelihood function is multimodal, then the maximum likelihood estimator is the $\hat{\theta}$ yielding the global maximum. Finally, it is to be noted that, since $y = \ln x$ is a monotonic transformation, the $\hat{\theta}$ which maximizes $L(\theta)$ also maximizes $\ln L(\theta)$. This fact is extremely useful and, as we shall see, is usually employed.

If the population density function has k unknown parameters $\theta_1, \ldots, \theta_k$ instead of just a single θ, then the likelihood function of a sample of size n drawn at random will be given by

$$L(x_1, \ldots, x_n; \theta_1, \ldots, \theta_k) = f(x_1; \theta_1, \ldots, \theta_k) \cdots f(x_n; \theta_1, \ldots, \theta_k). \qquad (17.4)$$

If L is regular in its first derivatives, the maximum likelihood estimators, $\hat{\theta}_1, \ldots, \hat{\theta}_k$, of $\theta_1, \ldots, \theta_k$ will then be given by the solutions to the system of equations

$$\frac{\partial L(\theta_1, \ldots, \theta_k)}{\partial \theta_i} = 0, \qquad i = 1, k. \qquad (17.5)$$

The second-order condition in this case takes the form (assuming that L is twice differentiable)

$$\sum_{i=1}^{k} \sum_{j=1}^{k} h_i h_j \frac{\partial^2 L}{\partial \theta_i \, \partial \theta_j} < 0, \qquad (17.6)$$

for any set of numbers $h_1, \ldots, h_k$, not all equal to zero.

17.2 EXAMPLES OF MAXIMUM LIKELIHOOD ESTIMATORS

The Binomial Parameter p

The binomial random variable has the density function

$$f(x; p) = p^x(1 - p)^{1-x}, \qquad x = 0, 1, \qquad (17.7)$$

so that the likelihood function for sample drawn at random of size n is

$$L(p) = \binom{n}{x} p^x (1-p)^{n-x}, \qquad x = 0, 1, \ldots, n. \tag{17.8}$$

Taking logarithms,

$$\ln L(p) = \ln \binom{n}{x} + x \ln p + (n-x) \ln (1-p), \tag{17.9}$$

and then differentiating with respect to p and setting the result equal to zero,

$$\begin{aligned} \frac{d \ln L}{dp} &= \frac{x}{p} - \frac{n-x}{1-p} \\ &= 0, \end{aligned} \tag{17.10}$$

we find

$$\hat{p} = \frac{x}{n}. \tag{17.11}$$

A quick check shows that

$$\begin{aligned} \frac{d^2 \ln L}{2} &= -\frac{x}{p^2} - \frac{n-x}{(1-p)^2} \\ &< 0, \end{aligned} \tag{17.12}$$

so that $\hat{p}$, the proportion of successes in the sample, is indeed the maximum likelihood estimator for p.

Since $E(X) = np$, $\hat{p}$ is unbiased and, since $\lim_{n\to\infty} \operatorname{var}(\hat{p}) = \lim_{n\to\infty} p(1-p)/n = 0$, in view of Theorem 16.1, $\hat{p}$ is also consistent. Finally, since from the first line of (17.10), $d \ln L/dp$ can be written in the form

$$\frac{d \ln L}{dp} = A(p)(\hat{p} - p), \tag{17.13}$$

where $A(p) = n/p(1-p)$, it follows from Theorem 16.7 that $\hat{p}$ attains the Cramer-Rao lower bound for the variance of an unbiased estimator.

The Poisson Parameter λ

The Poisson random variable, it will be recalled, has the density function

$$f(x; \lambda) = \frac{\lambda^x e^{-\lambda}}{x!}, \qquad \lambda > 0, \qquad x = 0, 1, 2, \ldots. \tag{17.14}$$

Hence, the likelihood function for a sample of size n drawn at random is

$$L(\lambda) = \frac{1}{x_1! \cdots x_n!} e^{-n\lambda} = \lambda^{\Sigma x_i}. \tag{17.15}$$

Again, taking logarithms,

$$\ln L(\lambda) = -n\lambda + (\sum x_i) \ln \lambda - \ln (x_1! \cdots x_n!), \tag{17.16}$$

differentiating with respect to λ,

$$\frac{dL(\lambda)}{d\lambda} = -n + \frac{\sum x_i}{\lambda}, \tag{17.17}$$

and then equating to zero, we finally find for $\hat{\lambda}$,

$$\begin{aligned} \hat{\lambda} &= \frac{\sum x_i}{n} \\ &= \bar{x}. \end{aligned} \tag{17.18}$$

Checking,

$$\begin{aligned} \frac{d^2 \ln L(\lambda)}{d\lambda^2} &= -\frac{\sum x_i}{\lambda^2} \\ &< 0, \end{aligned} \tag{17.19}$$

so that $\bar{x}$ is in fact the maximum likelihood estimator of λ. We already know $\bar{x}$ to be an unbiased and consistent estimator of λ, and, since (17.17) can be written as

$$\frac{d \ln L}{d\lambda} = \frac{n}{\lambda}(\bar{x} - \lambda), \tag{17.20}$$

its variance, in addition, attains the Cramer-Rao lower bound.

The Parameter of the Uniform Distribution

Assume that X is a continuous random variable with density function

$$f(x; \alpha, \beta) = \frac{1}{\beta - \alpha}, \qquad \alpha \leq x \leq \beta. \tag{17.21}$$

A sample of size n drawn at random from this distribution will have for its likelihood function

$$L(\beta - \alpha) = \frac{1}{(\beta - \alpha)^n}. \tag{17.22}$$

However, in this case, the usual procedure is a loss, since

$$\frac{dL(\beta - \alpha)}{d(\beta - \alpha)} = -\frac{n}{\beta - \alpha} \tag{17.23}$$

will equal zero only when $\beta - \alpha$ is infinite. While this, of course, makes no sense, it nevertheless is evident from (17.22) that $L(\beta - \alpha)$ will be maximized when $\beta - \alpha$ is as large as possible (but finite). Accordingly, suppose that we order the n sample values, $x_1, \ldots, x_n$, from smallest to largest, denoting the smallest value of x' and the largest by x''. Then α can be no larger than x' and β can be no smaller than x''. The maximum likelihood estimator of $\beta - \alpha$, therefore, will be given by $x'' - x'$—that is, by the sample range.

The Parameters of the Laplace Distribution

The Laplace distribution, to recall, has the density function

$$f(x; \alpha, \theta) = \frac{\alpha}{2} e^{-\alpha|x-\theta|}, \qquad \alpha > 0, \qquad -\infty < \theta < \infty, \qquad -\infty < x < \infty. \tag{17.24}$$

We shall assume both α and θ to be unknown. The likelihood function for a sample of size n drawn at random from this distribution is therefore

$$L(\alpha, \theta) = \left(\frac{\alpha}{2}\right)^n e^{-\alpha\Sigma|x_i-\theta|}, \tag{17.25}$$

or in logarithms,

$$\ln L(\alpha, \theta) = n \ln \alpha - n \ln 2 - \alpha\sum|x_i - \theta|. \tag{17.26}$$

Because of the absolute value function in the right-hand side of (17.26), the estimators of α and θ must be obtained in two stages. The first stage involves the fixing of the value of α, at $\alpha = \alpha_0$ say, and then maximizing with respect to θ, while the second stage consists of taking the estimate of θ thus obtained and then maximizing with respect to α.

On the assumption that $\alpha = \alpha_0$, the logarithm of the likelihood function in (17.26) becomes

$$\ln L^*(\theta) = a^* - \alpha_0\sum|x_i - \theta|, \tag{17.27}$$

where $a^* = \text{n} \ln \alpha_0 - n \ln 2$. Since α_0 must be positive, it is clear that this expression will be at a maximum at the value of θ which minimizes $\sum|x_i - \theta|$. To find this θ, let us arrange the n sample observations in ascending order of magnitude, so that $x_1, x_2, \ldots, x_k$ are below θ in magnitude and $x_{k+1}, \ldots, x_n$ above. At this point, the number k is, of course, not known, so that finding the value of k that minimizes

$$W = -(x_1 - \theta) - (x_2 - \theta) - \cdots - (x_k - \theta) + (x_{k+1} - \theta) + \cdots + (x_n - \theta) \qquad (17.28)$$

is equivalent to finding the θ that minimizes $\sum |x_i - \theta|$. Proceeding in conventional fashion, we find

$$\frac{dW}{d\theta} = k - (n - k), \qquad (17.29)$$

which, when equated to zero, yields

$$k = \frac{n}{2}, \qquad (17.30)$$

thus indicating the maximum likelihood estimator of θ to be the sample median. But, since

$$\frac{d^2W}{d\theta^2} = 0, \qquad (17.31)$$

we do not know whether the sample median minimizes or maximizes $\sum |x_i - \theta|$.

However, consider the expression

$$M = |x_1 - \theta'| + |x_2 - \theta'| + \cdots + |x_k - \theta'| + \cdots + |x_n - \theta'|, \qquad (17.32)$$

where θ' is the value of θ which minimizes $\sum |x_i - \theta|$ and where, as in (17.28), the n sample observations are arranged in increasing order of magnitude, with k and $k + 1$ being the watershed between $x_i - \theta'$ negative and $x_i - \theta'$ positive. Let δ be a small positive number. Then, in the expression

$$M' = |x_1 - \theta' - \delta| + \cdots + |x_k - \theta' - \delta| + |x_{k+1} - \theta' - \delta| + \cdots + |x_n - \theta' - \delta|, \qquad (17.33)$$

we will have each term in (17.32) increased by δ for $x_i - \theta' < 0$ and decreased by δ for each $x_i - \theta' > 0$. However, since $\theta' + \delta > \theta'$ and since θ' minimizes M, the number of terms increased by δ can never be fewer than those reduced by δ. Similarly in the expression,

$$M'' = |x_1 - \theta' + \delta| + \cdots + |x_k - \theta' + \delta| + x_{k+1} - \theta' + \delta| + \cdots + |x_n - \theta' + \delta|, \qquad (17.34)$$

the number of terms increased by δ can never be fewer than the number reduced by δ. Consequently, it follows that $\sum|x_i - \theta|$ is minimized when the number of terms below θ is equal to the number above, but this is precisely the condition that defines the sample median. Thus, for given α, the sample median does indeed provide the maximum likelihood estimator of θ.

Let us now return to (17.26) and replace θ with $\hat{\theta}$. Since α is not an argument in the absolute value function, we can differentiate with respect to α in the usual way, obtaining

$$\frac{\partial \ln L(\alpha; \hat{\theta})}{\partial \alpha} = \frac{n}{\alpha} - \sum|x_i - \hat{\theta}|. \tag{17.35}$$

Equating this to zero, we then find for $\hat{\alpha}$,

$$\hat{\alpha} = \frac{\sum|x_i - \hat{\theta}|}{n}. \tag{17.36}$$

To establish that the second-order conditions for a maximum of ln L are satisfied by $\hat{\alpha}$ and $\hat{\theta}$ involves a tedious noncalculus argument and will not be undertaken.

The Variance of the Normal Distribution with Known Mean

The likelihood function for a sample of size n drawn at random from the normal distribution with known mean μ and unknown variance σ^2 is

$$L(\sigma^2) = (2\pi\sigma^2)^{-n/2} \exp\left[-\frac{\sum(x_i - \mu)^2}{2\sigma^2}\right], \tag{17.37}$$

or in logarithms,

$$\ln L(\sigma^2) = -\frac{n}{2}\ln 2\pi - \frac{n}{2}\ln \sigma^2 - \frac{\sum(x_i - \mu)^2}{2\sigma^2}. \tag{17.38}$$

Differentiating with respect to σ^2,

$$\frac{d \ln L(\sigma^2)}{d(\sigma^2)} = -\frac{n}{2\sigma^2} + \frac{\sum(x_i - \mu)^2}{2\sigma^4}, \tag{17.39}$$

and then equating the result to zero, we find the maximum likelihood estimator of σ^2 to be

$$\hat{\sigma}^2 = \frac{\sum(X_i - \mu)^2}{n}. \tag{17.40}$$

Checking that the second-order condition is satisfied is left to the reader.

17.3 THE INVARIANCE OF MAXIMUM LIKELIHOOD ESTIMATORS

In the last example suppose that instead of σ^2 we were interested in the maximum likelihood estimator of σ. Returning, therefore, to (17.38), we have

$$\frac{d \ln L(\sigma)}{d\sigma} = -\frac{2n\sigma}{2\sigma^2} + \frac{2\sigma\sum(x_i - \mu)^2}{2\sigma^4}. \tag{17.41}$$

Hence,

$$\hat{\sigma} = \left[\frac{\sum(x_i - \mu)^2}{n}\right]^{1/2}. \tag{17.42}$$

The maximum likelihood estimator of σ thus turns out to be the square root of the maximum likelihood estimator of σ^2.

This last result is not just a coincidence, but rather reflects the invariance property of maximum likelihood estimators summarized in the following theorem.

THEOREM 17.1
Let $\hat{\theta}$ be the maximum likelihood estimator of a parameter θ, and let $\theta^* = u(\theta)$ be a monotonic function of θ with single-valued inverse v, $\theta = v(\theta^*)$. Then $u(\hat{\theta})$ is the maximum likelihood estimator of $u(\theta)$.

PROOF. Since $\hat{\theta}$ is the maximum likelihood estimator of θ, $L(\theta)$ will be maximized at $\theta = \hat{\theta}$. However, since v, the inverse of the function u, is single-valued, the value of $L(\theta) = L[v(\theta^*)]$ will accordingly be the same at $\theta^* = \hat{\theta}^*$ as at $\theta = \hat{\theta}$.

17.4 MAXIMUM LIKELIHOOD AND THE CRAMER-RAO LOWER BOUND

From Theorem 16.7, we know that if the Cramer-Rao lower bound to the variance of an unbiased estimator is to be attained, the derivative of the logarithm of the likelihood function must be able to be written as

$$\frac{\partial \ln L(\theta)}{d\theta} = A(\theta)(\tilde{\theta} - \theta), \tag{17.43}$$

where $\tilde{\theta}$ is a statistic (in fact, the minimum variance unbiased estimator of θ) and $A(\theta)$ is independent of the sample. Equating (17.43) to zero, we then find that the maximum likelihood estimator $\hat{\theta}$ coincides with $\tilde{\theta}$.[1] Thus, it follows that in the cases where the Cramer-Rao bound can be reached, the maximum likelihood estimator is minimum variance unbiased. The first example in Section 17.2 above is a case in point.

17.5 MAXIMUM LIKELIHOOD AND SUFFICIENCY

If a sufficient statistic for a parameter θ exists, it is an easy step to show that the maximum likelihood estimator must be a function of it. That this is the case follows from the fact that, in accordance with Theorem 16.2, sufficiency implies that the likelihood function can be factored,

$$L(x_1, \ldots, x_n; \theta) = g(y; \theta)h(x_1, \ldots, x_n), \tag{17.44}$$

$h(x_1, \ldots, x_n)$ being independent of θ. Choice of $\hat{\theta}$ to maximize $L(x_1, \ldots, x_n; \theta)$ is therefore equivalent to the choosing of $\hat{\theta}$ to maximize $g(y; \theta)$, with the result that $\hat{\theta}$ will be a function of the sufficient statistic Y alone. This being the case, it follows from Theorem 16.4 that all of the optimum properties of sufficient statistics are conferred upon maximum likelihood estimators. Consequently, in a situation where a sufficient statistic for a parameter is known to exist, we can take the maximum likelihood estimator, find the function of it which is unbiased for the parameter (assuming that the maximum likelihood estimator is not already unbiased), and thereby end up with an estimator, which, in accordance with Theorem 16.10, is unique minimum variance unbiased.

17.6 OTHER PROPERTIES OF MAXIMUM LIKELIHOOD ESTIMATORS

Maximum likelihood estimators possess a number of other attractive properties, but which we shall only mention briefly in passing. These include consistency under very general conditions on the likelihood function and asymptotic normality under the conditions specified in the following theorem [a proof can be found in Cramer (1946, pp. 501–504)].

[1] That $\hat{\theta}$ is unique in this case follows from the assumption implicit in the derivation of the Cramer-Rao lower bound that L is regular in its first derivatives. Checking that the second-order condition is satisfied at $\hat{\theta} = \tilde{\theta}$ is left to the reader.

THEOREM 17.2

Let $X_1, \ldots, X_n$ denote a sample of size n drawn at random from the distribution with density function $f(x; \theta)$. Assume that the likelihood function $L(\theta)$ is twice differentiable with respect to θ in an interval of θ including the true value θ_0, and assume

$$E\left(\frac{\partial \ln L(\theta)}{\partial \theta}\right) = 0 \tag{17.45}$$

and that

$$\begin{aligned} R^2(\theta) &= -E\left(\frac{\partial^2 \ln L(\theta)}{\partial \theta^2}\right) \\ &= E\left[\left(\frac{\partial \ln L}{\partial \theta}\right)^2\right] \end{aligned} \tag{17.46}$$

exists and is nonzero for all θ in the interval. Then the maximum likelihood estimator $\hat{\theta}$ is asymptotically normally distributed with mean θ_0 and variance $1/R^2(\theta)$.

Thus, under the conditions postulated on the likelihood function, the maximum likelihood estimator, besides being asymptotically normal, is asymptotically efficient in that [see expressions (16.57) and (16.58)] the asymptotic variance reaches the Cramer-Rao lower bound.

Finally, it is useful to note that, while for finite samples maximum likelihood estimators are often biased, they can frequently be altered, if this is desired, to an unbiased estimator by a simple transformation. A case in point is the maximum likelihood estimator of the variance of the normal distribution when both the mean and variance are unknown. The maximum likelihood estimator of σ^2 in this case is given by

$$\hat{\sigma}^2 = \frac{\sum(X_i - \bar{X})^2}{n}. \tag{17.47}$$

However, the reader can check that the expected value of $\hat{\sigma}^2$ is

$$E(\hat{\sigma}^2) = \frac{n-1}{n}\sigma^2. \tag{17.48}$$

Consequently, multiplication of $\hat{\sigma}^2$ by $n/(n-1)$ will provide an unbiased estimator for σ^2.

REFERENCES

Cramer, H., *Mathematical Methods of Statistics*, Princeton University Press, 1946, chap. 33, pp. 498–506.

Freeman, H., *An Introduction to Statistical Inference*, Addison-Wesley, 1963, chap. 26.

Kendall, M. G. and Stuart, A., *The Advanced Theory of Statistics*, vol. II, Charles Griffin and Co., 1961, chap. 18.

CHAPTER 18

HYPOTHESIS TESTING

In Chapters 16 and 17, we were concerned with the problem of using sample information to estimate an unknown parameter (or parameters) in the population. This, as was noted at the time, is one of the central problems in statistical inference. In this chapter, we approach the problem of unknown population characteristics from a different direction, namely, that of using sample information to test *hypotheses* about these unknown characteristics. Although hypothesis testing and estimation are not logically divorced from one another, they are traditionally introduced as separate concepts, and we shall maintain this tradition. The approach to hypothesis testing followed in this chapter is the one that was developed by J. Neyman and E. S. Pearson in a series of papers published in the late 1920s and 1930s, and has come to be known as the *classical method of hypothesis testing*.

In general, our frame of reference will be the following: We have a population described by the density function $f(x; \theta)$, where θ is a parameter, and we wish to use the informaion given in a sample of size n drawn at random from this density to test hypotheses about the value of θ. The set of all values that θ can take on will be denoted by Ω and shall be referred to as the *parameter space*. Usually, the form of $f(x; \theta)$ is assumed to be known, which in many cases may put a rather severe restriction on the analysis, but procedures exist for testing hypotheses about the shape of $f(x; \theta)$ as well. The population density function may have several parameters instead of only one, in which case the concern may be with the entire set of parameters or possibly just a subset.

18.1 THE NEYMAN-PEARSON THEORY OF HYPOTHESIS TESTING

Let us begin by noting that there are two types of hypotheses, simple and composite. A *simple* hypothesis completely specifies the population distribution and therefore refers to a single point in the parameter space. For example, if the parent distribution is normal with unknown mean μ and variance equal to 1, then the hypothesis H: $\mu = 0$ completely specifies the population distribution and is therefore simple. On the other hand, the hypothesis may refer to a region in the sample space, instead of a point, in which case the hypothesis is *composite*. In the preceding example, if the hypothesis were H: $\mu > 0$, the hypothesis would be composite. If the population density were $f(x; \theta_1, \ldots, \theta_k)$, then another example of a composite hypothesis would be H: $\theta_1 = 1$, $\theta_2 = 2$, a composite hypothesis because $\theta_3, \ldots, \theta_k$ are unspecified and accordingly can take on any set of values.

In the classical theory of hypothesis testing, a hypothesis, whether simple or composite, is always tested against an alternative hypothesis, which itself can be simple or composite. The first hypothesis is usually referred to as the *null* hypothesis and the second as the *alternative* hypothesis. These will be denoted by H_0 and H_a, respectively. With reference to the example in the preceding paragraph, an illustration of two simple hypotheses is

$$H_0\colon \mu = 0,$$
$$H_a\colon \mu = 10,$$

while an instance of a simple hypothesis being tested against a composite alternative is

$$H_0\colon \mu = 0,$$
$$H_a\colon \mu > 0.$$

Types of Errors Involved in Hypothesis Testing

Once a hypothesis and its alternative is set up, we must proceed to set up a test based on sample information which allows the null hypothesis to be accepted or else rejected in favor of its alternative. Suppose that the hypothesis H_0: $\theta = \theta_0$ is to be tested against the alternative H_a: $\theta = \theta_1$. Then, depending on the true value of θ (and for the moment we assume that one of θ_0, θ_1 is the true value), four possibilities can arise:

1. H_0 can be accepted when θ_0 is correct.
2. H_0 can be rejected when θ_0 is correct.
3. H_0 can be accepted when θ_1 is correct.
4. H_0 can be rejected when θ_1 is correct.

Correct decisions are made in cases 1 and 4, but errors are committed in cases 2 and 3. In case 2, we reject H_0 when it is correct, while, in case 3, we accept H_0 when it is wrong. These errors are usually referred to as *Type I* and *Type II*, respectively, and are summarized in Table 18.1.

TABLE 18.1 Type of Errors in Testing Hypotheses

Hypothesis Accepted	True Value of θ	
	θ_0	θ_1
H_0	Correct	II
H_a	I	Correct

Construction of Tests: Critical Regions

We now proceed to the basic theory underlying hypothesis testing. Assume that the population is described by the distribution with density function $f(x;\,\theta)$, where the form of f is known, but θ is unknown, and suppose that we wish to test the hypothesis H_0: $\theta = \theta_0$ against the alternative hypothesis H_a: $\theta = \theta_1$. Assume that we draw a sample of size n at random from $f(x;\,\theta)$ and formulate the statistic

$$\hat{\theta} = \hat{\theta}(x_1, \ldots, x_n;\, \theta). \tag{18.1}$$

[θ is included in (18.1) only to remind us that the shape of the function $\hat{\theta}$ depends on θ.]

Let us now divide the sample space of $\hat{\theta}$ into two mutually exclusive and exhaustive regions, A_1 and A_2. If $\hat{\theta}$ falls in A_1, H_0 is to be rejected in favor of H_a, while if $\hat{\theta}$ falls in A_2, H_0 is to be accepted. Figure 18.1 depicts such a situation for $\hat{\theta}$ based on a sample of size 2. The region labeled A_1 is the rejection region, while the rest of the sample space composes the acceptance region. H_0 is accepted if $\hat{\theta}$ falls in A_2, and rejected if it falls in A_1.

It is to be noted, however, that even if θ_0 is the true value of θ, $\hat{\theta}$ will fall in A_1 a certain proportion of the time, say α, in which cases H_0 will be

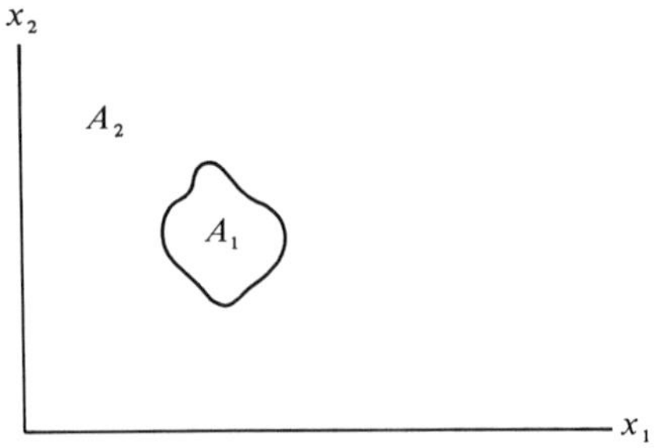

FIGURE 18.1 A_1 the rejection region; A_2 the acceptance region.

rejected even though it is true. Therefore, a Type I error would be committed α proportion of the time; alternatively, we can say that the probability of a Type I error is α. On the other hand, if θ_1 is the true state of nature, $\hat{\theta}$ will still fall in A_2 a certain proportion of the time, say β, in which cases a Type II error will be committed, since H_0 would be accepted even though H_a is true. Thus, if θ_1 is the true value of θ_1, we clearly want $\hat{\theta}$ to fall in A_1 in order to be able to reject H_0. This will occur $1 - \beta$ proportion of the time. Usually, the number $1 - \beta$ is called the *power* of the test, since it describes the test's ability to reject H_0 when H_a is true. Therefore, to sum up, we have

$$P(\hat{\theta} \in A_1 | H_0) = \alpha \qquad \text{(the probability of a Type I error)},$$
$$P(\hat{\theta} \in A_2 | H_a) = \beta \qquad \text{(the probability of a Type II error)},$$
$$P(\hat{\theta} \in A_1 | H_a) = 1 - \beta \qquad \text{(the power of the test)}.$$

Suppose, for the moment, that $\hat{\theta}$ has the distribution shown in Figure 18.2, and let α be a fixed number. Then it is evident that an indefinitely large number of areas beneath $f(\hat{\theta})$ exist with size equal to α. Two such areas are shown in the figure.

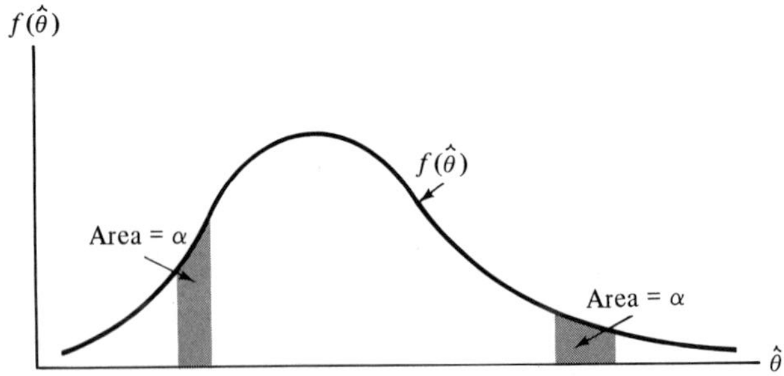

FIGURE 18.2 Two areas beneath $f(\hat{\theta})$ of size α.

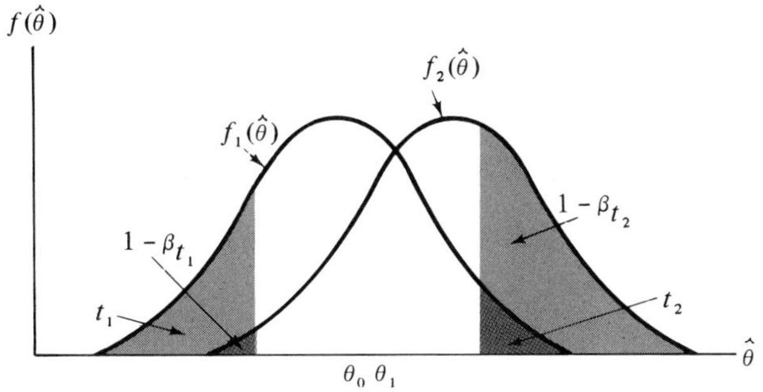

FIGURE 18.3 Distributions of $\hat{\theta}$ for $\theta = \theta_1$ and $\theta = \theta_2$.

Let us now refer to Figure 18.3, which shows the distribution of $\hat{\theta}$, first for $\theta = \theta_1$, and then for $\theta = \theta_2$. The two horizontally lined areas under $f_1(\hat{\theta})$, denoted t_1 and t_2, are both of size α, while the diagonally lined areas under $f_2(\hat{\theta})$ are of size $1 - \beta_{t_1}$ and $1 - \beta_{t_2}$, respectively, and give the powers of the two tests on the assumption that the true value of θ is θ_1. If the true value of θ is in fact θ_1, which test, t_1 or t_2, is to be preferred? It is quickly seen that, in this case, it is t_2, since, for the same probability of a Type I error, the probability of correctly rejecting H_0: $\theta = \theta_1$ if H_a: $\theta = \theta_2$ is true is much greater with t_2—that is, the power of the test is much larger with t_2. This immediately suggests the procedure to be followed in choosing the critical region of rejection to be used in testing a hypothesis:

> Choose the test (i.e., the critical region of rejection) in such a way that for a given size of α the power of the test is maximized—that is, for a given probability of Type I error, minimize the probability of Type II error.

Such a rejection region is called a *best critical region* (BCR). We shall now discuss a procedure for finding a BCR when the null hypothesis and its alternative are both simple.

The Neyman-Pearson Lemma

To begin with, let $L(x|H)$ denote the likelihood function of the sample under the assumption that hypothesis H regarding the parameter θ is true. Assume that H_0: $\theta = \theta_0$ is the hypothesis to be tested, and let H_1: $\theta = \theta_1$ be the alternative. For a given rejection region R, the power of the test, $1 - \beta$, is then given by

$$1 - \beta = \int_R L(x|H_1)\,dx, \tag{18.2}$$

where a single integral is used to represent the n-fold integration over the rejection region R, while the probability of a Type I error is equal to

$$\alpha = \int_R L(x \mid H_0)\, dx. \tag{18.3}$$

The best critical region will be given by the rejection region R which maximizes (18.2) subject to (18.3).

The following theorem, due to Neyman and Pearson and known as the Neyman–Pearson lemma, provides a systematic method for determining R.

THEOREM 18.1 (Neyman-Pearson Lemma)
Let $X_1, \ldots, X_n$ denote a sample size n drawn at random from the distribution with density function $f(x; \theta)$. Let θ_0 and θ_1 be distinct values of θ and let k be a positive number. Let $H_0\colon \theta = \theta_0$ and $H_1\colon \theta = \theta_1$ be two simple hypotheses. Finally, let C and C' be a subset of the sample space and its complement, respectively, such that

$$\frac{L(x \mid H_0)}{L(x \mid H_1)} \le k, \qquad \text{for each } x \in C, \tag{18.4}$$

$$\frac{L(x \mid H_0)}{L(x \mid H_1)} \ge k, \qquad \text{for each } x \in C', \tag{18.5}$$

$$\alpha = \int_C L(x \mid H_0)\, dx. \tag{18.6}$$

Then C is a best critical region of size α for testing the hypothesis $H_0\colon \theta = \theta_0$ against the alternative hypothesis $H_1\colon \theta = \theta_1$.

PROOF. If C is the only region of size α, there is nothing to prove so let us suppose that there is at least one other region of size α, which we shall denote by A. We now wish to show that

$$\int_C L(\theta_1) - \int_A L(\theta_1) \ge 0. \tag{18.7}$$

Note, first, that C can be written as the union of the disjoint sets $C \cap A$ and $C \cap A'$, and similarly for A as the union of $A \cap C$ and $A \cap C'$. Accordingly, we have

$$\begin{aligned}\int_C L(\theta_1) - \int_A L(\theta_1) &= \int_{C\cap A} L(\theta_1) + \int_{C\cap A'} L(\theta_1) - \int_{A\cap C} L(\theta_1)\\ &\quad - \int_{A\cap C'} L(\theta_1)\\ &= \int_{C\cap A'} L(\theta_1) - \int_{A\cap C'} L(\theta_1).\end{aligned} \tag{18.8}$$

In view of the hypothesis (18.4), $L(\theta_1) \geq (1/k)L(\theta_0)$ at each point of C, and hence at each point of $C \cap A'$. Consequently,

$$\int_{C \cap A'} L(\theta_1) \geq \frac{1}{k} \int_{C \cap A'} L(\theta_0). \tag{18.9}$$

However, by hypothesis (18.5) of the theorem ,$L(\theta_1) \leq (1/k)L(\theta_0)$ at each point of C', and hence at each point of $A \cap C'$, so that

$$\int_{A \cap C'} L(\theta_1) \leq \frac{1}{k} \int_{A \cap C'} L(\theta_0). \tag{18.10}$$

Inequalities (18.9) and (18.10) together then imply that

$$\int_{C \cap A'} L(\theta_1) - \int_{A \cap C'} L(\theta_1) \geq \frac{1}{k} \int_{C \cap A'} L(\theta_0) - \frac{1}{k} \int_{A \cap C'} L(\theta_0), \tag{18.11}$$

whence from Equation (18.8) we obtain

$$\int_C L(\theta_1) - \int_A L(\theta_1) \geq \frac{1}{k} \left[\int_{C \cap A'} L(\theta_0) - \int_{A \cap C'} L(\theta_0) \right]. \tag{18.12}$$

However,

$$\begin{aligned} \int_{C \cap A'} L(\theta_0) - \int_{A \cap C'} L(\theta_0) &= \int_{C \cap A'} L(\theta_0) + \int_{C \cap A} L(\theta_0) - \int_{A \cap C} L(\theta_0) \\ &\quad - \int_{A \cap C'} L(\theta_0) \\ &= \int_C L(\theta_0) - \int_A L(\theta_0) \\ &= \alpha - \alpha \\ &= 0. \end{aligned} \tag{18.13}$$

Upon substitution of this result in (18.12), we finally obtain

$$\int_C L(\theta_1) - \int_A L(\theta_1) \geq 0, \tag{18.7}$$

as was to be shown.[1]

We shall illustrate the foregoing with an example. Let the population be $N(\mu, 1)$, and let the two hypotheses be $H_0\colon \mu = \mu_0$ and $H_1\colon \mu = \mu_1$. Then

$$L(x \mid H_j) = (2\pi)^{-n/2} \exp[-\tfrac{1}{2} \sum_i (x_i - \mu_j)^2], \qquad j = 0, 1, \tag{18.14}$$

[1] If x is discrete instead of continuous, the proof is the same, with integration replaced by summation.

so that

$$\frac{L(x|H_0)}{L(x|H_1)} = \exp\left[-\frac{1}{2}\sum(x_i - \mu_0)^2 + \frac{1}{2}\sum(x_i - \mu_1)^2\right] \qquad (18.15)$$
$$\leq k_\alpha$$

for some positive number k which depends on α. Taking logarithms, we have from (18.15),

$$-\sum(x_i - \mu_0)^2 + \sum(x_i - \mu_1)^2 \leq 2 \ln k. \qquad (18.16)$$

However, since

$$\sum(x_i - \mu_0)^2 = \sum(x_i - \bar{x})^2 + n(\bar{x} - \mu_0)^2, \qquad (18.17)$$

and similarly for $\sum(x_i - \mu_1)^2$, (18.16) becomes

$$-\sum(x_i - \bar{x})^2 - n(\bar{x} - \mu_0)^2 + \sum(x_i - \bar{x})^2 + n(\bar{x} - \mu_1)^2 \leq 2 \ln k_\alpha \qquad (18.18)$$

or

$$n[(\bar{x} - \mu_1)^2 - (\bar{x} - \mu_0)^2] \leq 2 \ln k_\alpha. \qquad (18.19)$$

Finally, after squaring the terms within brackets and rearranging, we obtain

$$\bar{x}(\mu_0 - \mu_1) \leq \frac{1}{2}(\mu_0{}^2 - \mu_1{}^2) + \frac{1}{n}\ln k_\alpha. \qquad (18.20)$$

If $\mu_1 > \mu_0$, the BCR is then defined by

$$\bar{x} \geq \frac{1}{2}(\mu_0 + \mu_1) - \frac{\ln k_\alpha}{n(\mu_1 - \mu_0)}, \qquad (18.21)$$

while if $\mu_0 > \mu_1$, the BCR is given by

$$\bar{x} \leq \frac{1}{2}(\mu_0 + \mu_1) + \frac{\ln k_\alpha}{n(\mu_0 - \mu_1)}, \qquad (18.22)$$

Before going on to derive the exact values of the right-hand sides of (18.21) and (18.22), we should note that, in this example, the test statistic is the sample mean, which, is, of course, what we might have expected given that $\bar{X}$ is a minimum variance unbiased and sufficient statistic for μ. Moreover, the two BCR's, (18.21) and (18.22), are eminently sensible, since for $\mu_1 > \mu_0$ we should expect to reject H_0 in favor of H_1 if the sample mean exceeds a certain value, while for $\mu_0 > \mu_1$ we should expect to reject H_0 if $\bar{x}$ is below a certain value.

Turning now to the derivation of the values of the right-hand sides of (18.21) and (18.22), we know that, whatever the sample size, the mean of a sample drawn from $N(\mu, 1)$ will be distributed normally with mean μ and variance $1/n$. Therefore, to obtain a BCR of size α for testing H_0: $\mu = \mu_0$ against H_1: $\mu = \mu_1$, $\mu_1 > \mu_0$, we must determine $\bar{x}_\alpha$ so that

$$\int_{\bar{x}_\alpha}^{\infty} \left(\frac{n}{2\pi}\right)^{1/2} \exp\left[-\frac{n}{2}(\bar{x} - \mu_0)^2\right] d\bar{x} = \alpha, \tag{18.23}$$

which is easily done with the aid of a table for the standardized normal integral. Let

$$z = \frac{\bar{x} - \mu_0}{\sqrt{1/n}}. \tag{18.24}$$

The distribution of Z is then

$$G(z) = \int_{-\infty}^{z} (2\pi)^{-1/2} e^{-(1/2)y^2}\, dy, \tag{18.25}$$

so that the task is to find the value of z, say z_α, such that

$$G(z_\alpha) = 1 - \alpha. \tag{18.26}$$

However, since

$$z = \frac{\bar{x} - \mu_0}{\sqrt{1/n}}, \tag{18.27}$$

we have

$$\bar{x}_\alpha = \mu_0 + \frac{z_\alpha}{\sqrt{n}}. \tag{18.28}$$

The BCR in this case is thus the set of sample values for which

$$\bar{x} \geq \mu_0 + \frac{z_\alpha}{\sqrt{n}}. \tag{18.29}$$

That is, if the sample mean is greater than or equal to $\mu_0 + z_\alpha/\sqrt{n}$, H_0: $\mu = \mu_0$ is rejected in favor of H_1: $\mu = \mu_1$.

Since

$$G(z_\alpha) = 1 - G(-z_\alpha), \tag{18.30}$$

a similar argument will show that the BCR for testing H_0 against H_1 with $\mu_0 > \mu_1$ is given by the set of sample values for which

$$\bar{x} \leq \mu_0 - \frac{z_\alpha}{\sqrt{n}}. \tag{18.31}$$

In this case, if $\bar{x}$ is less than or equal to $\mu_0 - z_\alpha/\sqrt{n}$, H_0 is rejected in favor of H_1.

For example, if $\mu_0 = 5$, $n = 25$, and $\alpha = 0.05$ (how α is determined will be discussed later), we have from the table of the standardized normal integral, $z_{0.05} = 1.6449$, so that from (18.29),

$$\bar{x}_{0.05} = 5 + \frac{1.6449}{5}$$

$$= 5.3290. \tag{18.32}$$

The power of the test for $\mu_1 > \mu_0$ can be written directly, since it is simply

$$1 - \beta = \int_{\bar{x}_\alpha}^{\infty} \left(\frac{n}{2\pi}\right)^{1/2} \exp\left[-\frac{n}{2}(\bar{x} - \mu_1)^2\right] d\bar{x}. \tag{18.33}$$

For $\mu = \mu_0$, the standardized variable, as we know,

$$u = \frac{\bar{x} - \mu_0}{\sqrt{1/n}}, \tag{18.34}$$

is normal with mean 0 and variance 1, and the same for

$$v = \frac{\bar{x} - \mu_1}{\sqrt{1/n}} \tag{18.35}$$

for $\mu = \mu_1$. However,

$$v = u - \sqrt{n}(\mu_1 - \mu_0), \tag{18.36}$$

so that in terms of v, the power of the test is given by

$$1 - \beta = \int_{z_\alpha}^{\infty} (2\pi)^{-1/2} e^{-(1/2)[u - \sqrt{n}(\mu_1 - \mu_0)]^2} \, du. \tag{18.37}$$

It is evident from (18.37) that $1 - \beta$ is a monotonic increasing function both of the sample size n and the difference $\mu_1 - \mu_0$ between the hypothesized values between which the test has to choose. Finally, an argument similar to that just given will show that for $u_0 > u_1$ the power of the test is equal to

$$1 - \beta = \int_{-\infty}^{z_\alpha} (2\pi)^{-1/2} e^{-(1/2)[u - \sqrt{n}(\mu_0 - \mu_1)]^2} \, du. \tag{18.38}$$

Best Critical Regions and Sufficient Statistics

In the example just discussed, we noted that the Neyman-Pearson criterion leads to a test involving a sufficient statistic for the parameter in question. This is no idle coincidence, for if a sufficient statistic for the parameter exists, the BCR is a function of this statistic. To see this, note that, if $\hat{\theta}$ is a sufficient statistic for θ, then the likelihood function for x given H_0 factors into

$$L(x \mid H_0) = g(\hat{\theta}; \theta_0)h(x), \tag{18.39}$$

and similarly for $L(x \mid H_1)$,

$$L(x \mid H_1) = g(\hat{\theta}; \theta_1)h(x). \tag{18.40}$$

Consequently, (18.4) becomes

$$\begin{aligned} \frac{L(x \mid H_0)}{L(x \mid H_1)} &= \frac{g(\hat{\theta}; \theta_0)}{g(\hat{\theta}; \theta_1)} \\ &\leq k_\alpha, \end{aligned} \tag{18.41}$$

which shows that the BCR is a function of the sufficient statistic.

18.2 TESTING A SIMPLE HYPOTHESIS AGAINST A COMPOSITE ALTERNATIVE

Thus far, we have been considering the most elementary situation, that where the competing hypotheses are both simple. In terms of the theory (but not in terms of scientific procedure), it is a matter of convenience or convention in this case which hypothesis is considered "under test" and which is the "alternative" since they are interchangeable. Once we begin to generalize the problem, however, this symmetry disappears, as in the case we now consider where H_0 is simple, but H_a is composite and consists of a class of simple alternatives. The most common case of this kind is where H_0 contains one point in the sample space and H_a includes all the rest, namely,

$$H_0\colon\ \theta = \theta_0,$$
$$H_a\colon\ \theta \neq \theta_0.$$

Almost as common are the cases

$$\begin{aligned} H_0&\colon\ \theta = \theta_0 \\ H_a&\colon\ \theta > \theta_0 \end{aligned} \qquad \text{or} \qquad \begin{aligned} H_0&\colon\ \theta = \theta_0, \\ H_a&\colon\ \theta < \theta_0. \end{aligned}$$

From the discussion just concluded, we know that application of the Neyman-Pearson lemma leads to a BCR for the case where H_0 and H_a are both simple. Consequently, we can apply that result to any particular member of the composite H_a, say H_i, and obtain a BCR for that H_i.

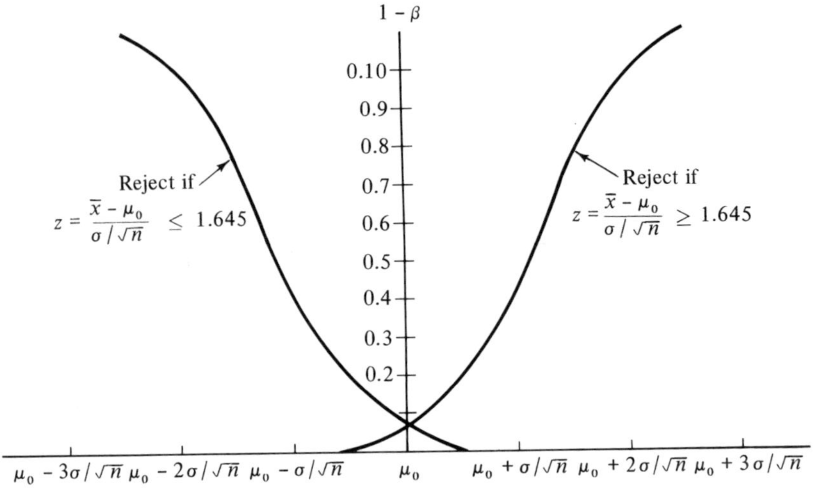

FIGURE 18.4 Power functions for testing H_0: $\mu = \mu_0$ against H_1: $\mu > \mu_0$ and H_1: $\mu < \mu_0$ for $\alpha = 0.05$.

However, this region will in general vary from one H_i to another, and it is obviously infeasible to determine a different BCR for each of the unspecified alternatives. Accordingly, we are led to enquire whether there is a single region which is best for all members H_i of H_a. Such a region, if it exists, is called *uniformly most powerful* (UMP) and the test based on it a UMP test.

Let us refer again to the testing of hypotheses about the population mean when the population distribution is $N(\mu, 1)$. Careful study of Figure 18.2 should convince the reader that not only does the test defined by (18.30) give the BCR for testing H_0: $\mu = \mu_0$ against H_a: $\mu = \mu_1$, $\mu_1 > \mu_0$, but also gives the BCR for testing μ_0 against *any* value of μ greater than μ_0. This test, therefore, is UMP for testing H_0: $\mu = \mu_0$ against H_a: $\mu > \mu_0$. Similarly, the BCR defined by (18.31) is UMP for testing H_0: $\mu = \mu_0$ against H_a: $\mu < \mu_0$. The power functions for these two tests for $\alpha = 0.05$ are given in Figure 18.4.

Figure 18.4 shows that, for this example, a UMP test for testing μ_0 against *all* values of μ does not exist, since for $\mu > \mu_0$ the best test is $\bar{x} \geq \mu_0 + z_\alpha/\sqrt{n}$, while for $\mu < \mu_0$ the best test is $\bar{x} \leq \mu_0 - z_\alpha/\sqrt{n}$. Unfortunately, this is typical, for UMP tests seldom exist for hypotheses of the type H_0: $\mu = \mu_0$ against H_a: $\mu \neq \mu_0$.

As a practical matter, what this means is that we must seek tests that are most powerful within a restricted class of tests. One such restricted class is the class of *unbiased* tests. This class of tests has the desirable property that the Type I error is controlled by requiring its probability not to exceed α and at the same time we are assured that the probability of rejecting H_0 if it is not true exceeds α. An unbiased test is thus defined by the condition that the test has minimum power at $\theta = \theta_0$. A *uniformly most powerful unbiased* test (UMPU) for testing the hypothesis H_0: $\theta = \theta_0$ against H_a: $\theta \neq \theta_0$, therefore, is defined by the rejection region R of size α such that for R', any other region of size α,

$$P(\hat{\theta} \in R \mid H_a) \geq P(\hat{\theta} \in R' \mid H_a), \tag{18.42}$$

where $\hat{\theta}$ is the test statistic. A UMPU test for the example just considered will be derived in Section 18.4.

18.3 AN OPTIMUM TEST PROPERTY OF SUFFICIENT STATISTICS

This is a convenient place to prove a property of sufficient statistics regarding the testing of hypotheses which is analogous to the property stated in Theorem 16.10.

THEOREM 18.2
Let R be a critical region of size α for testing H_0, a hypothesis concerning θ in $L(x \mid \theta)$, against some alternative hypothesis H_a, and let Y be a sufficient statistic, both on H_0 and H_a, for θ. Then there exists a critical region, also of size α, based on a function of Y, which has the same power as R.

PROOF. We begin by defining the function

$$c(R) = \begin{cases} 1, & \text{if } x \in R, \\ 0, & \text{otherwise.} \end{cases} \tag{18.43}$$

Then, the integral

$$E[c(R)] = \int_x c(R)L(x \mid \theta)\, dx \tag{18.44}$$

gives the probability that the sample point falls in R, and hence is equal to the size α of the test when H_0 is true and to the power of the test when H_1 is true. Using the factorization property of the likelihood function in the presence of a sufficient statistic (Theorem 16.2), (18.44) becomes

$$\begin{aligned} E[c(R)] &= \int_x c(R)h(x \mid y)g(y \mid \theta)\, dx \\ &= E[c(R) \mid y]g(y \mid \theta) \\ &= E\{E[c(R) \mid Y]\}, \end{aligned} \tag{18.45}$$

where the expectation outside of the braces is with respect to the distribution of Y. ($E[c(R)]$ is a constant and is therefore not affected by this last expectation.) Thus, we have that the expected value of $E[c(R)|Y)]$, which is a function of Y and which is not dependent on θ since Y is a sufficient statistic, is the same as the expectation of $c(R)$. This being the case, there is therefore a critical region based on the sufficient statistic Y which has the same size and power as the original region R.

The importance of this theorem is that we can, without any loss of power, confine our attention to test procedures which are based on functions of sufficient statistics (assuming that the latter exist). Moreover, it is also to be noted that the theorem places no restrictions on the hypotheses involved. Either of them may be simple or composite.

18.4 TESTS BASED ON THE LIKELIHOOD RATIO

We now turn to a practical method for testing a simple or composite hypothesis against a simple or composite alternative, which is based on the *likelihood ratio*

$$L = \frac{L(x|H_0)}{L(x|H_a)}. \tag{18.46}$$

This procedure yields a test that is based on a sufficient statistic, if one exists, and if a UMP or UMPU test exists, this test often leads to it.

The basic idea of the likelihood ratio test is as follows. Let Ω represent the entire parameter space; H_0 and H_a thus denote certain regions of Ω. In particular, suppose that H_a refers to the entire parameter space Ω. Of the values of θ in H_0, choose the value which maximizes $L(x|H_0)$, and similarly for $\theta \in H_a$ and $L(x|H_a)$. This yields a value of L, say $\hat{L}$, which must lie between 0 and 1. (Values of $\hat{L}$ close to 1 favor H_0, while values of $\hat{L}$ close to 0 favor H_a.) $\hat{L}$ is therefore a random variable with distribution, say, $h(\hat{L})$. On the assumption that H_0 is true, we can then find a critical region of size α by solving the equation

$$\int_0^{L_\alpha} h(\hat{L}|H_0)\,dh = \alpha \tag{18.47}$$

for L_α. The probability of a Type II error will then be given by

$$\beta = \int_{L_\alpha}^1 h(\hat{L}|H_a)\,dh, \tag{18.48}$$

and the power of the test by

$$1 - \beta = \int_0^{L_\alpha} h(\hat{L} \mid H_a)\, dh. \tag{18.49}$$

When the underlying population distribution contains several parameters (say k), $\theta_1, \ldots, \theta_k$, the procedure is the same except that the likelihood functions are defined over a space of k dimensions. The hypotheses H_0 and H_a continue to restrict the parameter space, and the maximum likelihood estimates of the k parameters are inserted into (18.46). From here on, the procedure is the same as in the case of a single parameter.

The likelihood ratio test, it will be noted, is a natural extension of the Neyman-Pearson criterion to composite hypotheses via the principle of maximum likelihood. This is most evident when both hypotheses are simple, in which case the likelihood ratio (18.46) reduces to the Neyman-Pearson expression (18.4).

As a first example, let us return to the case where the underlying population is $N(\mu, 1)$ and where the hypotheses are $H_0\colon \mu = \mu_0$ and $H_a\colon \mu \neq \mu_0$. Since H_0 is simple, the numerator of the likelihood ratio is

$$L(x \mid H_0) = (2\pi)^{-n/2} \exp[-\tfrac{1}{2} \textstyle\sum (x_i - \mu_0)^2], \tag{18.50}$$

while the denominator is

$$L(x \mid H_a) = (2\pi)^{-n/2} \exp[-\tfrac{1}{2} \textstyle\sum (x_i - \bar{x})^2], \tag{18.51}$$

$\bar{x}$ being the maximum likelihood estimate of μ. The likelihood ratio is therefore equal to

$$L = \exp[-\tfrac{1}{2} \textstyle\sum (x_i - \mu_0)^2 + \tfrac{1}{2} \textstyle\sum (x_i - \bar{x})^2]. \tag{18.52}$$

However,

$$\begin{aligned}
&-\tfrac{1}{2} \textstyle\sum (x_i - \mu_0)^2 + \tfrac{1}{2} \textstyle\sum (x_i - \bar{x})^2 \\
&\quad = -\tfrac{1}{2} \textstyle\sum (x_i - \bar{x} + \bar{x} - \mu_0)^2 + \tfrac{1}{2} \textstyle\sum (x_i - \bar{x})^2 \\
&\quad = -\frac{1}{2} \textstyle\sum (x_i - \bar{x})^2 - (\bar{x} - \mu_0) \textstyle\sum (x_i - \bar{x}) - \dfrac{n}{2} (\bar{x} - \mu_0)^2 + \dfrac{1}{2} \textstyle\sum (x_i - \bar{x})^2 \\
&\quad = -\frac{n}{2} (\bar{x} - \mu_0)^2,
\end{aligned} \tag{18.53}$$

so that (18.52) becomes

$$\hat{L} = \exp\left[-\frac{n}{2} (\bar{x} - \mu_0)^2\right], \tag{18.54}$$

which is a function only of the sample mean $\bar{x}$.

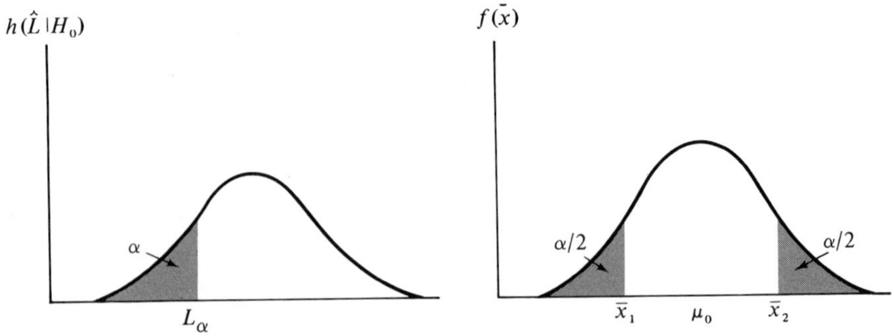

FIGURE 18.5 Correspondence between L_α and the critical regions for $\bar{x}$.

Fix α. Then, on the assumption that μ_0 is the true value of μ, the critical value of $\hat{L}$, L_α, is obtained from Equation (18.47),

$$\int_0^{L_\alpha} h(\hat{L}|H_0)\,dh = \alpha. \tag{18.47}$$

To determine L_α, it is thus necessary to know the shape of h. [Note that from (18.54) there will be two values of $\bar{x}$ that correspond to L_α.]

However, in the present case, knowledge of h can actually be put to the side. In view of the fact that the parent distribution is $N(\mu, 1)$, $\bar{x}$ is $N(\mu, 1/n)$, which means that in place of having to find L_α, a critical region of size α for $\bar{x}$ can be used instead. Since two values of $\bar{x}$ (say $\bar{x}_1$ and $\bar{x}_2$) correspond to L_α, two critical regions whose areas sum to α will be involved, one in each tail of the distribution, and from the symmetry of the normal distribution, it is evident that each of these will be of size $\alpha/2$. The correspondence between L_α and the critical regions for $\bar{x}$ is illustrated in Figure 18.5.

Although we shall not prove it here, this test on the population mean is UMPU.[2] Its power function is given in Figure 18.6, and is compared with

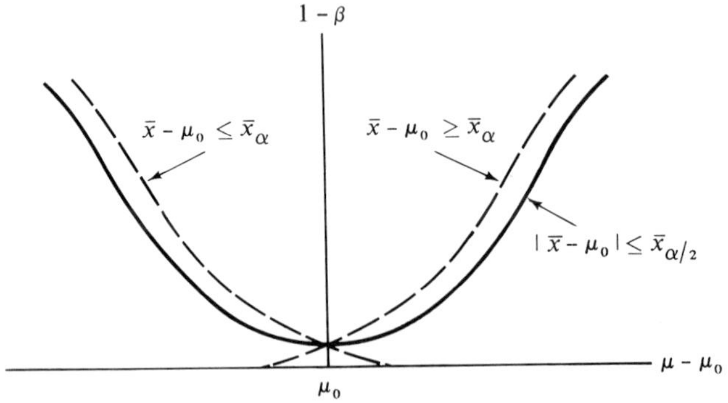

FIGURE 18.6 Power function of the likelihood ratio test.

[2] For a proof, see Kendall and Stuart (1961, pp. 212–213).

the power functions corresponding to the UMP tests obtained from the Neyman-Pearson lemma of the hypotheses $\mu = \mu_0$ against $\mu_1 > \mu_0$ and $\mu = \mu_0$ against $\mu_1 < \mu_0$.

As a second example, assume again that the underlying population is normal, but with both mean and variance unknown. As before, let the hypotheses to be tested be $H_0: \mu = \mu_0$ and $H_a: \mu \neq \mu_0$. Since σ^2 is unknown, we now must maximize the numerator in the likelihood ratio with respect to σ^2, μ being fixed at μ_0, and the denominator with respect to both μ and σ^2. As we already know, the maximum likelihood estimate for σ^2 with μ fixed at μ_0 is $\sum (x_i - \mu_0)^2/n$, while the maximum likelihood estimates for μ and σ^2 are $\bar{x}$ and $\sum (x_i - \bar{x})^2/n$. Hence, the likelihood ratio is equal to

$$L = \left[\frac{n}{2\pi \sum (x_i - \mu_0)^2}\right]^{n/2} e^{-n/2} \Bigg/ \left[\frac{n}{2\pi \sum (x_i - \bar{x})^2}\right]^{n/2} e^{-n/2}, \quad (18.55)$$

which reduces to

$$\hat{L} = \left[\frac{\sum (x_i - \bar{x})^2}{\sum (x_i - \mu_0)^2}\right]^{n/2}. \quad (18.56)$$

Writing the denominator as

$$\sum (x_i - \mu_0)^2 - \sum (x_i - \bar{x})^2 + n(\bar{x} - \mu_0)^2 \quad (18.57)$$

and letting

$$t = \frac{\bar{x} - \mu_0}{[\sum (x_i - \bar{x})^2/n(n-1)]^{1/2}}, \quad (18.58)$$

we can after some manipulation reduce (18.56) to

$$\hat{L} = \left[\frac{1}{1 + t^2/(n-1)}\right]^{n/2}, \quad (18.59)$$

However, as will be recalled from Section 13.2, t is distributed as "Student's" t-distribution with $n - 1$ degrees of freedom. Since $\hat{L}$ is a function only of t, instead of having to find a critical value of $\hat{L}$, L_α, corresponding to a critical region of size α, we can use the equivalent critical region of size α associated with the t-distribution. It is seen in (18.59) that each value of $\hat{L}$ corresponds to two values of t (say t_1 and t_2), so that there is again a critical region in each tail of the distribution. Finally, the symmetry of the t-distribution dictates that the size of the two regions will be $\alpha/2$ each. Thus, $H_0: \mu = \mu_0$ will be accepted if

$$-t_{\alpha/2} \leq t \leq t_{\alpha/2}, \quad (18.60)$$

where $t_{\alpha/2}$ for $n - 1$ degrees of freedom is obtained from a table of the t-distribution. With this example, as with the first, the likelihood ratio leads to a test which is UMPU.[3]

In these two examples, we have been able to sidestep the problem of actually having to evaluate the value of L_α by making use of the fact that $\hat{L}$ in each case is a function of a statistic whose exact distribution is known. Unfortunately, however, this does not always occur, which means that we then must face the problem of deriving explicitly the distribution of $\hat{L}$. Since this is nearly always a formidable task, it is usually necessary to resort to an approximation.

Since $\hat{L}$ is distributed on the interval (0, 1), it follows that the quantity $-2 \ln \hat{L}$ will be distributed on the interval $(0, \infty)$. Since it, too, is defined over the interval $(0, \infty)$, the usual practice is to approximate the distribution of $-2 \ln \hat{L}$ by the chi-square distribution. Such a procedure finds justification in the fact that as n increases, the distribution of $-2 \ln \hat{L}$ when H_0 holds tends to a χ^2 distribution with k degrees of freedom, k being the number of parameters treated explicitly in the hypotheses H_0 and H_a.[4] With this approximation, L_α can then be found by solving the equation

$$\int_0^{L_\alpha} \chi^2_{(k)}\, d\chi^2 = \alpha, \tag{18.61}$$

with the aid of a table for the chi-square integral. The reader must be cautioned, however, that this procedure has only asymptotic validity, and then only under the conditions on the likelihood function specified in Theorem 17.2.

As was mentioned at the beginning of this section, the likelihood ratio leads to a test based on a sufficient statistic in situations where the latter exists. That this is so follows from the fact that when a sufficient statistic is available, the maximum likelihood estimator is a function of it.

18.5 THE ORIGIN OF HYPOTHESES AND CHOICE OF α

Before turning to some examples of the tests that are most frequently encountered in statistical inference, two rather problematic questions need to be faced: From where do hypotheses come? And what determines the size of α?

[3] For a proof, see the reference in footnote 2.

[4] For this convergence in distribution to occur, it is necessary that the likelihood function be such that the maximum likelihood estimators are asymptotically normal and efficient (see Theorem 17.2). For a proof of the theorem involved, see Kendall and Stuart (1961, pp. 230–231).

The Origin of Hypotheses

This question is one of the most difficult in all of statistical inference to give a satisfactory answer, and the author freely confesses to an uneasy feeling about the discussion that follows. The problem is, of course, that we do not know what the truth is—if we did, then there would be no need for statistical inference—hence there is a large element of groping in the dark. Fortunately, however, we are seldom completely in the dark, in that there is usually some prior knowledge to guide us. The prior knowledge takes the form (1) of accumulated results from past research, (2) deductions from theories formulated to explain the particular phenomena at hand, or (3) simply questions that can only be answered by looking at the data. It will be useful to note an example from each of these sources.

Results from past research. In most cases, this is the most obvious source of prior information. A good example in economics is the following. Suppose that you are an expert sent out by the United Nations to study the distribution of personal income in Colombia. You find waiting for you upon your arrival in Bogota an extensive set of data obtained from a recent household budget survey which gives, among other things, the frequency distribution of income earners for, for example, 20 income classes. Assume that your first question concerns the integrity of the data—that is, does it really measure what it purports to? In approaching this question, you make use of the following fact: One of the best-established empirical laws in all of economics is that for high incomes the proportion of households z having an income greater than some value x is very accurately described by the relation

$$P(Z > x) = ax^{-\alpha}, \tag{18.62}$$

where a and α are positive constants. Usually α is found to be about 1.7. This is the celebrated law of Pareto, named after its discoverer, the economist Vilfredo Pareto.[5]

Taking the logarithm of (18.62), we obtain

$$\ln p(X > x) = \ln a - \alpha \ln x, \tag{18.63}$$

which shows that $P(Z > x)$ (the *upper* tail of the distribution function) is linear in logarithms, with slope equal to $-\alpha$. A test of the integrity of the Colombian income data, therefore, would be to plot the proportion of income receivers in the sample with incomes

[5] The properties of this distribution were discussed in Section 8.5.

greater than x against x on double logarithmic paper. If the upper tail of the distribution is not nearly a straight line, then you could conclude with some assurance that the income data are faulty.[6]

Deduction from theory. As an example of a hypothesis that is derived from theoretical considerations, we shall take the shape of the relationship between consumption spending and income that is implied by the *permanent income hypothesis* of Milton Friedman. This theory implies that, in the long run, consumption is proportional to income, or that

$$c_p = ky_p \tag{18.64}$$

where c_p and y_p are "permanent" consumption and "permanent" income, respectively, and k is a constant. Assuming that we have adequate measures of c_p and y_p, the permanent income hypothesis then suggests that in the equation

$$c_p = a + by_p + u, \tag{18.65}$$

where u is an unobservable error term assumed to be random with mean zero and variance σ^2, a should be equal to zero. This can be tested as a hypothesis.

Questions raised in the course of events. Suppose that an agronomist has developed a new strand of wheat and he is interested in whether this new variety will give better yields than an existing one. In this situation, the natural procedure would be to plant several test plots of the two varieties under identical conditions and then to test the hypothesis that the yields are the same against the alternative hypothesis that the yield of the new variety is greater than that of the old.

The situation just described, where hypotheses emerge more or less naturally in the course of investigation, is very common, not only in agronomy, but in the physical, biological, and social sciences in general.

Perhaps the only general statement that should be made about the origin of hypotheses is that none should be made. The important consideration is not where hypotheses come from, but whether they are relevant and lend themselves to possible refutation. The question of relevance is a difficult

[6] On the other hand, if the line were too straight, you would want to check up on how the data were collected in order to make sure that the information had not been "manufactured" to begin with using the law of Pareto.

one and is something that sincere investigators can disagree about; however, a hypothesis is useless in advancing scientific knowledge if there is no possibility of it being refuted by empirical evidence.

So let us assume that the problem of the origin of hypotheses is taken care of and continue from there. In the classical theory of hypothesis testing, we have two hypotheses, one to be tested H_0, and its alternative H_a. We begin by assuming H_0 to be true, and then draw a sample, supposedly at random, from the underlying population. We formulate a statistic whose distribution is known, or a least approximately so. Next, we isolate a region beneath the density function of this statistic that is small *if the hypothesis H_0 is true.* In fact, this probability is viewed as being so small that it is practically certain that the statistic will not fall in the region when only a single sample is drawn. Therefore, if the statistic does fall in the region for our particular sample, we conclude that the event of this occurring is so unlikely on the assumption that H_0 is true that we repudiate H_0 in favor of H_a. However, in doing so, we remain fully aware that if H_0 were true and if we were able to sample repeatedly, testing H_0 each time, then we would incorrectly reject H_0 a proportion of the time equal to the area of the critical region. This is the Type I error.

On the other hand, we also face the danger of accepting H_0 when it is not true—the Type II error—and the idea is that we take care in selecting the critical region so that the probability of this kind of error is as small as possible. However, a criticism that is often directed at the Neyman-Pearson method of hypothesis testing is that, for practical purposes, the probability of a Type II error is $1 - \alpha$ (α being the probability of a Type I error), since it is virtually impossible to hit exactly the true value of θ in H_0.

Strictly speaking, this is a valid criticism. However, two things should be kept in mind. First, when we fail to reject H_0, we are doing just that and nothing more. In accepting H_0, we are not saying that H_0 is true, but only that the statistical evidence is *consistent* with it being true. Additional evidence may lead to a rejection of H_0, and we must never foreclose this possibility. In short, a law is a law only until further notice. Secondly, if θ is not equal to θ_0, but to $\theta_0 + \varepsilon$, where ε is a small number, there will, in most situations, be little harm in accepting the hypothesis that θ is equal to θ_0. As the true θ moves farther and farther from θ_0, the consequences of accepting $\theta = \theta_0$ undoubtedly will become more serious, but if the test has been selected with care, its power to reject H_0 will increase *pari passu.* In other words, the probability of making a wrong decision decreases as the "loss" that would result from the decision increases.

Thus, we come to what in the author's opinion is the crux of the problem in the Neyman-Pearson theory of hypothesis testing, and this is to be able to set up the alternative hypothesis in such a way that it is meaningful *vis-à-vis* the hypothesis being tested, yet allows the sample information maximum scope to discriminate between the two hypotheses. For example, if the task is to test the hypothesis that the long-run marginal propensity to consume out of income in Canada is 0.92, then taking for the alternative hypothesis that the marginal propensity to consume is -1.0, although it would allow the data a great deal of scope to discriminate between the two hypotheses, would be meaningless since a long-run marginal propensity to consume of -1.0 is inconceivable. On the other hand, taking 0.90 as the alternative hypothesis would be a meaningful value, but it is unlikely that the sample information would be sharp enough to distinguish between the two hypotheses. In this case, the best procedure would be to take $\theta \neq 0.92$ as the alternative hypothesis.

If hypothesis testing is to be a productive exercise, it is clear that one of H_0 or H_a must embody existing knowledge, even if it is little more than impression or judgment, about the parameter (or parameters) in question. The logical place to include this information is in H_0, the reasons for this being twofold:

1. Including the prior knowledge in H_0 forces investigators to be specific about their beliefs and then keeps these beliefs in sharp focus.
2. Since, with α a small number, rejection of H_0 only occurs in the face of strong evidence to the contrary, H_0 is much more likely to be questioned when it is rejected than when it is accepted. On the other hand, scientific knowledge generally progresses most when existing beliefs are called into question, and usually this is only done when these are confronted with strong conflicting evidence. The parallel is obvious.

We shall return to the questions raised in these paragraphs in Chapter 22.

Choosing the Size of α

It should be evident from our discussion of Type I and Type II errors that α, β, and n, the sample size, are not independent of one another. Once α and the sample size are established, β is determined by the location of the critical region. Thus, there is a tradeoff between the probability of Type I error and the probability of Type II error (and the sample size also, although for the moment we shall assume this to be given). In particular, the probability of Type II error can be reduced only at the expense of increasing the probability of Type I error, and conversely.

How, then, are α and β determined? In principle, we should have measures of the losses associated with each type of error, and then α and β could be determined so as to minimize the total loss. In some cases, particularly where industrial processes are involved, this can actually be done because good estimates of the losses involved exist. In most situations, however, and especially in the social sciences, it is impossible to obtain objective estimates of these costs. As a result, α is usually set by convention at a low level, the most frequent values being 0.10, 0.05, or 0.01. The value of α thus determined is usually referred to as the *significance level* of the test. For example, if the null hypothesis H_0: $\theta = \theta_0$ is rejected in favor of the alternative hypothesis H_a: $\theta \neq \theta_0$ at $\alpha = 0.05$, we say that θ is significantly different from θ_0 at the 5 percent level of significance. Finally, it should be evident that once set, α *cannot be changed during the remainder of the test.*

Although we have been assuming that the sample size is fixed, α should not be determined independently of n. For, if the test is based on a statistic which is a consistent estimator of the parameter in question (as most of the tests most frequently used are), the null hypothesis will almost invariably be rejected (since it is virtually impossible that the true value of θ will be that specified in H_0) for large sample sizes. Thus, when the test involved is consistent, it is necessary that α be reduced in size as the sample size increases. Since the power of the test in this case varies directly with n, this can be done without any loss of power.

Finally, it should be mentioned that even though it may not be possible to specify with any accuracy the losses associated with Type I and Type II errors, we are usually in a position to say which is the more important and α can then be set accordingly. For example, if it is felt that the consequences of Type I error are relatively minor, than α can be set at a relatively high level, say 0.10, and thereby reduce the probability of Type II error.

18.6 EXAMPLES OF THE MOST FREQUENTLY EMPLOYED TESTS

In this section, we shall illustrate several of the most commonly employed tests in hypothesis testing with examples drawn from data on personal consumption expenditure by British and Scottish working-class households in 1938. The data analyzed are presented in Table 18.2. The following tests will be discussed:

1. The mean in a normal distribution.
2. The equality of means in two normal distributions.
3. The equality of variances in two normal distributions.

TABLE 18.2 Total Expenditure per Person, Working-Class Families, in London and Scotland in 1938 (pence per week)

Income Class	London	Scotland
1	95.00	102.31
2	140.71	132.78
3	172.94	168.18
4	198.40	195.38
5	228.97	232.92
6	263.17	264.81
7	301.11	301.58
8	343.33	334.50
9	392.82	393.89
10	460.36	458.70
11	552.96	545.55
12	761.33	889.00
	$n = 12$	$n = 12$
	$\Sigma x = 3911.10$	$\Sigma x = 4019.55$
	$\bar{x} = 325.92$	$\bar{x} = 334.96$
	$\Sigma x^2 = 1{,}679{,}950.55$	$\Sigma x^2 = 1{,}875{,}218.80$
	$\Sigma(x - \bar{x})^2 = 405{,}225.28$	$\Sigma(x - \bar{x})^2 = 528{,}820.29$

SOURCE: Prais, S. J. and Houthakker, H. S., *The Analysis of Family Budgets, Second Impression*, Cambridge University Press, 1955, 1971, Table B2, p. 189.

Although the data in Table 18.2 are actually group means for 12 total expenditure classes, we shall, for purposes of illustration, treat each entry as though it were the expenditure of a single individual.

The Mean in a Normal Distribution

To begin with, we shall assume that total expenditure per working-class person (pence per week) in both the London area and in Scotland are distributed normally with means μ_1 and μ_2 and variances $\sigma_1{}^2$ and $\sigma_2{}^2$, respectively.[7] The first test will be relative to the hypotheses

$$H_0:\ \mu_1 = 300 \text{ pence per week}$$
$$H_1:\ \mu_1 \neq 300 \text{ pence per week}$$
$$\alpha = 0.05.$$

Variance known. We shall assume, first, that the variance for London is known and equal to 33,768. On the assumption that H_0 is true, we know from previous discussion that the quantity,

$$Z = \frac{\bar{X}_1 - \mu_1}{\sigma_1/\sqrt{n}} \tag{18.66}$$

[7] The assumption of underlying normality with income and total expenditure data is, strictly speaking, unwarranted. However, the purpose of this section is illustration, rather than serious research.

is normally distributed with mean 0 and variance 1. And from the discussion earlier in this chapter, we know that

$$\bar{X}_1 \le \mu_1 - z_{\alpha/2}\left(\frac{\sigma_1}{\sqrt{n}}\right),$$

$$\bar{X}_1 \ge \mu_1 + z_{\alpha/2}\left(\frac{\sigma_1}{\sqrt{n}}\right), \tag{18.67}$$

provide a critical region that is UMPU for a test of the form: $\mu = \mu_1$ against $\mu \ne \mu_1$. Therefore, we need only to calculate (18.67). We have

$$\frac{\sigma_1}{\sqrt{n}} = \left(\frac{33{,}768}{12}\right)^{1/2}$$

$$= 53.03, \tag{18.68}$$

so that

$$z = \frac{325.92 - 300.00}{53.05}$$

$$= 0.4886. \tag{18.69}$$

For $\alpha = 0.05$, the table for the unit normal distribution[8] gives $z_{0.025} = 1.96$ (since half of the critical region of size 0.05 lies in each tail); hence,

$$z \le -1.96,$$

$$z \ge 1.96, \tag{18.70}$$

provide the critical regions for z. Since $z = 0.4886$ does not lie in either of these, we accordingly cannot reject the hypothesis that μ_1 is equal to 300.

Variance unknown. We now assume that ${\sigma_1}^2$ is unknown, which means that we can no longer use (18.66) to test the hypothesis on μ_1. However, from earlier chapters we know that on the assumption that $E(\bar{X}) = \mu_1$, the quantity

$$T = \frac{\bar{X} - \mu_1}{S_{\bar{x}}}, \tag{18.71}$$

where

$$S_{\bar{x}} = \left[\frac{\sum (X_i - \bar{X})^2}{n(n-1)}\right]^{1/2}, \tag{18.72}$$

[8] See, for example, Mood and Graybill (1963, Table II, p. 431).

has the t-distribution with $n-1$ degrees of freedom. Moreover, earlier in this chapter we saw that the UMPU test of $\mu = \mu_1$ against $\mu \neq \mu_1$ is given by the critical regions for $\bar{x}$:

$$\bar{x} \leq \mu_1 - t_{\alpha/2} S_{\bar{x}},$$
$$\bar{x} \geq \mu_1 + t_{\alpha/2} S_{\bar{x}}. \tag{18.73}$$

Thus, we must compute the quantity in (18.71). Since

$$s_{\bar{x}} = \left(\frac{405{,}225}{11 \cdot 12}\right)^{1/2}$$
$$= 55.41, \tag{18.74}$$

we have

$$t = \frac{325.92 - 300.00}{55.41}$$
$$= 0.4678. \tag{18.75}$$

From the table for the t-distribution,[9] we find that for $\alpha = 0.05$ and $k = n - 1 = 11$, $t_{0.025} = 2.201$. Hence, the critical regions for t are given by

$$t \leq -2.202,$$
$$t \geq 2.201. \tag{18.76}$$

Since $t = 0.4678$ is outside of either of the critical regions, we once again cannot reject the hypothesis that μ_1 is equal to 300.

Equality of Means in Two Normal Populations

An often encountered situation in scientific research is where we have samples drawn from two populations and wish to test the hypothesis that the populations have a common mean. With reference to the data in Table 18.2, we may want to know whether the mean total expenditure for individuals is the same in London as in Scotland. Therefore, we will test the hypotheses:

$$H_0\colon \mu_1 = \mu_2,$$
$$H_1\colon \mu_1 \neq \mu_2,$$
$$\alpha = 0.10.$$

We shall assume that the variances of the populations are equal but unknown.

[9] See Mood and Graybill (1963, Table IV, p. 433).

On the assumption that μ_1 does equal μ_2, it follows from our earlier discussion of the t-distribution that the quantity

$$T = \frac{\bar{X}_1 - \bar{X}_2}{S_{\bar{x}_1 - \bar{x}_2}}, \tag{18.77}$$

where

$$S_{\bar{x}_1 - \bar{x}_2} = [S_{\bar{x}_1}{}^2 + S_{\bar{x}_2}{}^2]^{1/2}, \tag{18.78}$$

has the t-distribution with $2n - 2$ degrees of freedom. Moreover, an argument similar to the one used in deriving the test via the likelihood ratio for the population mean when the variance is unknown leads to a test on $\bar{x}_1 - \bar{x}_2$ whose critical regions are given by

$$\begin{aligned} \bar{x}_1 - \bar{x}_2 &\le \mu_1 - \mu_2 - t_{\alpha/2}\, S_{\bar{x}_1 - \bar{x}_2}, \\ \bar{x}_1 - \bar{x}_2 &\ge \mu_1 - \mu_2 + t_{\alpha/2}\, S_{\bar{x}_1 - \bar{x}_2}. \end{aligned} \tag{18.79}$$

This test is UMPU for the test of $\mu_1 = \mu_2$ against $\mu_1 \ne \mu_2$. Thus, we must calculate (18.77), and, since

$$\begin{aligned} s_{\bar{x}_1 - \bar{x}_2} &= \left[\frac{405{,}225.28}{(11 \cdot 12)} + \frac{528{,}820.29}{(11 \cdot 12)}\right]^{1/2} \\ &= 84.12, \end{aligned}$$

we have

$$\begin{aligned} t &= \frac{325.92 - 334.96}{84.12} \\ &= -0.1075. \end{aligned} \tag{18.80}$$

From the table for the t-distribution with $\alpha = 0.10$ and $k = 2n - 2 = 22$ degrees of freedom, we find $t_{0.05} = 1.717$. Therefore, the critical regions on t are given by

$$\begin{aligned} t &\le -1.717, \\ t &\ge 1.717. \end{aligned} \tag{18.81}$$

Hence since $t = -0.1075$ does not fall in either of the critical regions, we cannot reject the hypothesis that $\mu_1 = \mu_2$.

We should note before leaving this example that were the assumed common variance of the two populations to be known, the normal distribution, rather than the t-distribution, would be used in the test. The rest of the procedure, however, would be exactly the same.

The Equality of Variances in Two Normal Populations

We now turn to a test of the equality of the variances in the two populations and test the hypotheses:

$$H_0\colon \sigma_1^2 = \sigma_2^2,$$
$$H_1\colon \sigma_1^2 \neq \sigma_2^2,$$
$$\alpha = 0.05.$$

From our earlier discussion of the F-distribution, we know that the quantity

$$F = \frac{n_1 S_1^2/\sigma_1^2}{n_1 - 1} \bigg/ \frac{n_2^2 S_2^2/\sigma_2^2}{n_2 - 1} \tag{18.82}$$

has the F-distribution with $n_1 - 1$ and $n_2 - 1$ degrees of freedom. Therefore, on the assumption that $\sigma_1^2 = \sigma_2^2$ and with $n_1 = n_2 = n$, we will have that

$$F = \frac{\sum (X_{1i} - \overline{X}_1)^2}{\sum (X_{2i} - \overline{X}_2)^2} \tag{18.83}$$

has the F-distribution with $n - 1$ and $n - 1$ degrees of freedom. Application of the likelihood ratio would then lead to a test using the F-distribution with two critical regions in the tails whose sum is equal to α. This test is UMPU for the above hypotheses. Since the F-distribution is not symmetrical the critical regions in the tails are not of equal size. However, since the regions are generally difficult to find in the F-table, they are usually made equal for convenience. (This being the case, the test is no longer strictly UMPU.)

Therefore, it remains to calculate (18.82), namely,

$$F = \frac{405{,}225.28}{528{,}820.29} = 0.7663. \tag{18.84}$$

From the table of the F-distribution,[10] we find that for $\alpha = 0.10$ and $n - 1$ and $n - 1$ degrees of freedom,

$$F_{0.05}(11, 11) = 0.355,$$
$$F_{0.95}(11, 11) = 2.82. \tag{18.85}$$

Since $F = 0.7663$ does not lie in either of the critical regions, we cannot reject the hypothesis that $\sigma_1^2 = \sigma_2^2$.

[10] See Mood and Graybill (1963, Table V, p. 434).

18.7 GOODNESS OF FIT

So far we have only considered situations where the population distribution is assumed to be known and the tests involve only parameters of the distribution. Often, however, it will be the underlying distribution itself that is of primary concern. Hence, in concluding this chapter, it is appropriate to outline a procedure for testing the goodness of fit of an empirical distribution with an *assumed* population distribution.

Let the empirical distribution be given by $g(x)$ and the theoretical distribution by $f(x)$. Although x may be either discrete or continuous, if it is continuous, $g(x)$ must be calculated in the discussion that follows at some point (usually the midpoint) within a range of x. Suppose that we have k classes for x—that is, if x is discrete, we have values of $g(x)$ for k values of x, while if x is continuous, we have values of $g(x)$ for k nonoverlapping intervals of x. The number of classes is immaterial, although as a general rule k should be at least 5. The only actual requirement is that the frequency of x be at least 1 in each class.

Denote the frequency of x within class i by g_i for the observed distribution and by f_i for the hypothesized distribution ($i = 1, k$). It follows that quantity

$$\phi = \sum_{i=1}^{k} \frac{(g_i - f_i)^2}{f_i} \tag{18.86}$$

is distributed approximately as chi-square with $k - m - 1$ degrees of freedom,[11] where m is the number of parameters in $f(x)$ that it is necessary to estimate from the sample information. It is evident from the form of (18.86) that low values of ϕ will favor the hypothesis that the observed frequencies were generated by $f(x)$, while high values will favor the contrary view. Therefore, a critical region of size α for the test of the hypotheses

$$H_0\colon\ g(x) = f(x),$$
$$H_1\colon\ g(x) \neq f(x),$$

will be given by

$$\int_t^{\infty} \chi^2_{(z)}\, d\chi^2 = \alpha, \tag{18.87}$$

where $z = k - m - 1$. In other words, the critical region of size α for the rejection of H_0 will be given by

$$\phi \geq t = \chi^2_{\alpha(k-m-1)}. \tag{18.88}$$

[11] For a derivation of this result, see Mood and Graybill (1963, pp. 309–310).

Where the value of $\chi^2_{\alpha(k-m-1)}$ is obtained from the table for the chi-square distribution.

For an example, we will take the data relating to the connections to wrong numbers that are tabulated in Table 7.1. In this case, $g(x)$ will refer to the numbers appearing under N_x, while $f(z)$ will refer to the numbers under $N_p(x, 8.74)$. The latter numbers refer to the frequencies that would arise from a Poisson distribution with $\lambda = 8.74$. Calculating (18.86) from these two columns of numbers, we have

$$\phi = 10.29. \tag{18.89}$$

From the table for the chi-square distribution,[12] we have for $k - m - 1 = 15 - 1 - 1 = 13$ degrees of freedom and $1 - \alpha = 0.95$, $\chi^2_{0.95(13)} = 22.4$. Since $\phi = 10.29 < 22.4$, we cannot reject the hypothesis that the connections to wrong numbers follow a Poisson distribution.

REFERENCES

Freeman, H., *An Introduction to Statistical Inference*, Addison-Wesley, 1963, chap. 28.

Kendall, M. G. and Stuart, A., *The Advanced Theory of Statistics*, vol. II, Charles Griffin and Co., 1961, chaps. 22–24.

Mood, A. M. and Graybill, F. A., *Introduction to the Theory of Statistics*, 2nd ed., McGraw-Hill, 1963, chap. 12.

[12] See Mood and Graybill (1963, Table III, p. 432).

CHAPTER 19

INTERVAL ESTIMATION

In this chapter, we shall look at the problem of estimation from a different vantage point. We shall continue to be concerned with the estimation of the parameter θ in the distribution $f(x;\ \theta)$, but rather than focusing on a single estimate of θ, as was the case in Chapters 16 and 17, our efforts will now be directed to the construction of an interval that is likely to include the true value of the parameter. Such a procedure is referred to as *interval estimation*, and can be viewed as the attempt to make explicit the uncertainty that is implicit in point estimation.

19.1 INTERVAL ESTIMATION DEFINED

By way of introduction, we might begin with a point estimate of θ say $\hat{\theta}$, which, for the sake of argument, we shall assume to be unbiased and whose variance is given by $\sigma^2(\hat{\theta})$. Then we may say that it is probable that the true value of θ lies in the range $\hat{\theta} \pm \sigma(\hat{\theta})$, very probable that it lies in the range $\hat{\theta} \pm 2\sigma(\hat{\theta})$, and so on. This is in essence the idea behind interval estimation.

To make this more explicit, let $X_1, \ldots, X_n$ denote a sample of size n drawn at random from $f(x;\ \theta)$ and let Z be a function of the X's and of θ, but whose distribution $g(z)$ is independent of θ. Then, given any probability $1 - \alpha$, where α is small, we can find a value z_1 of z, such that for all values of θ,

$$\int_{z_1}^{\infty} g(u)\, du = 1 - \alpha, \qquad (19.1)$$

which in probability notation is equivalent to

$$P(Z \geq z_1) = 1 - \alpha. \tag{19.2}$$

If it happens that the inequality $z \geq z_1$ in (19.2) can be written in the form $\theta \leq t_1$ or $\theta \geq t_1$, where t_1 is a function of z_1 and the X's, but not of θ, it follows that (19.2) can be written in the form

$$P(\theta \leq t_1) = 1 - \alpha. \tag{19.3}$$

For example, if $Z = \bar{X} - \theta$, we would have $\bar{X} - \theta \geq z_1$, or $\theta \leq \bar{X} - z_1$, so that t_1 would be equal to $\bar{X} - z_1$.

More generally, if a statistic t_1 can be found which depends on the X's and $1 - \alpha$, but not on θ, and which satisfies (19.3), then even if the distribution of Z were to depend on θ, Equation (19.3) could be used to make probability statements about θ. However, it is important to emphasize that it is t_1, and not θ, that is the random variable. That is, it is not being asserted that the probability is $1 - \alpha$ that θ is greater than t_1; on the contrary, since θ is an unknown constant, this probability is either 0 or 1. What (19.3) does assert is that with repeated sampling the probability is $1 - \alpha$ that the random variable t_1 will be greater than or equal to θ, *no matter what the value of* θ. Therefore, if we make the statement $\theta \leq t_1$ for any sample $X_1, \ldots, X_n$ that arises, we shall be correct $1 - \alpha$ percent of the time in the long run.

Although the argument with a single quantity t_1 suffices to bring out the basic notion underlying the theory of interval estimation, in practice we are usually concerned with finding two quantities t_1 and t_2 such that

$$P(t_1 \leq \theta \leq t_2) = 1 - \alpha, \tag{19.4}$$

so that we can make the assertion that the probability is $1 - \alpha$ that the interval from t_1 to t_2 will cover the true value of θ. Such an interval is called a *confidence interval* of size $1 - \alpha$ for θ. However, at the risk of being redundant, we should again remind ourselves that it is the interval t_1 to t_2 that is random, and not θ.

It should be evident that for a given size of α we should prefer a shorter over a longer interval, which leaves us with the task of finding the optimal values of t_1 and t_2. Fortunately, since there is an intimate connection between confidence intervals and hypothesis testing, this task is not as formidable as it might seem. First, however, we must establish this connection.

19.2 RELATION BETWEEN CONFIDENCE INTERVALS AND HYPOTHESIS TESTING

In the preceding chapter, we considered tests of the hypothesis $H_0: \theta = \theta_0$ (the alternative hypothesis, though always implicit, need not be made explicit for the argument that follows). Let $\hat{\theta}$ be the test statistic, and let $R(\theta_0)$ be the rejection region and $A(\theta_0)$ $[= R'(\theta_0)]$ be the acceptance region; that is, we

$$\text{reject } H_0: \ \theta = \theta_0 \text{ if } \hat{\theta} \text{ falls in } R(\theta_0),$$
$$\text{accept } H_0: \ \theta = \theta_0 \text{ if } \hat{\theta} \text{ falls in } A(\theta_0).$$

$R(\theta)$ and $A(\theta)$ obviously exhaust the sample space of $\hat{\theta}$. The probabilities associated with these operational rules are

$$P[\hat{\theta} \in R(\theta_0) \mid \theta_0] = \alpha,$$
$$P[\hat{\theta} \in A(\theta_0) \mid \theta_0] = 1 - \alpha, \tag{19.5}$$

for any fixed $\theta = \theta_0$.

We now assume that a region $A(\theta)$ exists for every value of θ, so that we can write

$$P[\hat{\theta} \in A(\theta) \mid \theta] = 1 - \alpha. \tag{19.6}$$

Next, we look at the parameter space for θ (say Ω), which we shall assume to be the entire real line. For each value of $\hat{\theta} \in A(\theta)$, there exists a region ω in Ω such that any $\theta \in \omega$ will give rise to $\hat{\theta} \in A(\theta)$ at least $1 - \alpha$ proportion of the time. Since ω depends on $\hat{\theta}$, we write it as $\omega(\hat{\theta})$. We can illustrate with a couple of hypothetical values of $\hat{\theta}$. (See Figure 19.1.) In other words, we have

$$\hat{\theta} \in A(\theta) \langle = \rangle \omega(\hat{\theta}) \supset \theta, \tag{19.7}$$

so that we have a correspondence between the probabilities on the random variable $\hat{\theta}$ and probabilities on the random interval $\omega(\hat{\theta})$, that is,

$$P[\hat{\theta} \in A(\theta) \mid \theta] = P[\omega(\hat{\theta}) \supset \theta \mid \theta] = 1 - \alpha. \tag{19.8}$$

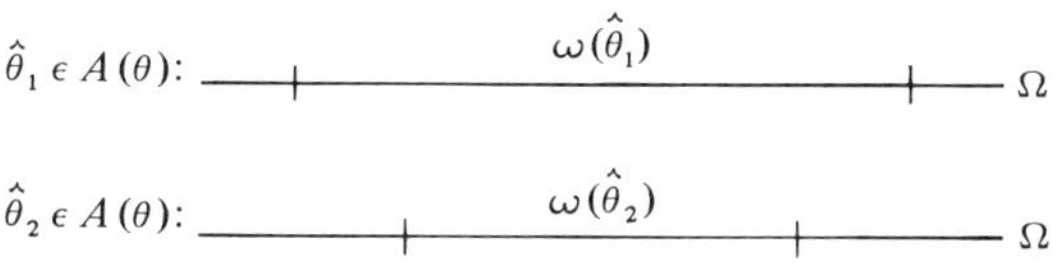

FIGURE 19.1

If the lower limit of $\omega(\hat{\theta})$ is written t_1 and the upper limit t_2, (19.8) implies that

$$P(t_1 \leq \theta \leq t_2) = 1 - \alpha, \tag{19.9}$$

which is (19.4). If for every $\hat{\theta}$ the set $\omega(\hat{\theta})$ is nonempty and closed, then t_1 and t_2 form a confidence interval on θ of size α. Thus, we see that the relationship between confidence intervals and hypothesis testing is direct.

We now exploit this result in establishing the confidence intervals which correspond to the tests which were discussed in the preceding chapter, namely, the uniformly most powerful test (UMP) and the uniformly most powerful unbiased test (UMPU).

For the UMP test, it will be recalled that we required a region R of size α equal to that of any other region R^* and with power at least as great as R^*, that is,

$$\begin{aligned} P(\hat{\theta} \in R \mid \theta_0) &= P(\hat{\theta} \in R^* \mid \theta_0) = \alpha, \\ P(\hat{\theta} \in R \mid \theta) &\geq P(\hat{\theta} \in R^* \mid \theta). \end{aligned} \tag{19.10}$$

Such regions, as we saw, seldom exist. Assuming for the moment, however, that R does exist for a particular problem, the confidence interval of size $1 - \alpha$ that corresponds to it is obtained as follows.

From (19.10), we have

$$\begin{aligned} P(\hat{\theta} \in A \mid \theta_0) &= P(\hat{\theta} \in A^* \mid \theta_0) = 1 - \alpha, \\ P(\hat{\theta} \in A \mid \theta) &\leq P(\hat{\theta} \in A^* \mid \theta), \end{aligned} \tag{19.11}$$

which, in view of (19.8) and (19.9), is equivalent to

$$\begin{aligned} P(t_1 \leq \theta_0 \leq t_2 \mid \theta_0) &= P(t_1' \leq \theta_0 \leq t_2' \mid \theta_0) = 1 - \alpha, \\ P(t_1 \leq \theta_0 \leq t_2 \mid \theta) &\leq P(t_1' \leq \theta_0 \leq t_2' \mid \theta), \end{aligned} \tag{19.12}$$

for each possible θ_0 and θ. Thus, the optimal interval corresponding to the UMP test is given by t_1 to t_2. It is worth noting that this interval is optimal not in the sense that $t_2 - t_1$ is the shortest interval possible, but rather in the sense that it is least likely to cover a false value of θ. For the same reasons that UMP tests seldom exist, optimal intervals in this sense also seldom exist.

For the UMPU test we require an unbiased region R of size α that is at least as powerful as any other unbiased region R^* also of size α—that is, we require R such that R has minimum power at $\theta = \theta_0$ and for which

$$\begin{aligned} P(\hat{\theta} \in R \mid \theta_0) &= P(\hat{\theta} \in R^* \mid \theta_0) = \alpha, \\ P(\hat{\theta} \in R \mid \theta) &\geq P(\hat{\theta} \in R^* \mid \theta). \end{aligned} \tag{19.13}$$

A transformation similar to that used in finding the confidence interval corresponding to a UMP test will again yield an interval of the form

$$P(t_1 \le \theta_0 \le t_2 \mid \theta_0) = P(t_1' \le \theta_0 \le t_2' \mid \theta_0) = 1 - \alpha,$$
$$P(t_1 \le \theta_0 \le t_2 \mid \theta) \ge P(t_1' \le \theta_0 \le t_2' \mid \theta). \tag{19.14}$$

However, since t_1, t_2, t_1', and t_2' now refer to lower and upper bounds on intervals corresponding to unbiased tests, we can no longer interpret the interval t_1 to t_2 as shortest in the sense of the interval t_1 to t_2 in (19.12).

19.3 A GENERAL METHOD FOR OBTAINING CONFIDENCE INTERVALS

Before moving on to consider examples of the most frequently used confidence intervals, it will be useful to outline briefly a general method for obtaining them.

Assume that we have a population described by $f(x;\, \theta)$, where θ is unknown, and suppose that we have an estimator of θ, $\hat{\theta} = \hat{\theta}(X_1, \ldots, X_n)$, based on a sample of size n from $f(x;\, \theta)$. Let the distribution of $\hat{\theta}$ be $g(\hat{\theta};\, \theta)$, and assume that we want a confidence interval of size $1 - \alpha$ for θ. If an arbitrary number, say θ', is substituted for θ in $g(\hat{\theta}, \theta)$, the distribution of $\hat{\theta}$ will be completely specified and we can make probability statements about $\hat{\theta}$.

In particular, we can find two numbers h_1 and h_2, which depend on θ', for which

$$P(\hat{\theta} < h_1) = \int_{-\infty}^{h_1} g(z;\, \theta')\, dz = \frac{\alpha}{2},$$

$$P(\hat{\theta} > h_2) = \int_{h_2}^{\infty} g(z;\, \theta')\, dz = \frac{\alpha}{2}. \tag{19.15}$$

Since h_1 and h_2 are functions of θ', they are also functions of θ (since θ' is arbitrary), $h_1(\theta)$ and $h_2(\theta)$, so that from (19.15) we can write

$$P[h_1(\theta) < \hat{\theta} < h_2(\theta)] = \int_{h_1(\theta)}^{h_2(\theta)} g(z;\, \theta)\, dz = 1 - \alpha. \tag{19.16}$$

The functions $h_1(\theta)$ and $h_2(\theta)$ may be plotted against θ as in Figure 19.2. It then follows that a vertical line through any point θ' will intersect the two curves, which, when projected into the vertical axis, will give limits between which $\hat{\theta}$ will fall with probability $1 - \alpha$ when $\theta = \theta'$.

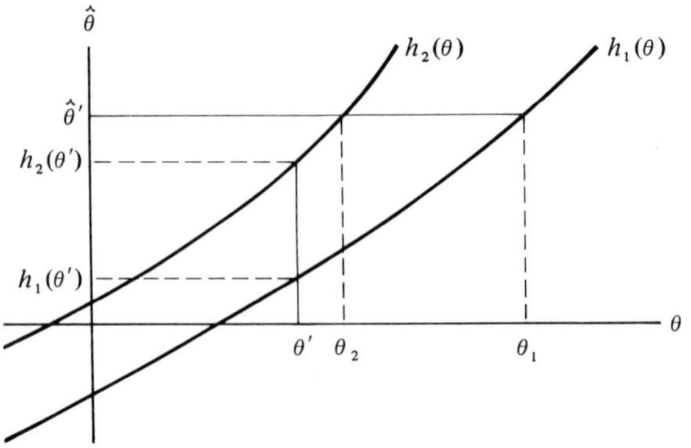

FIGURE 19.2 Confidence interval for θ.

Thus, in order to construct a $1 - \alpha$ confidence interval for θ, we draw a sample of size n from $f(x; \theta)$ and compute the value of $\hat{\theta}$, say $\hat{\theta}'$. Next, we draw a horizontal line through $\hat{\theta}'$ intersecting the curves $h_1(\theta)$ and $h_2(\theta)$, and then project the two points of intersection onto the θ-axis. These are labeled θ_1' and θ_2' in Figure 19.2. Since from the method of construction,

$$P(\theta_2' < \theta < \theta_1') = 1 - \alpha, \tag{19.17}$$

no matter what the value of θ, it follows that the interval θ_2' to θ_1' provides a $1 - \alpha$ percent confidence interval for θ.

As a simple example, consider the estimation of θ in the exponential distribution,

$$f(x; \theta) = \frac{1}{\theta} e^{-x/\theta}, \tag{19.18}$$

for samples of size 1. We shall use the maximum likelihood estimator as the estimator of θ, which turns out to be simply the observation x'. Therefore,

$$g(\hat{\theta}; x') = \frac{1}{x'} e^{-x/x'}. \tag{19.19}$$

Hence, we must solve the equations

$$\frac{1}{x'} \int_0^{h_1(x')} e^{-x/x'}\, dx = \frac{\alpha}{2},$$

$$\frac{1}{x'} \int_{h_2(x')}^{\infty} e^{-x/x'}\, dx = \frac{\alpha}{2}, \tag{19.20}$$

for $h_1(x')$ and $h_2(x')$. Performing the integrations, we find

$$-e^{-x/x'}\Big]_0^{h_1(x')} = \frac{\alpha}{2},$$

$$-e^{-x/x'}\Big]_{h_2(x')}^{\infty} = \frac{\alpha}{2}, \tag{19.21}$$

or

$$-e^{-h_1(x')/x'} + 1 = \frac{\alpha}{2}$$

$$-e^{-h_2(x')/x} = \frac{\alpha}{2} \tag{19.22}$$

which finally reduces to

$$h_1(x') = -x' \ln\left(1 - \frac{\alpha}{2}\right),$$

$$h_2(x') = -x' \ln\left(\frac{\alpha}{2}\right). \tag{19.23}$$

In other words, a confidence interval of size $1 - \alpha$ for θ is provided by the interval $-x' \ln(\alpha/2)$ to $-x' \ln(1 - \alpha/2)$.

In this example, admittedly a contrived one, we were able to solve the equations in (19.15) directly in terms of $\hat{\theta}$, and thus did not have to determine explicitly the functions $h_1(\theta)$ and $h_2(\theta)$. Fortunately, this occurs in a number of important cases, and as a second illustration, we shall construct a confidence interval for the mean in a normal population with (known) variance σ^2 based on a sample of size n.

The maximum likelihood estimator of μ, as we already know, is the sample mean $\bar{X}$, which, as we also know, is distributed normally with mean μ and variance σ^2/n. Therefore, if our particular sample value is $\bar{x}'$, we can find a confidence interval of size $1 - \alpha$ for μ by solving the two equations

$$\frac{1}{\sqrt{2\pi}} \int_{-\infty}^{t_1'} e^{-(1/2)z^2}\, dz = \frac{\alpha}{2},$$

$$\frac{1}{\sqrt{2\pi}} \int_{t_2'}^{\infty} e^{-(1/2)z^2}\, dz = \frac{\alpha}{2}, \tag{19.24}$$

for t_1' and t_2', where we have made the usual transformation

$$z = \frac{\bar{x} - \bar{x}'}{\sigma/\sqrt{n}}. \tag{19.25}$$

The values of t_1' and t_2' are easily found with the aid of a table for the unit normal distribution. Thus, we will have

$$P\left(t_1' < \frac{\bar{X} - \bar{x}'}{\sigma/\sqrt{n}} < t_2'\right) = 1 - \alpha, \tag{19.26}$$

or rearranging and using $-z_{\alpha/2}$ and $z_{\alpha/2}$ for t_1' and t_2', respectively,

$$P\left[\bar{X} - \frac{\sigma}{\sqrt{n}}(z_{\alpha/2}) < \bar{x}' < \bar{X} + \frac{\sigma}{\sqrt{n}}(z_{\alpha/2})\right] = 1 - \alpha. \tag{19.27}$$

Therefore, a confidence interval constructed according to formula (19.16) of size $1 - \alpha$ for μ will be given by the interval $\bar{x} - z_{\alpha/2}(\sigma/\sqrt{n})$ to $\bar{x} + z_{\alpha/2}(\sigma/\sqrt{n})$ obtained when we substitute μ for $\bar{x}'$.

An alternative way of deriving (19.27) is simply to proceed from the fact that $\bar{X}$ from a normal parent is distributed normally with mean μ and variance σ^2/n. The quantity,

$$Z = \frac{\bar{X} - \mu}{\sigma/\sqrt{n}}, \tag{19.28}$$

is therefore normal with mean 0 and variance 1. From the unit normal table, we then find the value $-z_{\alpha/2}$ and $z_{\alpha/2}$ for which

$$P(-z_{\alpha/2} < Z < z_{\alpha/2}) = 1 - \alpha. \tag{19.29}$$

But in view of (19.28), (19.29) is equal to

$$P\left(-z_{\alpha/2} < \frac{\bar{X} - \mu}{\sigma/\sqrt{n}} < z_{\alpha/2}\right) = 1 - \alpha, \tag{19.30}$$

which after some manipulation reduces to

$$P\left[\bar{X} - z_{\alpha/2}\left(\frac{\sigma}{\sqrt{n}}\right) < \mu < \bar{X} + z_{\alpha/2}\left(\frac{\sigma}{\sqrt{n}}\right)\right] = 1 - \alpha. \tag{19.31}$$

Since (19.31) holds no matter what the value of μ, $\bar{x} - z_{\alpha/2}(\sigma/\sqrt{n})$ to $\bar{x} + z_{\alpha/2}(\sigma/\sqrt{n})$ therefore provides a $1 - \alpha$ percent confidence interval for μ.

REFERENCES

Freeman, H., *An Introduction to Statistical Inference*, Addison-Wesley, 1963, chap. 29.

Kendall, M. G. and Stuart, A., *The Advanced Theory of Statistics*, vol. II, Charles Griffin & Co., 1961, chap. 20.

CHAPTER 20

THE LINEAR REGRESSION MODEL AND THE PRINCIPLE OF LEAST SQUARES

In this chapter, we shall analyze the statistical model that underlies much of present-day research in the nonexperimental sciences. This is the linear regression model. In the process, we shall examine in detail the properties of yet another method of estimation, namely, the one based on the celebrated principle of least squares.

20.1 THE LINEAR REGRESSION MODEL AND ITS ASSUMPTIONS

We begin by assuming that we are interested in studying the behavior of a quantity y which we postulate to be related to n other quantities $x_1, \ldots, x_n$ and a random variable u according to

$$y = \beta_1 x_1 + \cdots + \beta_n x_n + u. \tag{20.1}$$

We shall refer to y as the *dependent* variable and to $x_1, \ldots, x_n$ as *independent* variables. The parameters $\beta_1, \ldots, \beta_n$ are assumed to be unknown constants, and our task is to estimate them on the basis of a sample of observations on y and $x_1, \ldots, x_n$.

Assuming that our sample consists of T observations on y and each of the nx's, we shall write the model in matrix notation as[1]

[1] The same convention with regard to notation will be used in this chapter as in Chapter 12. See footnote 1 in that chapter.

$$y = \mathbf{X\beta} + u, \tag{20.2}$$

where

$$y = \begin{bmatrix} y_1 \\ y_2 \\ \vdots \\ y_T \end{bmatrix}, \quad \mathbf{X} = \begin{bmatrix} x_{11} & x_{12} & \cdots & x_{1n} \\ x_{21} & x_{21} & \cdots & x_{2n} \\ \vdots & \vdots & & \vdots \\ x_{T1} & x_{T2} & \cdots & x_{Tn} \end{bmatrix}, \quad u = \begin{bmatrix} u_1 \\ u_2 \\ \vdots \\ u_T \end{bmatrix}, \quad \mathbf{\beta} = \begin{bmatrix} \beta_1 \\ \beta_2 \\ \vdots \\ \beta_n \end{bmatrix}. \tag{20.3}$$

We can allow for an intercept in the model by assuming $x_{t1} = 1$ for all t.

We now make the following assumptions about **u** and **X**:

(i) $E(u_t) = 0,$ for all t;
(ii) $E(u_t^2) = \sigma^2 < \infty,$ for all t;
(iii) $E(u_i u_j) = 0,$ for $i \neq j$;
(iv) **X** is a fixed set of numbers;
(v) the rank of **X** is n.

The first assumption is an important one, but we can guarantee its being fulfilled. For suppose that $E(u_t) = \mu \neq 0$. Then we can define a new constant term for the equation, $\beta_1{}^* = \beta_1 + \mu$, and a new error term, $u^* = u - \mu$, the latter having expected value equal to zero. Assumption (ii) states that the variance of the error term is finite and the same for every observation, while assumption (iii) asserts that the n elements of u are independent. Property (ii) is referred to as *homoscedasticity*. (An error term which has $E(u_t^2) = \sigma_t^2$ is said to be *heteroscedastic*.) Properties (ii) and (iii) taken together assert that the covariance matrix of u has the form

$$E(uu') = \sigma^2\mathbf{I}, \qquad \sigma^2 < \infty. \tag{20.4}$$

The fourth assumption, which states that the value of **X** is fixed in repeated sampling, establishes that the error term u in no way depends on the values taken by the independent variables. Alternatively, this assumption can be relaxed to allow **X** itself to be random, but in this case **X** and u must be distributed independently. Finally, assumption (v) asserts that there are no exact linear relationships connecting the independent variables.

Let us now turn to the estimation of the parameters in $\mathbf{\beta}$. The conventional approach to this task is to choose the estimate of $\mathbf{\beta}$ in such a way that the sum of squared deviations of the elements of y from the estimated regression hyperplane is at a minimum. That is to say, we select $\mathbf{\beta}$ so as to minimize the quantity

$$\begin{aligned} \phi &= u'u \\ &= (y - \mathbf{X\beta})'(y - \mathbf{X\beta}). \end{aligned} \tag{20.5}$$

This is the famous principle of least squares, and its estimator, which we shall now derive, has enjoyed immense popularity since its invention by Gauss, DeMoivre, and Laplace in the early 1800s.

Multiplying out the right-hand side of the second line in (20.5), we have for ϕ,

$$\begin{aligned}\phi &= \boldsymbol{y}'\boldsymbol{y} - \boldsymbol{\beta}'\mathbf{X}'\boldsymbol{y} - \boldsymbol{y}'\mathbf{X}\boldsymbol{\beta} + \boldsymbol{\beta}'\mathbf{X}'\mathbf{X}\boldsymbol{\beta} \\ &= \boldsymbol{y}'\boldsymbol{y} - 2\boldsymbol{\beta}'\mathbf{X}'\boldsymbol{y} + \boldsymbol{\beta}'\mathbf{X}'\mathbf{X}\boldsymbol{\beta}, \end{aligned} \tag{20.6}$$

since $\boldsymbol{y}'\mathbf{X}\boldsymbol{\beta} = \boldsymbol{\beta}'\mathbf{X}'\boldsymbol{y}$ (ϕ being a quadratic form, and therefore a scalar). Differentiating ϕ with respect to β, we have

$$\frac{\partial\phi}{\partial\boldsymbol{\beta}} = -2\mathbf{X}'\boldsymbol{y} + 2\mathbf{X}'\mathbf{X}\boldsymbol{\beta}. \tag{20.7}$$

Equating $\partial\phi/\partial\beta$ to zero,

$$-2\mathbf{X}'\boldsymbol{y} + 2\mathbf{X}'\mathbf{X}\boldsymbol{\beta} = \mathbf{0}, \tag{20.8}$$

we then find for the minimizing $\boldsymbol{\beta}$,

$$\hat{\beta} = (\mathbf{X}'\mathbf{X})^{-1}\mathbf{X}'\boldsymbol{y}. \tag{20.9}$$

[That the matrix $\mathbf{X}'\mathbf{X}$ is nonsingular is guaranteed by assumption (v).] For ϕ to be a minimum at $\hat{\beta}$, it is necessary and sufficient that the matrix of cross partial derivatives, $\partial^2\phi/\partial\boldsymbol{\beta}\partial\boldsymbol{\beta}'$, be positive definite. From (20.7),

$$\frac{\partial^2\phi}{\partial\boldsymbol{\beta}\partial\boldsymbol{\beta}'} = 2\mathbf{X}'\mathbf{X}, \tag{20.10}$$

so that this is in fact the case, since $\mathbf{X}'\mathbf{X}$ is necessarily positive definite.[2]

The estimator $\hat{\beta}$ defined in (20.9) is referred to as the *least squares estimator* of $\boldsymbol{\beta}$. We shall now undertake a study of its properties.

20.2 PROPERTIES OF THE LEAST SQUARES ESTIMATOR

Using (20.2), we can write $\hat{\beta}$ in (20.9) as

$$\begin{aligned}\hat{\beta} &= (\mathbf{X}'\mathbf{X})^{-1}\mathbf{X}'(\mathbf{X}\boldsymbol{\beta} + \boldsymbol{u}) \\ &= \boldsymbol{\beta} + (\mathbf{X}'\mathbf{X})^{-1}\mathbf{X}'\boldsymbol{u}. \end{aligned} \tag{20.11}$$

Taking the expected value of $\hat{\beta}$, we then have

$$\begin{aligned}E(\hat{\beta}) &= \boldsymbol{\beta} + E[(\mathbf{X}'\mathbf{X})^{-1}\mathbf{X}'\boldsymbol{u}] \\ &= \boldsymbol{\beta} + (\mathbf{X}'\mathbf{X})^{-1}\mathbf{X}'E(\boldsymbol{u}) \qquad \text{[from assumption (iv)]} \\ &= \boldsymbol{\beta}, \end{aligned} \tag{20.12}$$

[2] See Theorem A.5.8 of Appendix 5.

since from assumption (i) $E(\boldsymbol{u}) = \mathbf{0}$. Thus, we find $\hat{\boldsymbol{\beta}}$ to be unbiased. For the covariance matrix of $\hat{\boldsymbol{\beta}}$, we have from (20.12), (20.11), and assumptions (ii) and (iii) [as represented in expression (20.4)],

$$\begin{aligned} E[(\hat{\boldsymbol{\beta}} - \boldsymbol{\beta})(\hat{\boldsymbol{\beta}} - \boldsymbol{\beta})'] &= E[(\mathbf{X}'\mathbf{X})^{-1}\mathbf{X}'\boldsymbol{u}\boldsymbol{u}'\mathbf{X}(\mathbf{X}'\mathbf{X})^{-1}] \\ &= (\mathbf{X}'\mathbf{X})^{-1}\mathbf{X}'E(\boldsymbol{u}\boldsymbol{u}')\mathbf{X}(\mathbf{X}'\mathbf{X})^{-1} \\ &= \sigma^2(\mathbf{X}'\mathbf{X})^{-1}, \end{aligned} \tag{20.13}$$

which we shall henceforth denote by $V(\hat{\boldsymbol{\beta}})$.

Let us now focus for a moment on the vector of least squares residuals, defined by

$$\hat{\boldsymbol{u}} = \boldsymbol{y} - \mathbf{X}\hat{\boldsymbol{\beta}}, \tag{20.14}$$

which, in view of (20.2) and (20.11), can be written as

$$\hat{\boldsymbol{u}} = [\mathbf{I} - \mathbf{X}(\mathbf{X}'\mathbf{X})^{-1}\mathbf{X}']\boldsymbol{u}. \tag{20.15}$$

Note first that, since $E(\boldsymbol{u}) = \mathbf{0}$,

$$E(\hat{\boldsymbol{u}}) = \mathbf{0}. \tag{20.16}$$

Next, since

$$\begin{aligned} \mathbf{X}'\hat{\boldsymbol{u}} &= \mathbf{X}'[\mathbf{I} - \mathbf{X}(\mathbf{X}'\mathbf{X})^{-1}\mathbf{X}']\boldsymbol{u} \\ &= \mathbf{X}'\boldsymbol{u} - \mathbf{X}'\mathbf{X}(\mathbf{X}'\mathbf{X})^{-1}\mathbf{X}'\boldsymbol{u} \\ &= \mathbf{X}'\boldsymbol{u} - \mathbf{X}'\boldsymbol{u} \\ &= \mathbf{0}, \end{aligned} \tag{20.17}$$

it follows that the residuals are orthogonal to the independent variables. Since we are assuming the first vector in $\mathbf{X}$ to be a vector of 1's, this condition also implies that the residuals sum to zero. This last result implies, in turn, that the estimated regression hyperplane contains the mean of $\boldsymbol{y}$, and, therefore, the point of means of the sample. It is to be noted, however, that both of these results depend on the equation being estimated having an intercept. If the equation is homogeneous, then, in general, the residuals will not necessarily sum to zero and neither will the fitted hyperplane necessarily contain the point of sample means.

We turn next to the expected value of the sum of squared residuals, which, in view of (20.15) and the fact that the matrix $[\mathbf{I} - \mathbf{X}(\mathbf{X}'\mathbf{X})^{-1}\mathbf{X}']$ is symmetrical and idempotent,[3] can be written as

$$\begin{aligned} \hat{\boldsymbol{u}}'\hat{\boldsymbol{u}} &= \boldsymbol{u}'[\mathbf{I} - \mathbf{X}(\mathbf{X}'\mathbf{X})^{-1}\mathbf{X}']\boldsymbol{u} \\ &= \boldsymbol{u}'\boldsymbol{u} - \boldsymbol{u}'\mathbf{X}(\mathbf{X}'\mathbf{X})^{-1}\mathbf{X}'\boldsymbol{u}. \end{aligned} \tag{20.18}$$

[3] For the definition of an idempotent matrix, see Appendix 5.

Taking the expected value, we then have

$$E(\hat{\boldsymbol{u}}'\hat{\boldsymbol{u}}) = E(\boldsymbol{u}'\boldsymbol{u}) - E[\boldsymbol{u}'\mathbf{X}(\mathbf{X}'\mathbf{X})^{-1}\mathbf{X}'\boldsymbol{u}]. \tag{20.19}$$

Since $\boldsymbol{u}'\boldsymbol{u} = \sum u_t^2$, the first term on the right is simply $T\sigma^2$; evaluation of the second term, however, is less straightforward. To do this, observe that, since $\hat{\boldsymbol{u}}'\hat{\boldsymbol{u}}$ is a scalar, and, therefore a 1×1 matrix, $\boldsymbol{u}'\mathbf{X}(\mathbf{X}'\mathbf{X})^{-1}\mathbf{X}'\boldsymbol{u} = \mathrm{Tr}[\boldsymbol{u}'\mathbf{X}(\mathbf{X}'\mathbf{X})^{-1}\mathbf{X}'\boldsymbol{u}]$, where Tr denotes "the trace of."[4] Therefore,

$$\begin{aligned}
E[\boldsymbol{u}'\mathbf{X}(\mathbf{X}'\mathbf{X})^{-1}\mathbf{X}'\boldsymbol{u}] &= E\{\mathrm{Tr}[\boldsymbol{u}'\mathbf{X}(\mathbf{X}'\mathbf{X})^{-1}\mathbf{X}'\boldsymbol{u}]\}\\
&= E\{\mathrm{Tr}[\mathbf{X}'\boldsymbol{u}\boldsymbol{u}'\mathbf{X}(\mathbf{X}'\mathbf{X})^{-1}]\}\\
&= \mathrm{Tr}\{E[\mathbf{X}'\boldsymbol{u}\boldsymbol{u}'\mathbf{X}(\mathbf{X}'\mathbf{X})^{-1}]\}\\
&= \mathrm{Tr}\{\mathbf{X}'E(\boldsymbol{u}\boldsymbol{u}')\mathbf{X}(\mathbf{X}'\mathbf{X})^{-1}\}\\
&= \mathrm{Tr}\{\sigma^2\mathbf{X}'\mathbf{X}(\mathbf{X}'\mathbf{X})^{-1}\}\\
&= \sigma^2\mathrm{Tr}(\mathbf{I}_n)\\
&= n\sigma^2.
\end{aligned} \tag{20.20}$$

Hence, returning to (20.19), we have for the expected value of the residual sum of squares,

$$\begin{aligned}
E(\hat{\boldsymbol{u}}'\hat{\boldsymbol{u}}) &= T\sigma^2 - n\sigma^2\\
&= (T - n)\sigma^2.
\end{aligned} \tag{20.21}$$

Having considered the residual sum of squares, let us turn for a moment to the total (or original) sum of squares, defined as $\sum y_t^2 = \boldsymbol{y}'\boldsymbol{y}$. In terms of $\hat{\boldsymbol{\beta}}$ and $\hat{\boldsymbol{u}}$, this quantity can be written as

$$\begin{aligned}
\boldsymbol{y}'\boldsymbol{y} &= (\mathbf{X}\hat{\boldsymbol{\beta}} + \hat{\boldsymbol{u}})'(\mathbf{X}\hat{\boldsymbol{\beta}} + \hat{\boldsymbol{u}})\\
&= \hat{\boldsymbol{\beta}}\mathbf{X}'\mathbf{X}\hat{\boldsymbol{\beta}} + \hat{\boldsymbol{\beta}}'\mathbf{X}'\hat{\boldsymbol{u}} + \hat{\boldsymbol{u}}'\mathbf{X}\hat{\boldsymbol{\beta}} + \hat{\boldsymbol{u}}'\hat{\boldsymbol{u}}\\
&= \hat{\boldsymbol{\beta}}'\mathbf{X}'\mathbf{X}\hat{\boldsymbol{\beta}} + \hat{\boldsymbol{u}}'\hat{\boldsymbol{u}} \qquad (\text{since } \mathbf{X}'\hat{\boldsymbol{u}} = \mathbf{0})\\
&= \hat{\boldsymbol{y}}'\hat{\boldsymbol{y}} + \hat{\boldsymbol{u}}'\hat{\boldsymbol{u}},
\end{aligned} \tag{20.22}$$

where

$$\hat{\boldsymbol{y}} = \mathbf{X}\hat{\boldsymbol{\beta}}. \tag{20.23}$$

The last line shows $\boldsymbol{y}'\boldsymbol{y}$ to be broken down into the sum of $\hat{\boldsymbol{y}}'\hat{\boldsymbol{y}}$, usually referred to as the explained sum of squares, and the residual sum of squares $\hat{\boldsymbol{u}}'\hat{\boldsymbol{u}}$. However, a decomposition which is usually of more interest than the one given in (20.22) is the one based on the variation about the mean

$$\sum (y_t - \bar{y})^2 = \sum y_t^2 - \frac{(\sum y_t^2)}{T}, \tag{20.24}$$

[4] For a discussion of the trace of a matrix, see Appendix 5.

Since least squares has the property, assuming the model to contain an intercept, that $\bar{\mathbf{y}} = \bar{\hat{\mathbf{y}}}$, we have

$$\sum (y_t - \bar{\hat{\mathbf{y}}})^2 = \sum \hat{y}_t^2 - \frac{(\sum y_t)^2}{T}, \tag{20.25}$$

so that subtracting $(\sum y_t)^2/T$ from both sides of (20.22) it follows that

$$\sum (y_t - \bar{\mathbf{y}})^2 = \sum (\hat{y}_t - \bar{\hat{\mathbf{y}}})^2 + \sum \hat{u}_t^2. \tag{20.26}$$

The first term on the right in this expression is referred to as the explained variation or, alternatively, as that part of the variation in the dependent variable explained by the independent variables.

A quantity which finds frequent use as a descriptive statistic is given by the ratio of explained to total variation and is called the *multiple determination coefficient*. It is usually denoted by $\boldsymbol{R}^2$:

$$\begin{aligned} \boldsymbol{R}^2 &= \frac{\sum (\hat{y}_t - \bar{\hat{\mathbf{y}}})^2}{\sum (y_t - \bar{\mathbf{y}})^2} \\ &= 1 - \frac{\sum \hat{u}_t^2}{\sum (y_t - \bar{\mathbf{y}})^2}. \end{aligned} \tag{20.27}$$

The square root of the multiple determination coefficient defines the *multiple correlation coefficient*. Clearly, both $\boldsymbol{R}$ and $\boldsymbol{R}^2$ must lie between 0 and 1. An $\boldsymbol{R}^2$ close to 1 means, of course, that the model fits the data very well, while an $\boldsymbol{R}^2$ close to 0 means that there is little in the way of a linear relationship connecting $\mathbf{y}$ to the variables in $\mathbf{X}$.[5]

Since in most applications σ^2 is unknown, the statistic,

$$S_e^2 = \frac{\hat{\mathbf{u}}'\hat{\mathbf{u}}}{T - n}, \tag{20.28}$$

will provide an unbiased estimator. The quantity $T - n$, the number of observations less the number of parameters estimated, is referred to as the *number of degrees of freedom*. With σ^2 unknown, an unbiased estimator of the covariance matrix of $\hat{\boldsymbol{\beta}}$ will be provided by

$$\hat{V}(\hat{\boldsymbol{\beta}}) = S_e^2(\mathbf{X}'\mathbf{X})^{-1}. \tag{20.29}$$

We now define an unbiased estimator $\tilde{\boldsymbol{\beta}}$ to be *linear* if it can be written as a linear function of $\mathbf{y}$,

$$\tilde{\boldsymbol{\beta}} = \mathbf{A}\mathbf{y}, \tag{20.30}$$

[5] In the case where the model does not contain an intercept, the quantity defined by (20.27) is no longer bounded by 0 and 1. In this situation, a preferable alternative is to define the coefficient of multiple determination as the square of the simple correlation of $\mathbf{y}$ with $\hat{\mathbf{y}}$. This quantity is bounded by 0 and 1, and, moreover, coincides with (20.27) when the model has an intercept.

$\mathbf{A}$ being some $n \times T$ matrix which is of rank n and independent of $\boldsymbol{u}$. Since

$$\begin{aligned}\tilde{\beta} &= \mathbf{A}y \\ &= \mathbf{A}(\mathbf{X}\boldsymbol{\beta} + \boldsymbol{u}) \\ &= \mathbf{A}\mathbf{X}\boldsymbol{\beta} + \mathbf{A}\boldsymbol{u},\end{aligned} \tag{20.31}$$

unbiasedness requires that

$$\mathbf{AX} = \mathbf{I}. \tag{20.32}$$

We shall call the class of estimators whose typical element is $\tilde{\beta}$ to be the class of linear unbiased estimators of $\boldsymbol{\beta}$. Note that the least squares estimator is a member of this class, with $\mathbf{A} = (\mathbf{X}'\mathbf{X})^{-1}\mathbf{X}'$. We shall now show that, within this class of estimators, the least squares estimator has minimum variance.

THEOREM 20.1 (Gauss-Markov)
With reference to the model, $\boldsymbol{y} = \mathbf{X}\boldsymbol{\beta} + \boldsymbol{u}$, suppose that assumptions (i)–(v) with respect to $\boldsymbol{u}$ and $\mathbf{X}$ are fulfilled. Then, within the class of linear unbiased estimators of $\boldsymbol{\beta}$, the least squares estimator, $\hat{\beta} = (\mathbf{X}'\mathbf{X})^{-1}\mathbf{X}'\boldsymbol{y}$, is the estimator with minimum variance.

PROOF. Let us begin by noting that since

$$\tilde{\beta} - \boldsymbol{\beta} = \mathbf{A}\boldsymbol{u}, \tag{20.33}$$

the covariance matrix of $\tilde{\beta}$ is given by

$$V(\tilde{\beta}) = \sigma^2\mathbf{AA}'. \tag{20.34}$$

We now want to compare this covariance matrix with the one for least squares,

$$V(\hat{\beta}) = \sigma^2(\mathbf{X}'\mathbf{X})^{-1}, \tag{20.13}$$

Since $\mathbf{A}$ is an $n \times T$ matrix of rank n, we can always write it as

$$\mathbf{A} = (\mathbf{X}'\mathbf{X})^{-1}\mathbf{X}' + \mathbf{D}, \tag{20.35}$$

for some $n \times T$ matrix $\mathbf{D}$. Note that, since unbiasedness requires $\mathbf{AX} = \mathbf{I}$, we must have $\mathbf{DX} = \mathbf{0}$. The covariance matrix of $\tilde{\beta}$ can then be written as

$$\begin{aligned}V(\hat{\beta}) &= \sigma^2\{[(\mathbf{X}'\mathbf{X})^{-1}\mathbf{X}' + \mathbf{D}][(\mathbf{X}'\mathbf{X})^{-1}\mathbf{X}' + \mathbf{D}]'\} \\ &= \sigma^2[(\mathbf{X}'\mathbf{X})^{-1} + \mathbf{DX}(\mathbf{X}'\mathbf{X})^{-1} + (\mathbf{X}'\mathbf{X})^{-1}\mathbf{X}'\mathbf{D} + \mathbf{DD}'] \\ &= \sigma^2[(\mathbf{X}'\mathbf{X}^{-1} + \mathbf{DD}'],\end{aligned} \tag{20.36}$$

since $\mathbf{DX} = \mathbf{0}$. Thus, we find the covariance matrix of $\tilde{\boldsymbol{\beta}}$ to be equal to the covariance matrix of $\hat{\boldsymbol{\beta}}$ plus another matrix which is positive semidefinite.

Since the diagonal elements of a positive semidefinite matrix are necessarily nonnegative, it follows that the variances of the elements of $\tilde{\boldsymbol{\beta}}$, which appear on the diagonal of $V(\tilde{\boldsymbol{\beta}})$, can never be smaller than the variances of the elements of $\tilde{\boldsymbol{\beta}}$; that is, we have

$$\text{var}(\tilde{\beta}_i) \geq \text{var}(\hat{\beta}_i), \qquad i = 1, n. \tag{20.37}$$

But this is only part of the story. Consider $\tilde{\alpha}$, an arbitrary linear combination of the $\tilde{\beta}_i$,

$$\tilde{\alpha} = \mathbf{a}'\tilde{\boldsymbol{\beta}}, \tag{20.38}$$

for any nonzero $n \times 1$ vector $\mathbf{a}$. Then

$$\begin{aligned} E(\tilde{\alpha}) &= \mathbf{a}'E(\tilde{\boldsymbol{\beta}}), \\ &= \mathbf{a}'\boldsymbol{\beta} \end{aligned} \tag{20.39}$$

and

$$\begin{aligned} \text{var}(\tilde{\alpha}) &= \sigma^2\mathbf{a}'\mathbf{AA}'\mathbf{a} \\ &= \sigma^2\mathbf{a}'[(\mathbf{X}'\mathbf{X})^{-1} + \mathbf{DD}']\mathbf{a}. \end{aligned} \tag{20.40}$$

Let $\hat{\alpha} = \mathbf{a}'\hat{\boldsymbol{\beta}}$ be the least squares estimator of α. Clearly,

$$E(\hat{\alpha}) = \mathbf{a}'\boldsymbol{\beta} \tag{20.41}$$

and

$$\text{var}(\hat{\alpha}) = \sigma^2\mathbf{a}'(\mathbf{X}'\mathbf{X})^{-1}\mathbf{a}. \tag{20.42}$$

Whence, from (20.34),

$$\text{var}(\tilde{\alpha}) \geq \text{var}(\hat{\alpha}), \tag{20.43}$$

since $\mathbf{a}'\mathbf{DD}'\mathbf{a}$ is necessarily nonnegative.

From (20.37) and (20.43), we thus conclude that, within the class of linear unbiased estimators of $\boldsymbol{\beta}$, the least squares estimator, $\hat{\boldsymbol{\beta}} = (\mathbf{X}'\mathbf{X})^{-1}\mathbf{X}'\boldsymbol{y}$, has minimum variance in the sense that, for any other linear unbiased estimator $\hat{\boldsymbol{\beta}}$ of $\boldsymbol{\beta}$:

(a) $\text{var}(\hat{\beta}_i) \leq \text{var}(\tilde{\beta}_i), \qquad i = 1, n;$
(b) $\text{var}(\mathbf{a}'\hat{\boldsymbol{\beta}}) \leq \text{var}(\mathbf{a}'\tilde{\boldsymbol{\beta}})$, where $\mathbf{a}$ is any nonzero $n \times 1$ vector of real numbers.

This proves the theorem.

Since the restrictions it places on the error term $\boldsymbol{u}$ are for the most part minimal, this theorem provides strong justification for the use of least squares, and its importance can accordingly hardly be overemphasized. Moreover, in situations where it can be assumed that the elements of $\boldsymbol{u}$ are multivariate normal, least squares is also justified, as the next theorem shows, by the principle of maximum likelihood.

THEOREM 20.2

In the linear regression model, $\boldsymbol{y} = \mathbf{X}\boldsymbol{\beta} + \boldsymbol{u}$, suppose that assumptions (i)–(v) with respect to $\boldsymbol{u}$ and $\mathbf{X}$ are fulfilled, and in addition, assume that $\boldsymbol{u}$ is $N(\mathbf{0}, \sigma^2\mathbf{I})$. Then the least squares estimator, $\hat{\beta} = (\mathbf{X}'\mathbf{X})^{-1}\mathbf{X}'\mathbf{y}$, is also the maximum likelihood estimator.

PROOF. With $\boldsymbol{u}$ $N(\mathbf{0}, \sigma^2\mathbf{I})$, the likelihood function for the sample of T observations is

$$L(\mathbf{u}; \mathbf{X}, \boldsymbol{\beta}, \sigma^2) = (2\pi\sigma^2)^{-T/2}\left[\exp\left(-\frac{1}{2\sigma^2}\mathbf{u}'\mathbf{u}\right)\right], \tag{20.44}$$

or, in terms of $\boldsymbol{y}$ and $\mathbf{X}\boldsymbol{\beta}$,[6]

$$L(\boldsymbol{y}; \mathbf{X}, \boldsymbol{\beta}, \sigma^2) = (2\pi\sigma^2)^{-T/2}\left[\exp\left[-\frac{1}{2\sigma^2}(\boldsymbol{y} - \mathbf{X}\boldsymbol{\beta})'(\boldsymbol{y} - \mathbf{X}\boldsymbol{\beta})\right]\right], \tag{20.45}$$

Since σ^2 and $(\boldsymbol{y} - \mathbf{X}\boldsymbol{\beta})'(\mathbf{y} - \mathbf{X}\boldsymbol{\beta})$ are both necessarily positive, it is clear that the likelihood function will be at a maximum with respect to β when $(\boldsymbol{y} - \mathbf{X}\boldsymbol{\beta})'(\boldsymbol{y} - \mathbf{X}\boldsymbol{\beta})$ is at a minimum. Hence, the maximum likelihood estimator of $\boldsymbol{\beta}$ is given by the least squares estimator, $\hat{\beta} = (\mathbf{X}'\mathbf{X})^{-1}\mathbf{X}'\boldsymbol{y}$.

Before leaving this section, we shall establish that, on the assumption that the values taken by the independent variables are bounded, the least squares estimator is consistent.

THEOREM 20.3

In the linear regression model, $\boldsymbol{y} = \mathbf{X}\boldsymbol{\beta} + \boldsymbol{u}$, retain assumptions (i)–(v) and in addition assume that the elements of $\mathbf{X}$ are bounded. Then $\hat{\beta} = (\mathbf{X}'\mathbf{X})^{-1}\mathbf{X}'\boldsymbol{y}$ provides a consistent estimator of $\boldsymbol{\beta}$.

PROOF. Recall from Theorem 16.1 that a sufficient condition for an unbiased estimator to be consistent is for its variance to converge

[6] Note that the Jacobian of the transformation from $\boldsymbol{u}$ to $\boldsymbol{y}$ is one.

asymptotically to zero. We already know $\hat{\beta}$ to be unbiased, and its covariance matrix we can rewrite as

$$V(\hat{\beta}) = \frac{\sigma^2}{T}(T^{-1}\mathbf{X}'\mathbf{X})^{-1}. \tag{20.46}$$

Let us now take the limit of this expression as $T \to \infty$,

$$\begin{aligned} \lim_{T\to\infty} V(\hat{\beta}) &= \lim_{T\to\infty}\left[\frac{\sigma^2}{T}(T^{-1}\mathbf{X}'\mathbf{X})^{-1}\right] \\ &= \lim_{T\to\infty}\frac{\sigma^2}{T}\lim_{T\to\infty}(T^{-1}\mathbf{X}'\mathbf{X})^{-1}. \end{aligned} \tag{20.47}$$

The assumption that the elements of $\mathbf{X}$ are bounded implies that $T^{-1}\mathbf{X}'\mathbf{X}$, and therefore $(T^{-1}\mathbf{X}'\mathbf{X})^{-1}$, is bounded for all T; hence, $\lim_{T\to\infty}(T^{-1}\mathbf{X}'\mathbf{X})^{-1}$ can be replaced by a matrix of finite constants. Then, since

$$\lim_{T\to\infty}\frac{\sigma^2}{T} = 0, \tag{20.48}$$

it follows that

$$\lim_{T\to\infty} V(\hat{\beta}) = 0. \tag{20.49}$$

In view of Theorem 16.1, the consistency of $\hat{\beta}$ is thus established.

20.3 STATISTICAL INFERENCE IN THE LINEAR REGRESSION MODEL

Let us now turn to the distribution theory which provides for statistical inference in the least squares regression model. From this point on we assume that the error term is distributed normally with mean 0 and variance σ^2. The groundwork for the results about to be derived was laid in Chapters 12 and 14.

We shall begin with the distribution of $\hat{\beta}$. From (20.11), we have

$$\hat{\beta} - \boldsymbol{\beta} = (\mathbf{X}'\mathbf{X})^{-1}\mathbf{X}'\boldsymbol{u}, \tag{20.50}$$

which shows $\hat{\beta} - \boldsymbol{\beta}$ to be a linear function of the elements of $\boldsymbol{u}$. Since $\boldsymbol{u}$ is multivariate normal with mean $\mathbf{0}$ and covariance matrix $\sigma^2\mathbf{I}$, it accordingly follows from Theorem 12.5 that $\hat{\beta} - \boldsymbol{\beta}$ will be distributed multivariate normally with mean $\mathbf{0}$ and covariance matrix $\sigma^2(\mathbf{X}'\mathbf{X})^{-1}$. The distribution of $\hat{\beta}$, therefore, will be $N[\boldsymbol{\beta}, \sigma^2(\mathbf{X}'\mathbf{X})^{-1}]$.

Consider, next, the residual sum of squares, which from (20.18) is given by

$$\hat{\boldsymbol{u}}'\hat{\boldsymbol{u}} = \boldsymbol{u}'[\mathbf{I} - \mathbf{X}(\mathbf{X}'\mathbf{X})^{-1}\mathbf{X}']\boldsymbol{u}.$$

As we have already noted, the matrix $\mathbf{I} - \mathbf{X}(\mathbf{X}'\mathbf{X})^{-1}\mathbf{X}'$ is idempotent and has rank $T - n$. This being the case, it then follows from Theorem 15.2 that the quantity $\hat{\boldsymbol{u}}'\hat{\boldsymbol{u}}/\sigma^2$ will be distributed as chi-square with $T - n$ degrees of freedom.

In Theorem 15.3, it was proven that if a vector $\mathbf{z}$ is $N(\mathbf{0}, \mathbf{I})$, $\mathbf{A}$ an idempotent matrix, and $\mathbf{B}$ another matrix such that $\mathbf{BA} = \mathbf{0}$, then the linear form $\mathbf{Bz}$ is distributed independently of the quadratic form $\mathbf{z}'\mathbf{Az}$. With this in mind, consider the quantities

$$\frac{1}{\sigma}(\hat{\boldsymbol{\beta}} - \boldsymbol{\beta}) = \frac{1}{\sigma}(\mathbf{X}'\mathbf{X})^{-1}\mathbf{X}'\boldsymbol{u} \tag{20.51}$$

and

$$\frac{1}{\sigma^2}\boldsymbol{u}'\boldsymbol{u} = \frac{1}{\sigma^2}\boldsymbol{u}'[\mathbf{I} - \mathbf{X}(\mathbf{X}'\mathbf{X})^{-1}\mathbf{X}']\boldsymbol{u}. \tag{20.52}$$

Now, $\mathbf{I} - \mathbf{X}(\mathbf{X}'\mathbf{X})^{-1}\mathbf{X}'$ is idempotent, and

$$\begin{aligned}(\mathbf{X}'\mathbf{X})^{-1}\mathbf{X}'[\mathbf{I} - \mathbf{X}(\mathbf{X}'\mathbf{X})^{-1}\mathbf{X}'] &= (\mathbf{X}'\mathbf{X})^{-1}\mathbf{X}' - (\mathbf{X}'\mathbf{X})^{-1}\mathbf{X}' \\ &= \mathbf{0}. \end{aligned} \tag{20.53}$$

Hence, we can conclude that $(\hat{\boldsymbol{\beta}} - \boldsymbol{\beta})$ and $\hat{\boldsymbol{u}}'\hat{\boldsymbol{u}}$ are distributed independently. As will become clear in the next paragraph, this fact opens the door to the testing of hypotheses about $\boldsymbol{\beta}$.

Consider now the quantity

$$\begin{aligned}\tau &= \frac{\hat{\beta}_i - \beta_i}{\sigma_{\hat{\beta}_i}} \bigg/ \frac{S_{\hat{\beta}_i}}{\sigma_{\hat{\beta}_i}} \\ &= \frac{\hat{\beta}_i - \beta_i}{S_{\hat{\beta}_i}}, \end{aligned} \tag{20.54}$$

where $\sigma_{\hat{\beta}_i}$ is the square root of the ith element on the diagonal in the covariance matrix of $\hat{\boldsymbol{\beta}}$, $\sigma^2(\mathbf{X}'\mathbf{X})^{-1}$, and $S_{\hat{\beta}_i}$ is the square root of the ith element on the diagonal in the estimated covariance matrix, $S_e^{\,2}(\mathbf{X}'\mathbf{X})^{-1}$. Now $(\hat{\beta}_i - \beta_i)/\sigma_{\hat{\beta}_i}$ is $N(0, 1)$, while $S_{\hat{\beta}_i}/\sigma_{\hat{\beta}_i}$ is the square root of a variable which is chi-square and divided by its degrees of freedom, which is $T - n$. Moreover, in view of the preceding paragraph, these two quantities are distributed independently. Consequently, it follows that

$$\tau = \frac{\hat{\beta}_i - \beta_i}{S_{\hat{\beta}_i}}$$

has the t-distribution with $T - n$ degrees of freedom. This is an important result, for it means that the t-distribution can be employed in testing hypotheses as to the population value of β_i and the construction of confidence intervals in the manner of Sections 18.6 and 19.3.

The preceding provides for testing hypotheses with regard to a specific β_i. Suppose, instead, that we are interested in testing hypotheses about linear combinations of the β_i of the form

$$\gamma = \mathbf{k}'\boldsymbol{\beta}, \tag{20.55}$$

for $\mathbf{k}$ any vector of real numbers. To do this, we require the distribution of $\hat{\gamma} = \mathbf{k}'\hat{\boldsymbol{\beta}}$. Once again, we use Theorem 12.5, and conclude that $\mathbf{k}'\hat{\boldsymbol{\beta}}$ will be normally distributed with mean γ and variance $\sigma^2\mathbf{k}'(\mathbf{X}'\mathbf{X})^{-1}\mathbf{k}$. Consequently, following the argument of the preceding paragraph, the quantity

$$\gamma = \frac{\mathbf{k}'\hat{\boldsymbol{\beta}} - \mathbf{k}'\boldsymbol{\beta}}{S_{\mathbf{k}'\hat{\boldsymbol{\beta}}}}, \tag{20.56}$$

where

$$S_{\mathbf{k}'\hat{\boldsymbol{\beta}}} = [S_e{}^2\mathbf{k}'(\mathbf{X}'\mathbf{X})^{-1}\mathbf{k}]^{1/2}, \tag{20.57}$$

will have the t-distribution with $T - n$ degrees of freedom. From here on, the procedure is the same as for the case of a single regression coefficient.

Finally, suppose that we are interested in testing a hypothesis of the form $\boldsymbol{\beta}^* = \boldsymbol{\beta}_0{}^*$, where $\boldsymbol{\beta}^*$ is some subvector of $\boldsymbol{\beta}$. Indeed, let us consider the case where $\boldsymbol{\beta}^*$ is equal to $\boldsymbol{\beta}$ itself. On the hypothesis that $\boldsymbol{\beta} = \boldsymbol{\beta}_0$, we will have from (20.50),

$$\hat{\boldsymbol{\beta}} - \boldsymbol{\beta}_0 = (\mathbf{X}'\mathbf{X})^{-1}\mathbf{X}'\boldsymbol{u}. \tag{20.58}$$

Accordingly, consider the quantity

$$Q = (\hat{\boldsymbol{\beta}} - \boldsymbol{\beta}_0)'\mathbf{X}'\mathbf{X}(\hat{\boldsymbol{\beta}} - \boldsymbol{\beta}_0), \tag{20.59}$$

which in view of (20.58) can be written as

$$Q = \boldsymbol{u}'\mathbf{X}(\mathbf{X}'\mathbf{X})^{-1}\mathbf{X}'\boldsymbol{u}. \tag{20.60}$$

Note that the matrix $\mathbf{X}(\mathbf{X}'\mathbf{X})^{-1}\mathbf{X}'$ is idempotent with rank n; hence, Q/σ^2 is chi-square with n degrees of freedom. Note, too, that the product of $\mathbf{X}(\mathbf{X}'\mathbf{X})^{-1}\mathbf{X}'$ with $\mathbf{I} - \mathbf{X}(\mathbf{X}'\mathbf{X})^{-1}\mathbf{X}'$ is 0. Consequently, it follows from Theorem 15.4 that Q/σ^2 and $\hat{\boldsymbol{u}}'\hat{\boldsymbol{u}}/\sigma^2$ will be distributed independently. Whence, from Theorem 15.5, the ratio

$$f = \frac{(\hat{\boldsymbol{\beta}} - \boldsymbol{\beta}_0)'\mathbf{X}'\mathbf{X}(\hat{\boldsymbol{\beta}} - \boldsymbol{\beta}_0)}{n} \bigg/ \frac{\hat{\boldsymbol{u}}'\hat{\boldsymbol{u}}}{T - n}, \tag{20.61}$$

will have the F-distribution with n and $T - n$ degrees of freedom. This being the case, hypotheses about the vector $\boldsymbol{\beta}$ can be tested using the F-distribution as in Section 18.6.

Let us now turn to the case where we are interested in testing a hypothesis about just a subset of the elements of $\boldsymbol{\beta}$, say the last m. Accordingly, we rewrite the model as

$$y = \mathbf{X}_1\boldsymbol{\beta}_1 + \mathbf{X}_2\boldsymbol{\beta}_2 + \boldsymbol{u}, \tag{20.62}$$

where $\mathbf{X}_1$ is a $T \times (n - m)$ matrix denoting the observations on the first $n - m$ independent variables and $\mathbf{X}_2$ a $T \times m$ matrix denoting the observations on the last m, and similarly for $\boldsymbol{\beta}_1$ and $\boldsymbol{\beta}_2$. With this partitioning, expression (20.9) for $\hat{\beta}$ becomes

$$\begin{bmatrix} \hat{\beta}_1 \\ \hat{\beta}_2 \end{bmatrix} = \begin{bmatrix} \mathbf{X}_1'\mathbf{X}_1 & \mathbf{X}'_1\mathbf{X}_2 \\ \mathbf{X}_2'\mathbf{X}_1 & \mathbf{X}_2'\mathbf{X}_2 \end{bmatrix}^{-1} \begin{bmatrix} \mathbf{X}_1'y \\ \mathbf{X}_2'y \end{bmatrix}, \tag{20.63}$$

or, employing the rules for inversion of a matrix by partitioning,[7]

$$\begin{aligned} \begin{bmatrix} \hat{\beta}_1 \\ \hat{\beta}_2 \end{bmatrix} &= \begin{bmatrix} (\mathbf{X}_1'\mathbf{X}_1)^{-1}[\mathbf{I} + \mathbf{X}_1'\mathbf{X}_2\mathbf{D}^{-1}\mathbf{X}_2'\mathbf{X}_1(\mathbf{X}_1'\mathbf{X}_1)^{-1}] & -(\mathbf{X}_1'\mathbf{X}_1)^{-1}\mathbf{X}_1'\mathbf{X}_2\mathbf{D}^{-1} \\ -\mathbf{D}^{-1}\mathbf{X}_2'\mathbf{X}_1(\mathbf{X}_1'\mathbf{X}_1)^{-1} & \mathbf{D}^{-1} \end{bmatrix} \begin{bmatrix} \mathbf{X}_1'y \\ \mathbf{X}_2'y \end{bmatrix} \\ &= \begin{bmatrix} (\mathbf{X}_1'\mathbf{X}_1)^{-1}\mathbf{X}_1 y - (\mathbf{X}_1'\mathbf{X}_1)^{-1}\mathbf{X}_1'\mathbf{X}_2\mathbf{D}^{-1}\mathbf{X}_2\mathbf{M}_1 y \\ \mathbf{D}^{-1}\mathbf{X}_2'\mathbf{M}_1 y \end{bmatrix}, \end{aligned} \tag{20.64}$$

where

$$\mathbf{M}_1 = \mathbf{I} - \mathbf{X}_1(\mathbf{X}_1'\mathbf{X}_1)^{-1}\mathbf{X}_1' \tag{20.65}$$

and

$$\begin{aligned} \mathbf{D} &= \mathbf{X}_2'\mathbf{X}_2 - \mathbf{X}_2'\mathbf{X}_1(\mathbf{X}_1'\mathbf{X}_1)^{-1}\mathbf{X}_1'\mathbf{X}_2 \\ &= \mathbf{X}_2'\mathbf{M}_1\mathbf{X}_2. \end{aligned} \tag{20.66}$$

From the second line in (20.64), we have for $\hat{\beta}_2$,

$$\begin{aligned} \hat{\beta}_2 &= \mathbf{D}^{-1}\mathbf{X}_2'\mathbf{M}_1 y \\ &= \mathbf{D}^{-1}\mathbf{X}_2'\mathbf{M}_1(\mathbf{X}_1\boldsymbol{\beta}_1 + \mathbf{X}_2\boldsymbol{\beta}_2 + \boldsymbol{u}) \\ &= \boldsymbol{\beta}_2 + \mathbf{D}^{-1}\mathbf{X}_2'\mathbf{M}_1\boldsymbol{u}, \end{aligned} \tag{20.67}$$

[7] See Appendix 5.

where we have used the facts that

$$\begin{aligned} \mathbf{M}_1\mathbf{X}_1 &= [\mathbf{I} - \mathbf{X}_1(\mathbf{X}_1'\mathbf{X}_1)^{-1}\mathbf{X}_1']\mathbf{X}_1 \\ &= \mathbf{X}_1 - \mathbf{X}_1 \\ &= \mathbf{0} \end{aligned} \tag{20.68}$$

and

$$\begin{aligned} \mathbf{D}^{-1}\mathbf{X}_2'\mathbf{M}_1\mathbf{X}_2 &= \mathbf{D}^{-1}\mathbf{D} \\ &= \mathbf{I}. \end{aligned} \tag{20.69}$$

Consequently, from (20.67), we have

$$\hat{\beta}_2 - \boldsymbol{\beta}_2 = \mathbf{D}^{-1}\mathbf{X}_2'\mathbf{M}_1\boldsymbol{u}. \tag{20.70}$$

Consider, now, the expression

$$Q_2 = (\hat{\beta}_2 - \boldsymbol{\beta}_2^*)'\mathbf{D}(\hat{\beta}_2 - \boldsymbol{\beta}^*), \tag{20.71}$$

which on the hypothesis that $\boldsymbol{\beta}_2 = \boldsymbol{\beta}_2^*$ can be written

$$\begin{aligned} Q_2 &= \boldsymbol{u}'\mathbf{M}_1\mathbf{X}_2\mathbf{D}^{-1}\mathbf{D}\mathbf{D}^{-1}\mathbf{X}_2'\mathbf{M}_1\boldsymbol{u} \\ &= \boldsymbol{u}'\mathbf{P}\boldsymbol{u}, \end{aligned} \tag{20.72}$$

where

$$\mathbf{P} = \mathbf{M}_1\mathbf{X}_2\mathbf{D}^{-1}\mathbf{X}_2'\mathbf{M}_1. \tag{20.73}$$

Since

$$\begin{aligned} \mathbf{P}\cdot\mathbf{P} &= \mathbf{M}_1\mathbf{X}_2\mathbf{D}^{-1}\mathbf{X}_2'\mathbf{M}_1 \cdot \mathbf{M}_1\mathbf{X}_2\mathbf{D}^{-1}\mathbf{X}_2'\mathbf{M}_1 \\ &= \mathbf{M}_1\mathbf{X}_2\mathbf{D}^{-1}\mathbf{X}_2'\mathbf{M}_1\mathbf{X}_2\mathbf{D}^{-1}\mathbf{X}_2'\mathbf{M}_1 \qquad (\text{since } \mathbf{M}_1\cdot\mathbf{M}_1 = \mathbf{M}_1) \\ &= \mathbf{M}_1\mathbf{X}_2\mathbf{D}^{-1}\mathbf{D}\mathbf{D}^{-1}\mathbf{X}_2'\mathbf{M}_1 \qquad (\text{since } \mathbf{X}_2'\mathbf{M}_1\mathbf{X}_2 = \mathbf{D}) \\ &= \mathbf{M}_1\mathbf{X}_2\mathbf{D}^{-1}\mathbf{X}_2'\mathbf{M}_1 \\ &= \mathbf{P}, \end{aligned} \tag{20.74}$$

P is idempotent. In addition,

$$\begin{aligned} \text{rank }(\mathbf{P}) &= \text{Tr}(\mathbf{P}) \\ &= \text{Tr}(\mathbf{M}_1\mathbf{X}_2\mathbf{D}^{-1}\mathbf{X}^{2\prime}\mathbf{M}_1) \\ &= \text{Tr}(\mathbf{X}_2'\mathbf{M}_1\mathbf{M}_1\mathbf{X}_2\mathbf{D}^{-1}) \\ &= \text{Tr}(\mathbf{D}\mathbf{D}^{-1}) \\ &= m. \end{aligned} \tag{20.75}$$

Consequently, the quantity Q_2/σ^2 is chi-square with m degrees of freedom.

Next, let $\mathbf{M} = \mathbf{I} - \mathbf{X}(\mathbf{X}'\mathbf{X})^{-1}\mathbf{X}'$. Then it can be checked that $\mathbf{PM}_1 = \mathbf{P}$ and that $\mathbf{M} = \mathbf{M}_1 - \mathbf{P}$, so that

$$\begin{aligned}\mathbf{PM} &= \mathbf{P}(\mathbf{M}_1 - \mathbf{P}) \\ &= \mathbf{P} - \mathbf{P} \\ &= \mathbf{0},\end{aligned} \tag{20.76}$$

which in view of Theorem 15.4 establishes that $\boldsymbol{u}'\mathbf{P}\boldsymbol{u}$ and $\boldsymbol{u}'\mathbf{M}\boldsymbol{u}$ $[= \hat{\boldsymbol{u}}'\hat{\boldsymbol{u}}]$ are independently distributed. Hence, we can conclude that on the hypothesis that $\boldsymbol{\beta}_2 = \boldsymbol{\beta}_2{}^*$ the ratio

$$f = \frac{(\hat{\boldsymbol{\beta}}_2 - \boldsymbol{\beta}_2{}^*)'\mathbf{D}(\hat{\boldsymbol{\beta}}_2 - \boldsymbol{\beta}_2{}^*)}{m} \Bigg/ \frac{\hat{\boldsymbol{u}}'\hat{\boldsymbol{u}}}{T - n}, \tag{20.77}$$

will have the F-distribution with m and $T - n$ degrees of freedom.

20.4 THE AITKEN EXTENSION OF THE GAUSS-MARKOV THEOREM: GENERALIZED LEAST SQUARES

In the proof of the Gauss-Markov theorem, it was assumed that the covariance matrix of $\boldsymbol{u}$ has the form $\sigma^2\mathbf{I}$. Let us now take up the case, which is frequently encountered in practice, where the covariance matrix of $\boldsymbol{u}$ is given by

$$E(\boldsymbol{u}\boldsymbol{u}') = \boldsymbol{\Omega}, \tag{20.78}$$

where $\boldsymbol{\Omega}$ is some $T \times T$ positive definite symmetric matrix. However, we retain the assumption that $E(\boldsymbol{u}) = \mathbf{0}$ and also the assumption that $\mathbf{X}$ is of rank n and fixed in repeated sampling.

Let us consider, now, the estimator defined by

$$\hat{\boldsymbol{\beta}} = (\mathbf{X}'\boldsymbol{\Omega}^{-1}\mathbf{X})^{-1}\mathbf{X}'\boldsymbol{\Omega}^{-1}\boldsymbol{y}. \tag{20.79}$$

Since

$$\begin{aligned}E(\hat{\boldsymbol{\beta}}) &= E[(\mathbf{X}'\boldsymbol{\Omega}^{-1}\mathbf{X})^{-1}\mathbf{X}'\boldsymbol{\Omega}^{-1}\boldsymbol{y}] \\ &= E[(\mathbf{X}'\boldsymbol{\Omega}^{-1}\mathbf{X})^{-1}\mathbf{X}'\boldsymbol{\Omega}^{-1}(\mathbf{X}\boldsymbol{\beta} + \boldsymbol{u})] \\ &= \boldsymbol{\beta} + (\mathbf{X}'\boldsymbol{\Omega}^{-1}\mathbf{X})^{-1}\mathbf{X}'\boldsymbol{\Omega}^{-1}E(\boldsymbol{u}) \\ &= \boldsymbol{\beta},\end{aligned} \tag{20.80}$$

$\hat{\boldsymbol{\beta}}$ is unbiased. Its covariance matrix is

$$\begin{aligned}E[(\hat{\boldsymbol{\beta}} - \boldsymbol{\beta})(\hat{\boldsymbol{\beta}} - \boldsymbol{\beta})'] &= E[(\mathbf{X}'\boldsymbol{\Omega}^{-1}\mathbf{X})^{-1}\mathbf{X}'\boldsymbol{\Omega}^{-1}\boldsymbol{u}\boldsymbol{u}'\boldsymbol{\Omega}^{-1}\mathbf{X}(\mathbf{X}'\boldsymbol{\Omega}^{-1}\mathbf{X})] \\ &= (\mathbf{X}'\boldsymbol{\Omega}^{-1}\mathbf{X})^{-1}\mathbf{X}'\boldsymbol{\Omega}^{-1}E(\boldsymbol{u}\boldsymbol{u}')\boldsymbol{\Omega}^{-1}\mathbf{X}(\mathbf{X}'\boldsymbol{\Omega}^{-1}\mathbf{X}) \\ &= (\mathbf{X}'\boldsymbol{\Omega}^{-1}\mathbf{X})^{-1}.\end{aligned} \tag{20.81}$$

Moreover, $\hat{\boldsymbol{\beta}}$ is linear in $\boldsymbol{y}$, and therefore is a linear estimator.

Next, consider the general linear unbiased estimator of $\boldsymbol{\beta}$,

$$\tilde{\boldsymbol{\beta}} = \mathbf{A}\boldsymbol{y}. \tag{20.82}$$

As before, unbiasedness requires that

$$\mathbf{AX} = \mathbf{I}. \tag{20.83}$$

The covariance matrix of $\tilde{\boldsymbol{\beta}}$ is

$$V(\tilde{\boldsymbol{\beta}}) = \mathbf{A\Omega A}'. \tag{20.84}$$

Since $\hat{\boldsymbol{\beta}}$ is a special case of $\tilde{\boldsymbol{\beta}}$, we can write $\mathbf{A}$ as

$$\mathbf{A} = (\mathbf{X}'\boldsymbol{\Omega}^{-1}\mathbf{X})^{-1}\mathbf{X}'\boldsymbol{\Omega}^{-1} + \mathbf{D}, \tag{20.85}$$

for some $n \times T$ matrix $\mathbf{D}$. $V(\tilde{\boldsymbol{\beta}})$ then becomes

$$\begin{aligned} V(\tilde{\boldsymbol{\beta}}) &= [(\mathbf{X}'\boldsymbol{\Omega}^{-1}\mathbf{X})^{-1}\mathbf{X}'\boldsymbol{\Omega}^{-1} + D]\boldsymbol{\Omega}[(\mathbf{X}'\boldsymbol{\Omega}^{-1}\mathbf{X})^{-1}\mathbf{X}'\boldsymbol{\Omega}^{-1} + \mathbf{D}]' \\ &= (\mathbf{X}'\boldsymbol{\Omega}^{-1}\mathbf{X})^{-1} + \mathbf{DD}', \end{aligned} \tag{20.86}$$

where we have made use of the fact that, from (20.83), $\mathbf{DX} = \mathbf{0}$.

From (20.86), we see that the covariance matrix of the general linear unbiased estimator is equal to the covariance matrix of $\hat{\boldsymbol{\beta}}$ plus another matrix which is positive semidefinite. Accordingly, we have proven the following.

THEOREM 20.4 (Aitken Generalization of Gauss-Markov Theorem) In the linear regression model, $\boldsymbol{y} = \mathbf{X}\boldsymbol{\beta} + \boldsymbol{u}$, assume that:

(i) $E(\boldsymbol{u}) = \mathbf{0}$;
(ii) $E(\boldsymbol{uu}') = \boldsymbol{\Omega}$ $\quad (|\boldsymbol{\Omega}| \neq 0)$;
(iii) $\mathbf{X}$ is a fixed set of numbers of rank n.

Let $\hat{\boldsymbol{\beta}} = (\mathbf{X}'\boldsymbol{\Omega}^{-1}\mathbf{X})^{-1}\mathbf{X}'\boldsymbol{\Omega}^{-1}\boldsymbol{y}$, and let $\hat{\gamma} = \mathbf{k}'\hat{\boldsymbol{\beta}}$, where $\mathbf{k}$ is any vector of real numbers, $\mathbf{k} \neq \mathbf{0}$. Then, within the class of linear unbiased estimators of $\boldsymbol{\beta}$, $\hat{\boldsymbol{\beta}}$ is the estimator for which

$$\text{var}(\hat{\gamma}) \leq \text{var}(\tilde{\gamma}), \tag{20.87}$$

where $\tilde{\gamma}$ is any other linear unbiased estimator of $\gamma = \mathbf{k}'\boldsymbol{\beta}$.

This theorem, which is seen to reduce to Theorem 20.1 in the case where $E(\boldsymbol{uu}') = \sigma^2\mathbf{I}$, is usually referred to as the Aitken generalization of the Gauss-Markov theorem.[8] The method is known as *generalized least*

[8] After A. C. Aitken (1935).

squares, and the estimator, $\hat{\beta} = (\mathbf{X}'\boldsymbol{\Omega}^{-1}\mathbf{X})^{-1}\mathbf{X}'\boldsymbol{\Omega}^{-1}\boldsymbol{y}$, is called the generalized least squares estimator.

It is left to the reader to check:

1. That the generalized least squares estimator is the one obtained by minimizing the *weighted* sum of squares,

$$\begin{aligned}\phi(\boldsymbol{\beta}) &= \boldsymbol{u}'\boldsymbol{\Omega}^{-1}\boldsymbol{u} \\ &= (\boldsymbol{y} - \mathbf{X}\boldsymbol{\beta})'\boldsymbol{\Omega}^{-1}(\boldsymbol{y} - \mathbf{X}\boldsymbol{\beta}).\end{aligned} \tag{20.88}$$

2. That, on the assumption that $\boldsymbol{u}$ is $N(\mathbf{0}, \boldsymbol{\Omega})$, the generalized least squares estimator is also the maximum likelihood estimator.
3. That, on the assumption that the elements of $\mathbf{X}$ are bounded, the generalized least squares estimator is consistent.

The principal drawback to the use of generalized least squares is that, in most applications, the covariance matrix $\boldsymbol{\Omega}$ is unknown and must be estimated. Fortunately, depending on the form that $\boldsymbol{\Omega}$ is thought to take, there now exist a number of ways of going about this.[9] It should be noted, however, that with an estimate of $\boldsymbol{\Omega}$, say $\hat{\boldsymbol{\Omega}}$, the estimator given by

$$\hat{\beta} = (\mathbf{X}'\hat{\boldsymbol{\Omega}}^{-1}\mathbf{X})^{-1}\mathbf{X}'\hat{\boldsymbol{\Omega}}^{-1}\boldsymbol{y} \tag{20.89}$$

is no longer the estimator of Theorem 20.4, although if $\hat{\boldsymbol{\Omega}}$ is consistent for $\boldsymbol{\Omega}$, it will achieve that estimator asymptotically.

Finally, it should be mentioned that, on the assumption that the error term $\boldsymbol{u}$ is $N(\mathbf{0}, \boldsymbol{\Omega})$, the distribution theory for undertaking statistical inference with generalized least squares parallels that presented for ordinary least squares in Section 20.3 above.

REFERENCES

Aitken, A. C., "On Least Squares and Linear Combination of Observations," *Proceedings of the Royal Society of Edinburgh*, vol. 55, 1935.

Goldberger, A. S., *Econometric Theory*, Wiley, 1964, chap. 4.

Malinvaud, E., *Statistical Methods of Econometrics*, 2nd ed., American Elsevier, 1970, chap. 5.

[9] See, for example, Malinvaud (1970, chap. 13).

CHAPTER 21

ORDER STATISTICS

The preceding five chapters have deviated very little from the assumption that the reference population is described by a normal distribution. Much of the time, this assumption receives justification either from actual evidence that the population is in fact normal (or at least approximately so) or else from operation of the classical central limit theorem. However, it is becoming more and more frequent that, for one reason or another, researchers do not feel the assumption of normality to be warranted and proceed instead under the assumption that the reference population is described by simply a general distribution. In these situations, it is usually necessary to employ procedures of inference that are based on distributions of certain order statistics. The purpose of this chapter is to provide the student with a brief introduction to this subject.

21.1 ORDER STATISTICS DEFINED

Let $X_1, X_2, \ldots, X_n$ be a sample of size n drawn at random from a population described by the density function $f(x)$. Let $X_{(i)}$ denote the ith smallest of the values taken by the n observations, so that

$$X_{(1)} \leq X_{(2)} \leq \cdots \leq X_{(n)}. \tag{21.1}$$

In view of these inequalities, $X_{(1)}, X_{(2)}, \ldots, X_{(n)}$ represents the n observations of the sample arranged in ascending order of magnitude. The quantity $X_{(i)}$ is referred to as the ith-*order statistic* of the sample.

The study of order statistics deals with the properties and applications of these ordered random variables and of functions involving them.

Examples include the *extremes* $X_{(1)}$ and $X_{(n)}$, the *range* $X_{(n)} - X_{(1)}$, the *extreme deviate* (from the mean) $X_{(n)} - \overline{X}$, and the *studentized range* $(X_{(n)} - X_{(1)})/S_v$, where S_v is an estimator of σ (for a normal distribution) based on v degrees of freedom. These statistics all have important applications. The extremes are useful in the study of floods and droughts,[1] as well as in the analysis of breaking strengths and fatigue failures. The range is well known to provide a quick estimator of σ for the normal distribution and, in addition, has found wide acceptance in the field of quality control. The extreme deviate is a basic tool in the detection of outliers, since large values of $(X_{(n)} - \overline{X})/\sigma$ will indicate the presence of one or a few excessively large observations. Finally, the studentized range is useful in the same context when outliers are not confined to a single direction. The studentized range also supplies the basis for many quick tests with small samples and is of key importance in ranking "treatment" means in the analysis of variance.

The extreme usefulness of order statistics derives from the following fact which shall be proven in Section 21.4: *The distribution of the area beneath the density function between any two ordered observations is independent of the form of the density function.* Among other things, this property frequently enables a researcher to employ techniques of statistical inference that are applicable regardless of the form of the density function of the population. Such techniques are referred to as *nonparametric* or *distribution-free* methods. Some examples will be considered in Sections 21.3 and 21.4.

21.2 DISTRIBUTION OF ORDER STATISTICS

Let $X_1, X_2, \ldots, X_n$ be a sample drawn at random from a continuous population with density function $f(x)$, and suppose that we want the joint distribution of the order statistics $X_{(1)}, X_{(2)}, \ldots, X_{(n)}$. For simplicity, we shall discuss the case where $n = 3$; however, it will become clear that the arguments employed are completely general.

Let X_1, X_2, X_3 be a sample of size 3 from $f(x)$, $-\infty < x < \infty$, and let

$$\begin{array}{llll} x_{(1)} = x_1; & x_{(2)} = x_2; & x_{(3)} = x_3, & \text{if } x_1 < x_2 < x_3, \\ x_{(1)} = x_1; & x_{(2)} = x_3; & x_{(3)} = x_2, & \text{if } x_1 < x_3 < x_2, \\ x_{(1)} = x_2; & x_{(2)} = x_1; & x_{(3)} = x_3, & \text{if } x_2 < x_1 < x_3, \\ x_{(1)} = x_2; & x_{(2)} = x_3; & x_{(3)} = x_1, & \text{if } x_2 > x_3 < x_1, \\ x_{(1)} = x_3; & x_{(2)} = x_1; & x_{(3)} = x_2, & \text{if } x_3 < x_1 < x_2, \\ x_{(1)} = x_3; & x_{(2)} = x_2; & x_{(3)} = x_1, & \text{if } x_3 < x_2 < x_1. \end{array} \tag{21.2}$$

[1] See Gumbel (1958).

These six regions $x_1 < x_2 < x_3$; $x_1 < x_3 < x_2$; ...; $x_3 < x_2 < x_1$ are disjoint and their union with certain sets that have probability zero, such as $x_1 = x_2 = x_3$, and so on, is the three-dimensional space $-\infty < x_1 < \infty$, $-\infty < x_2 < \infty$, $-\infty < x_3 < \infty$. We shall now find the joint density of $X_{(1)}, X_{(2)}, X_{(3)}$ for each of the six regions.

To begin with, note that the joint density of the X's is

$$g(x_1, x_2, x_3) = f(x_1)f(x_2)f(x_3), \qquad -\infty < x_1, x_2, x_3 < \infty. \tag{21.3}$$

For the region $x_1 < x_2 < x_3$, the density of the $X_{(.)}$'s is

$$h(x_{(1)}, x_{(2)}, x_{(3)}) = f(x_{(1)})f(x_{(2)})f(x_{(3)}), \qquad x_{(1)} < x_{(2)} < x_{(3)}, \tag{21.4}$$

since the Jacobian of the transformation from x_1, x_2, x_3 to $x_{(1)}, x_{(2)}, x_{(3)}$ is equal to 1. For the region $x_1 < x_3 < x_2$, the density of the $X_{(.)}$'s is

$$h(x_{(1)}, x_{(2)}, x_{(3)}) = f(x_{(1)})f(x_{(3)})f(x_{(2)}), \qquad x_{(1)} < x_{(2)} < x_{(3)}. \tag{21.5}$$

Similarly, for the other four regions:

$$h(x_{(1)}, x_{(2)}, x_{(3)}) = f(x_{(2)})f(x_{(1)})f(x_{(3)}), \qquad x_{(1)} < x_{(2)} < x_{(3)}, \tag{21.6}$$

$$h(x_{(1)}, x_{(2)}, x_{(3)}) = f(x_{(2)})f(x_{(3)})f(x_{(1)}), \qquad x_{(1)} < x_{(2)} < x_{(3)}, \tag{21.7}$$

$$h(x_{(1)}, x_{(2)}, x_{(3)}) = f(x_{(3)})f(x_{(1)})f(x_{(2)}), \qquad x_{(1)} < x_{(2)} < x_{(3)}, \tag{21.8}$$

$$h(x_{(1)}, x_{(2)}, x_{(3)}) = f(x_{(3)})f(x_{(2)})f(x_{(1)}), \qquad x_{(1)} < x_{(2)} < x_{(3)}. \tag{21.9}$$

Note that the portions of the density $h(x_{(1)}, x_{(2)}, x_{(3)})$ in (21.4)–(21.9) are identical and over the same region $x_{(1)} < x_{(2)} < x_{(3)}$. Since the six regions $x_1 < x_2 < x_3$, and so on, are disjoint and, except for sets having probability zero, exhaust the three-dimensional space $-\infty < x_1 < \infty$; $-\infty < x_2 < \infty$; $-\infty < x_3 < \infty$, it follows that $h(x_{(1)}, x_{(2)}, x_{(3)})$ is obtained by summation over the regions. Hence,

$$h(x_{(1)}, x_{(2)}, x_{(3)}) = 6f(x_{(1)})f(x_{(2)})f(x_{(3)}), \quad -\infty < x_{(1)} < x_{(2)} < x_{(3)} < \infty. \tag{21.10}$$

or

$$h(x_{(1)}, x_{(2)}, x_{(3)}) = 3!f(x_{(1)})f(x_{(2)})f(x_{(3)}), \quad -\infty < x_{(1)} < x_{(2)} < x_{(3)} < \infty. \tag{21.11}$$

To extend this result to a sample of size n, we need only to observe that there are $n!$ regions of the kind $x_1 < x_2 < \cdots < x_n$ obtained by permuting the x_i, $i = 1, \ldots, n$. Accordingly, we have proven the following theorem.

THEOREM 21.1

Let $X_1, \ldots, X_n$ be a sample drawn at random from the density $f(x)$, $-\infty > x < \infty$. The joint density of the order statistics $X_{(1)}, \ldots, X_{(n)}$ is

$$h(x_{(1)}, \ldots, x_{(n)}) = n!f(x_{(1)}) \cdots f(x_{(n)}),$$
$$-\infty < x_{(1)} < \cdots < x_{(n)} < \infty. \qquad (21.12)$$

The probability of the *ordered* sample is therefore seen to be given by $h(x_{(1)}, \ldots, x_{(n)})\, dx_{(1)} \cdots dx_{(n)}$.

We turn now to another result of interest and importance.

THEOREM 21.2

Let $X_1, \ldots, X_n$ be a sample of size n drawn at random from a continuous population with density function $f(x)$ and distribution function $F(x)$. Let $X_{(1)}, \ldots, X_{(n)}$ be the corresponding order statistics, and let

$$U_i = \int_{-\infty}^{X_{(i)}} f(z)\, dz$$
$$= F(X_{(i)}). \qquad (21.13)$$

Then the joint density function of the random variable $U_1, \ldots, U_n$ is

$$g(u_1, \ldots, u_n) = n!, \qquad 0 < u_1 < \cdots < u_n < 1, \qquad (21.14)$$

which does not depend on $f(x)$.

PROOF. We begin with the joint density of $X_{(1)}, \ldots, X_{(n)}$, which from the preceding theorem is equal to

$$h(x_{(1)}, \ldots, x_{(n)}) = n!f(x_{(1)}) \cdots f(x_{(n)}), \qquad (21.15)$$

and then transform the $x_{(i)}$ to u_i according to (21.13). From the formula for a change-in-variables, we will have for the joint density of $U_1, \ldots, U_n$,

$$g(u_1, \ldots, u_n) = h[\psi_1(u_1), \ldots, \psi_n(u_n)]\,|J|, \qquad (21.16)$$

where ψ_i is the inverse transformation from u_i to $x_{(i)}$ and J is the Jacobian of these transformations. However, in view of the fact that

$$\frac{\partial(x_{(1)}, \ldots, x_{(n)})}{\partial(u_1, \ldots, u_n)} = \left[\frac{\partial(u_1, \ldots, u_n)}{\partial(x_{(1)}, \ldots, x_{(n)})}\right]^{-1}, \qquad (21.17)$$

we have from (21.13),

$$J = \frac{1}{f[\psi(u_1)]} \cdots \frac{1}{f[\psi(u_n)]}. \tag{21.18}$$

Hence,

$$g(u_1, \ldots, u_n) = n!, \qquad 0 < u_1 < \cdots < u_n < 1, \tag{21.19}$$

as was to be shown.

Theorem 21.1 provides the joint density function of the ordered sample. We shall now derive a result that is more general than Theorem 21.1 and which will be used in the next section to obtain the distributions of several order statistics that are of frequent practical interest.

Let $r_1, r_2, \ldots, r_k$ be a set of integers with the property

$$1 \le r_1 < r_2 < \cdots < r_k \le n, \tag{21.20}$$

and let $X_{(r_1)}, X_{(r_2)}, \ldots, X_{(r_k)}$ be the order statistics corresponding to $r_1, r_2, \ldots, r_k$. We seek the joint density function, $h(x_{(r_1)}, x_{(r_2)}, \ldots, x_{(r_k)})$, of these order statistics. Of course, for $k = n$, h is as already obtained in Theorem 21.1. For illustration, assume that $n = 8$, $r_1 = 3$, $r_2 = r_k = 6$, so that we want $h(x_{(3)}, x_{(6)})$, which is the joint density function of the third and sixth ordered X_i. The necessary conditions on the remaining 6 $X_{(i)}$ in the sample are: $X_{(1)}, X_{(2)} < X_{(3)}$; $X_{(3)} < X_{(4)}, X_{(5)} < X_{(6)}$; $X_{(7)}, X_{(8)} > X_{(6)}$.

In general, $h(x_{(r_1)}, \ldots, x_{(r_k)})$ requires the following allocation of the $nX_{(i)}$: There is one $X_{(i)}$ at each of $x_{(r_1)}, x_{(r_2)}, \ldots, x_{(r_k)}$, $r_1 - 1$ less than $x_{(r_1)}$, $r_2 - 1$ less than $x_{(r_2)}, \ldots, r_k - 1$ less than $x_{(r_k)}$, and finally $n - r_k$ greater than $x_{(k)}$. The probability of *one* such allocation is as follows:

$$\left[\left\{\int_{-\infty}^{x(r1)} f(z)\,dz\right\}^{r_1 - 1}\left\{\int_{x(r1)}^{x(r2)} f(z)\,dz\right\}^{r_2 - r_1 - 1} \cdots \left\{\int_{(r_k)}^{\infty} f(z)\,dz\right\}^{n - r_k}\right]$$
$$\cdot f(x_{(r_1)})f(x_{(r_2)}) \cdots f(x_{(r_k)})\,dx_{(r_1)}\,dx_{(r_1)}\,dx_{(r_2)} \cdots dx_{(r_k)}. \tag{21.21}$$

The square bracket in the first line gives the probability of $r_1 - 1$ sample variates in the cell between $-\infty$ and $x_{(r_1)}$, $r_2 - r_1 - 1$ sample variates in the cell between $x_{(r_1)}$ and $x_{(r_2)}$, and so on, while the second line gives the probability of sample variates falling in intervals of length $dx_{(r_i)}$ around $x_{(r_1)}, \ldots, x_{(r_k)}$. However, n variates taken n at a time can be allocated among all of these cells in

$$\frac{n!}{1!1!\cdots 1!(r_1 - 1)!(r_2 - r_1 - 1)!\cdots(n - r_k)!}$$

different and equally likely ways. Hence, we have proven the following.

THEOREM 21.3
Let $X_1, \ldots, X_n$ be a sample drawn at random from the continuous density $f(x)$, let $r_1, r_2, \ldots, r_k$ be a set of integers such that

$$1 \leq r_1 < r_2 < \cdots < r_k \leq n,$$

and let $X_{(r_1)}, \ldots, X_{(r_k)}$ be the order statistics corresponding to the r_i. The joint density function of $X_{(r_1)}, \ldots, X_{(r_k)}$ is then

$$h(x_{(r_1)}), \ldots, x_{(r_k)} = \frac{n!}{(r_1 - 1)!(r_2 - r_1 - 1)! \cdots (n - r_k)!}$$

$$\times \left\{ \int_{-\infty}^{x(r_1)} f(z)\, dz \right\}^{r_1 - 1} \cdot \left\{ \int_{x(r_1)}^{x(r_2)} f(z)\, dz \right\}^{r_2 - r_1 - 1}$$

$$\cdots \left\{ \int_{x(r_k)}^{\infty} f(z)\, dz \right\}^{n - r_k} f(x_{(r_1)}) f(x_{(r_2)} \cdots f(x_{(r_k)}),$$

$$-\infty < x_{(r_1)} < \cdots < x_{(r_k)} < \infty. \tag{21.22}$$

We shall now consider some special cases. First, observe that, as was noted earlier, if $r_1 = 1$, $r_2 = 2$, $\ldots$, $r_k = n$, then the density function in (21.22) reduces to the density function (21.12) in Theorem 21.1. Next, suppose that we are interested in the distribution of the sample maximum, $X_{(n)}$. In this case, we will have $r_1 = r_k = n$, and (21.22) reduces to

$$h(x_{(n)}) = n \left\{ \int_{-\infty}^{x(n)} f(z)\, dz \right\}^{n-1} f(x_{(n)})\, dx_{(n)}, \tag{21.23}$$

which is a function of the sample size n and the population density $f(x)$.

It is useful to note that the expression in (21.23) can also be arrived at directly. The probability that any element of sample drawn at random from a population with density function $f(x)$ is less than or equal to a number $x_{(n)}$ is $F(x_{(n)})$, and since the elements of the sample are independent, the probability that all n elements of a sample are less than or equal to $x_{(n)}$ is $F^n(x_{(n)})$. However, this is identical with the probability $H(x_{(n)})$ that the largest element of the sample $X_{(n)}$ is less than or equal to $x_{(n)}$, so that

$$H(x_{(n)}) = F^n(x_{(n)}). \tag{21.24}$$

Differentiation of (21.25) with respect to $x_{(n)}$ yields (21.23).

As another example, we shall consider the distribution of the sample median, the middle value of the ordered observations, which we shall denote by X_d. For present purposes, we shall assume that n is odd, so that it can be written as $n = 2s + 1$. In this case, $r_1 = r_k = (n+1)/2 = s + 1$, and expression (21.22) accordingly reduces to

$$h(x_d) = \frac{(2s+1)!}{s!\,s!}\left\{\int_{-\infty}^{x_d} f(z)\,dz\right\}^s \left\{\int_{x_d}^{\infty} f(z)\,dz\right\}^s f(x_d). \qquad (21.25)$$

It can be shown[2] that as n gets large, the distribution of the sample median approaches a normal distribution with mean μ_d and variance

$$\frac{1}{4n[f(\mu_d)]^2},$$

where f is the density function of the population being sampled and μ_d is the population median. In the case where $f(x)$ is $N(\mu, \sigma^2)$, then the limiting distribution of x_d is normal with mean μ and variance $\pi\sigma^2/2n$.

The final distribution to be derived in this section is the distribution of the sample range $X_{(n)} - X_{(1)}$. We begin with the joint density function of $X_{(1)}$ and $X_{(n)}$, which is obtained from (21.22) by setting $r_1 = 1$ and $r_2 = r_k = n$. Thus,

$$h(x_{(1)}, x_{(n)}) = n(n-1)\left\{\int_{x_{(1)}}^{x_{(n)}} f(z)\,dz\right\}^{n-2} f(x_{(1)})f(x_{(n)}). \qquad (21.26)$$

However, we seek the distribution of a specific function of $X_{(1)}$ and $X_{(n)}$, namely, $X_{(n)} - X_{(1)}$. Accordingly, let $r = x_{(n)} - x_{(1)}$ and $v = x_{(1)}$. Since the Jacobian of this transformation is 1, the joint density function of R and V is therefore

$$g(r, v) = n(n-1)\left\{\int_{v}^{r+v} f(z)\,dz\right\}^{n-2} f(r+v)f(v),$$

$$-\infty < v < \infty; \qquad 0 < r < \infty. \qquad (21.27)$$

The density function of R is then obtained through integration over v. Practical application requires, of course, knowledge of the form of f, the density function of the population.

In the case where X is normal, the distribution of R has been derived by L. H. C. Tippett, and tables of the distribution function

$$G(w) = \int_0^w g(w)\,dw, \qquad (21.28)$$

where $w = r\sigma$, have been developed for $n = 2, \ldots, 20$ by Pearson and Hartley (1941–1942).

[2] See Wilks (1962, p. 273).

21.3 A DISTRIBUTION-FREE CONFIDENCE INTERVAL FOR THE MEDIAN

It was remarked earlier that order statistics find frequent use in developing procedures of statistical inference that do not depend on the distribution of the population being sampled. This section provides an illustration through the construction of a confidence interval for the median. The procedure is based on the binomial distribution.

Let μ_d be the median of the distribution that describes the population being sampled. By definition of the median, the probability that an observation falls to the left or to the right of μ_d is $\frac{1}{2}$ in each case. Accordingly, the probability that exactly i observations in a sample of size n drawn at random fall to the left of μ_d is just

$$\binom{n}{i}\left(\frac{1}{2}\right)^n.$$

Hence, the probability that $X_{(r)}$, the rth-order statistic, exceeds μ_d is given by

$$P(X_{(r)} > \mu_d) = \sum_{i=0}^{r-1}\binom{n}{i}\left(\frac{1}{2}\right)^n. \tag{21.29}$$

Similarly, the probability that the sth-order statistic is less than μ_d is given by

$$P(X_{(s)} < \mu_d) = \sum_{i=s}^{n}\binom{n}{i}\left(\frac{1}{2}\right)^n. \tag{21.30}$$

Let us now assume that $s > r$ and then add (21.29) and (21.30)

$$P(X_{(r)} > \mu_d) + P(X_{(s)} < \mu_d) = \sum_{i=0}^{r-1}\binom{n}{i}\left(\frac{1}{2}\right)^n + \sum_{i=s}^{n}\binom{n}{i}\left(\frac{1}{2}\right)^n. \tag{21.31}$$

Since necessarily $X_{(r)} \leq X_{(s)}$, this expression gives the probability that the interval $X_{(r)}$ to $X_{(s)}$ *does not* include μ_d. Consequently, subtraction of (21.31) from 1 yields the probability that the interval $X_{(r)}$ to $X_{(s)}$ *does* include μ_d,

$$\begin{aligned} P(X_{(r)} < \mu_d < X_{(s)}) &= 1 - \sum_{i=0}^{r-1}\binom{n}{i}\left(\frac{1}{2}\right)^n - \sum_{i=s}^{n}\binom{n}{i}\left(\frac{1}{2}\right)^n \\ &= \sum_{i=r}^{s=1}\binom{n}{i}\left(\frac{1}{2}\right)^n. \end{aligned} \tag{21.32}$$

Hence, we have proven the following.

THEOREM 21.4
Let $X_{(1)}, \ldots, X_{(n)}$ be the order statistics of a sample of size n drawn at random from a continuous population with density function $f(x)$. Let $s > r$. Then the probability that the random interval $X_{(r)}$ to $X_{(s)}$ covers the median μ_d of $f(x)$ is

$$P(X_{(r)} < \mu_d < X_{(s)}) = \sum_{i=r}^{s-1} \binom{n}{i} \left(\frac{1}{2}\right)^n. \tag{21.33}$$

In practice, s is ordinarily taken to be $n - r + 1$, so that the rth observations in order of magnitude from the bottom and from the top are used. In this case, (21.33) becomes

$$P(X_{(r)} < \mu_d < X_{(n-r)}) = \sum_{i=r}^{n-r} \binom{n}{i} \left(\frac{1}{2}\right)^n. \tag{21.34}$$

If a confidence interval on μ_d of size $1 - \alpha$ is desired, r is usually chosen as the largest integer for which the quantity in (27.34) is not less than $1 - \alpha$. Denoting this integer as r', the confidence interval $X_{(r')}$ to $X_{(n-r')}$ then satisfies

$$P(X_{(r')} < \mu < X_{(n-r')}) \geq 1 - \alpha. \tag{21.35}$$

21.4 TOLERANCE LIMITS

We shall now turn to a result that is due originally to S. S. Wilks and which has found widespread application in the formulation of tolerance limits for industrial processes. For illustration, suppose that an automatic machine in a ball-bearing factory is designed to manufacture bearings 0.25 in. in diameter. The bearings from the machine are considered acceptable if the diameter falls between the limits 0.249 and 0.251 in. Production from the machine is regularly monitored by measuring each day the diameters of a sample of bearings drawn at random and computing statistical tolerance limits, L_1 and L_2, from the sample. If L_1 is above 0.249 and L_2 below 0.251, the production is accepted. How large should the sample be in order that one can be assured with probability 0.90 that the statistical tolerance limits will contain at least 99 percent of the population of bearing diameters being turned out by the machine? The theorem to be proven shortly provides a solution to this problem.

More generally, let $f(x)$ be a density function describing a continuous population, and suppose that on the basis of a sample of n observations we wish to determine two numbers, L_1 and L_2, such that we can assert with some specified probability, $1 - \alpha$, that β, of the area beneath $f(x)$ lies between L_1 and L_2. In other words, we want to find two functions

$L_1 = L_1(X_1, \ldots, X_n)$ and $L_2 = L_2(X_1, \ldots, X_n)$, where $X_1, \ldots, X_n$ denotes a sample of size n drawn at random from $f(x)$, such that the probability that

$$\int_{L_1}^{L_2} f(x)\,dx > \beta \tag{21.36}$$

is at least $1 - \alpha$.

Although there are many functions L_1 and L_2 that could be used, we shall let $L_1 = X_{(1)}$, the smallest of the observations, and let $L_2 = X_{(n)}$, the largest of the observations. Solution of our problem thus requires obtaining the distribution of the area beneath $f(x)$ bounded by the order statistic $X_{(1)}$ and $X_{(n)}$.

THEOREM 21.5

Let $X_1, \ldots, X_n$ be a sample of size n from the continuous density $f(x)$ and let W be the area beneath $f(x)$ between the largest and the smallest observations, that is, between $X_{(n)}$ and $X_{(1)}$. Then the density function of W is given by

$$g(w) = n(n-1)w^{n-2}(1-w), \qquad 0 < w < 1. \tag{21.37}$$

PROOF. From (21.26), we have for the joint density of $X_{(1)}$ and $X_{(n)}$,

$$h(x_{(1)}, x_{(n)}) = n(n-1)\left\{\int_{x_{(1)}}^{x_{(n)}} f(z)\,dz\right\}^{n-2} f(x_{(1)})f(x_{(n)}). \tag{21.26}$$

We now undertake a change-of-variables from $X_{(1)}$ and $X_{(n)}$ to U and W, where

$$u = \int_{-\infty}^{x_{(1)}} f(z)\,dz, \qquad w = \int_{x_{(1)}}^{x_{(n)}} f(z)\,dz. \tag{21.38}$$

The Jacobian of this transformation is $f(x_{(1)})f(x_{(n)})$. However, since

$$\frac{\partial(x_{(1)}, x_{(n)})}{\partial(u, w)} = \left[\frac{\partial(u, w)}{\partial(x_{(1)}, x_{(n)})}\right]^{-1}, \tag{21.39}$$

the joint density of u and w reduces to

$$g(u, w) = n(n-1)w^{n-2}, \qquad 0 < u < (1-w) < 1. \tag{21.40}$$

Finally, integration over u yields the density function of W,

$$g(w) = n(n-1)w^{n-2}(1-w), \qquad 0 < w < 1, \tag{21.37}$$

as was to be shown.[3]

[3] The density function in (21.37) is a member of the well-known beta family of distributions. The beta distribution has two parameters and its density function is given by

$$F(x; \alpha, \beta) = \frac{(\alpha+\beta+1)!}{\alpha!\,\beta!}\,x^{\alpha}(1-x)^{\beta}, \qquad 0 < x < 1, \qquad \alpha, \beta > -1.$$

In the present case $\alpha = n-2$ and $\beta = 1$.

Let us now return to the question raised at the beginning of this section: How large a sample should be taken in order that the probability is 0.90 that 99 percent of a day's output of bearings will have diameters between the largest and smallest observations? We have $1 - \alpha = 0.90$ and $\beta = 0.99$, and we wish to determine n such that

$$P(W > \beta) = 1 - \alpha, \tag{21.41}$$

where the density of W is given by (21.37). Accordingly,

$$\begin{aligned}(1 - \alpha) &= P(W > \beta) \\ &= \int_\beta^1 n(n-1)w^{n-2}(1-w)\,dw \\ &= 1 - n\beta^{n-1} + (n-1)\beta^n.\end{aligned} \tag{21.42}$$

Substituting for α and β, we get the equation

$$0.90 = 1 - n(0.99)^{n-1} + (n-1)(0.99)^n, \tag{21.43}$$

which can be solved to determine n.

We shall close this brief introduction to order statistics by noting that Theorem 21.5 is in fact a special case of the following more general theorem:

THEOREM 21.6

Let $X_{(1)}, \ldots, X_{(n)}$ be the order statistics of a sample drawn at random from the continuous density function $f(x)$ and let W_{rs} denote the area beneath $f(x)$ between $X_{(r)}$ and $X_{(s)} (s > r)$. The density function of W_{rs} is then given by

$$g(w_{rs}) = \frac{n!}{(s-r-1)!\,(n-s+r)!}(w_{rs})^{s-r-1}(1-w_{rs})^{n-s+r}, \quad 0 < w_{rs} < 1, \tag{21.44}$$

which is independent of $f(x)$.

This theorem, it will be noted, is the theorem cited at the end of Section 21.1. Its proof is similar to the proof of Theorem 21.5, and is left to the interested reader.

REFERENCES

Freeman, H., *Introduction to Statistical Inference*, Addison-Wesley, 1963, chap. 22.

Gumbel, E. J., *Statistics of Extremes*, Columbia University Press, 1958.

Kendall, M. G. and Stuart, A., *The Advanced Theory of Statistics*, vol. 1, Charles Griffin and Co., 1958, chap. 14.

Mood, A. M. and Graybill, F. A., *Introduction to the Theory of Statistics*, 2nd ed., McGraw-Hill, 1963, chap. 16.

Pearson, E. S. and Hartley, H. O., "The Probability Integral of the Range in Samples of n Observations from a Normal Population," *Biometrika*, vol. 32 (1941–1942), pp. 301–310.

Wilks, S. S., *Mathematical Statistics*, Wiley, 1962, chap. 9.

CHAPTER 22

STATISTICAL INFERENCE IN PERSPECTIVE

The approach to statistical inference presented in the preceding six chapters has, with very little exception, followed a classical format. Minimum variance, maximum likelihood, and least squares have been advanced as underlying principles of estimation, while hypothesis testing has been discussed within the framework developed by Neyman and Pearson. The purpose of this final chapter is twofold: first, to provide the reader with a brief synopsis of statistical inference based on Bayesian principles, and then, with this as introduction, to acquaint him with the issues which fuel continual debate about the foundations of statistical inference. The goal is to present these issues in sufficient clarity that the reader can begin to form his own judgment as to the viewpoint he wishes to adopt.

22.1 A BRIEF SURVEY OF BAYESIAN INFERENCE

Thus far in our study of statistical inference, it has been assumed that the unknown parameters that we are interested in saying something about are nonrandom. In this section, we shall adopt the view that the parameters are themselves random variables with a joint distribution which, for the time being, we shall take to be known. As before, let the population of interest be described by the density function $f(x, \theta)$. We now assume that, prior to the availability of any sample evidence, θ is a random variable with known density function $h(\theta)$.[1] Let Ω be the space over which θ is defined, and let E be the sample space of X. Suppose, now, that a sample of size n is drawn at random from $f(x, \theta)$; how does the availability of this sample information affect the distribution of θ?

The answer to this question is given by the theorem of Bayes, Theorem 3.1. Let $h(\theta|x)$ be the conditional distribution of θ given the information provided by the sample $x' = (x_1, \ldots, x_n)$, and let $g(x|\theta)$ be the conditional density of $X_1, \ldots, X_n$ given θ. Then, from Theorem 3.1, we have

$$h(\theta|x) = \frac{f(x, \theta)}{\int_\Omega f(x, \theta)\, d\theta}$$

$$= \frac{g(x|\theta)h(\theta)}{\int_\Omega g(x|\theta)h(\theta)\, d\theta}. \tag{22.1}$$

Formula (22.1), which is simply Bayes theorem in terms of density functions, is the *sine qua non* of Bayesian inference. In Bayesian inference, we begin with a *prior distribution* of the parameter in question, and then proceed to draw a sample. The sample information is then combined via formula (22.1) with the prior distribution to produce a *posterior distribution* for the parameter. In this terminology, the prior distribution is represented by $h(\theta)$, while the posterior distribution is represented by $h(\theta|x)$.

In our discussion of Bayes theorem in Chapter 3, it was noted that its use in scientific endeavors is not uncontroversial because of the need to have a prior distribution for θ. In the final analysis, the attitude that we adopt in this controversy pretty much reduces to our interpretation of probability. If we hold the view that probability is an impersonal, objective quantity that is determined by structural characteristics of the phenomenon in question,[2] then Bayes theorem will find infrequent use, since in most situations an objective basis for determining the prior distribution will not be present. On the other hand, if we take the view that probability is essentially subjective in nature in that it measures the degree of belief on the part of the observer that the event in question is going to occur, then Bayes theorem can be used at will, since the observer is free to provide a prior distribution on the basis of his own personal judgment.

We shall go into the issues surrounding the different interpretations of probability in a later section of this chapter, but for now let us accept the existence of a prior distribution on θ, however arrived at, at face value and ask ourselves the question of what is thereby gained. Recall from Chapter

[1] Since θ is now viewed as a random variable, we must interpret $f(x, \theta)$ as the joint density of X and θ.

[2] The classical and relative frequency definitions of probability provide appropriate bases for this view.

18 that it was noted that a statistician working within the framework of classical inference cannot make a statement of the form:

The probability that θ lies within the interval θ_1 to θ_2 is equal to $1 - \alpha$. (22.2)

The best that the classical statistician can do in this regard is to say that if he were to sample repeatedly and to calculate θ_1 and θ_2 for each sample, then in $1 - \alpha$ proportion of the time his interval θ_1 to θ_2 would include the true value of θ. However, for the Bayesian statistician, the statement in (22.2) is straightforward, for the probability in question will be given by choosing θ_1 and θ_2 so as to satisfy

$$1 - \alpha = \int_{\theta_1}^{\theta_2} h(\theta \mid x)\, d\theta. \tag{22.3}$$

A second advantage conferred by a Bayesian framework is that it provides for a systematic assimilation of a priori knowledge. This is accomplished through the construction of the prior distribution on θ. The reader will recall our discussion in Chapter 18 of the problems connected with the selection of hypotheses in the Neyman-Pearson framework. The advice offered was to formulate the null hypothesis in such a way as to incorporate into it existing knowledge about the parameter in question. In a Bayesian format, there is ordinarily little interest in testing one hypothesis directly against another; the interest instead is usually in calculating the probability that such and such a hypothesis is true. Let us illustrate this with one of the examples considered in Chapter 18.

Suppose, once again, that we are interested in ascertaining the marginal propensity to consume in Canada. As was noted in Chapter 18, existing knowledge suggests that the marginal propensity to consume in an economy such as Canada's should be of the order of 0.92. In a Neyman-Pearson framework, 0.92 would therefore be the logical candidate for H_0. However, existing knowledge does not say that the Canadian marginal propensity to consume is 0.92 with certainty, but only that it is highly likely to be in the neighborhood of 0.92. Suppose, for the sake of argument, that we are willing to summarize our existing (i.e., a priori) beliefs as to the value of the Canadian marginal propensity to consume according to the following probability distribution (let β denote the MPC):

H_i	$p(H_i)$
$H_1: \beta < 0.85$	0.02
$H_2: 0.85 \leq \beta < 0.88$	0.03
$H_3: 0.88 \leq \beta < 0.91$	0.10
$H_4: 0.91 \leq \beta < 0.93$	0.70
$H_5: 0.93 \leq \beta < 0.96$	0.10
$H_6: 0.96 \leq \beta < 0.99$	0.03
$H_7: \beta \geq 0.99$	0.02

Our strong belief that β is in the neighborhood of 0.92 is expressed by assigning a prior probability of 0.70 to H_4.

Suppose, now, that we have a sample consisting of 128 quarterly time-series observations on consumption (c) and disposable income (y) covering the period 1947 through 1972 with which we proceed to estimate the model

$$c = a + by + u, \tag{22.4}$$

u being a random variable satisfying the Gauss-Markov assumptions for least squares. We then proceed to combine via fomula (21.1) the sample information on the distribution of β with the prior distribution given above.[3] The result, let us suppose, is the posterior distribution given below (x represents the sample):

H_i	$P(H_i \mid x)$
$H_1: \beta < 0.85$	0.01
$H_2: 0.85 \leq \beta < 0.88$	0.01
$H_3: 0.88 \leq \beta < 0.91$	0.05
$H_4: 0.91 \leq \beta < 0.93$	0.86
$H_5: 0.93 \leq \beta < 0.96$	0.05
$H_6: 0.96 \leq \beta < 0.99$	0.01
$H_7: \beta \geq 0.99$	0.01

On comparing this distribution with the prior distribution on β given above, we find the probability of β lying in the interval 0.91 to 0.93 to have increased, on the basis of the (hypothetical) sample evidence, to 0.86. In other words, the sample evidence in this case has strengthened our belief that the marginal propensity to consume in Canada lies in a small neighborhood of 0.92.

Although the Bayesian statistician assumes a prior probability distribution for θ over Ω, it does not follow that he necessarily views θ as a random variable. Like the classical statistician, the Bayesian can view θ as being unknown, but fixed. The prior distribution in this case summarizes his personal assessments (which may themselves be based on impersonal, objective evidence) as the likelihood that θ lies in different regions of Ω. The classical statistician for the most part is unwilling to make these assessments for it requires a nonobjective interpretation of probability. For him, the probability is either 1 or 0 that the true θ lies in any particular region in Ω. For either θ lies in the region in question or it does not, and accordingly it does not make any sense to him, since an objective basis is lacking, to say that the probability is such and such that θ lies in some particular region.

Before leaving this brief survey of Bayesian inference, let us return for a

[3] We pass over the mechanics of how this is done. The interested reader is referred to Zellner (1971, chap. 3).

moment to formula (22.1) and note that for x fixed and θ variable $g(x|\theta)$ in the numerator of the right-hand side defines the likelihood function for the sample $L(\theta)$. Accordingly, we can write $h(\theta|x)$ as

$$h(\theta|x) = kL(\theta)h(\theta), \tag{22.5}$$

where

$$k^{-1} = \int_{\Omega} g(x|\theta)h(\theta)\, d\theta. \tag{22.6}$$

The posterior distribution of θ is thus seen to be proportional to the product of the likelihood function and the prior distribution. If the prior distribution is "gentle" relative to the likelihood function, then the shape of the posterior distribution will be determined primarily by the sample. On the other hand, if the reverse is the case, then $h(\theta|x)$ will be determined primarily by the prior distribution. In the former case, the prior distribution can be said to be relatively uninformative, while in the latter case it is the sample that is uninformative.

From formula (22.5), it is clear that for the Bayesian statistician the sample information is summarized completely by the likelihood function. However, in situations where there exists a sufficient statistic, the sample information can be further reduced to the distribution of the sufficient statistic. That this is so follows from an application of Theorem 16.2. Let $t(X)$ be the sufficient statistic. Application of Theorem 16.2 then yields for $g(x|\theta)$ in expression (22.1),

$$g(x|\theta) = m[t(x)|\theta] \cdot q(x), \tag{22.7}$$

where the function $q(x)$ does not depend on θ. Formula (22.1) now becomes

$$\begin{aligned} h(\theta|x) &= \frac{m[t(x)|\theta]q(x)h(\theta)}{\int_{\Omega} m[t(x)|\theta]q(x)h(\theta)\, d\theta} \\ &= \frac{m[t(x)|\theta]h(\theta)}{\int_{\Omega} m[t(x)|\theta]h(\theta)\, d\theta} \qquad \text{[since } q(x) \text{ does not depend on } \theta\text{]} \\ &= k^* m[t(x)|\theta]h(\theta), \end{aligned} \tag{22.8}$$

where

$$(k^*)^{-1} = \int_{\Omega} m[t(x)|\theta]h(\theta)\, d\theta. \tag{22.9}$$

Since $m[t(x)|\theta]$ can be transformed into the density function for $t(x)$ (by dividing m by its integral over θ), we thus find that the density function of the posterior distribution can be expressed (apart from a constant) as the product of the density functions of the sufficient statistic and the prior distribution. This is an important result because in many cases the distribution of the sufficient statistic is simpler than the likelihood function.

22.2 SOME GUIDING PRINCIPLES OF STATISTICAL INFERENCE

Although Bayesian inference operates from very different principles than those provided by the classical framework of Neyman and Pearson, the reader should not get the impression that inferential statistics is composed of two, and only two, camps. There are many statisticians who do not find the Neyman-Pearson format to their liking, but who nevertheless fail to qualify as full-blown Bayesians. In this section, we shall discuss some of the principles which guide these practitioners. The discussion which follows is based primarily upon Raiffa (1968, pp. 286–289).[4]

The first principle that we shall discuss is the so-called *likelihood principle*. This principle, which should not be confused with the principle of maximum likelihood studied in Chapter 17, asserts that any information about an experiment and its outcome over and above the likelihood function is irrelevant for inferences or decisions about the population parameter. As was noted near the end of the preceding section, Bayesians accept this principle. As an example, let us consider the one discussed by Raiffa. Suppose that a medical researcher is interested in the unknown proportion p of patients who will recover from a new surgical operation, and assume that he has performed the operation three times with outcomes (S, S, F). The likelihood function $L(p)$ is then

$$L(p) = p^2(1 - p), \tag{22.10}$$

and it is irrelevant whether the experiment called for exactly three patients or whether he performed the operation until he encountered his first failure, or whether he was summoned away to some administrative post. In each case the likelihood function is the same.

Many basic statistical procedures do not satisfy the likelihood principle including tests of significance and the estimation of confidence intervals. Statisticians who are not convinced of the reasonableness of the likelihood principle feel that the experimental frame of reference cannot be ignored.

[4] From H. Raiffa, *Decision Analysis: Introductory Lectures on Choices under Uncertainty*, Addison-Wesley, 1968, chap. 10.

In particular, they feel that the experimental frame of reference is needed to supply conditional probability assessments for possible sample outcomes which, in fact, have not occurred. To illustrate the difference in viewpoints, suppose that a sociologist is convinced that students who live off campus date more frequently than those who live in dormitories and suppose that he samples students living in dormitories and off campus in equal proportions until he builds a favorable case for his point of view. He then stops and reports the results based on *all* of the students that he sampled. The likelihoodist would say that the fact that the sociologist stopped the experiment when he did makes no difference to the conclusions drawn from the sample, since any other stopping rule yielding the same size of sample would yield an identical likelihood function. The non-likelihoodist, however, would object violently to this procedure, since in his view significance tests, confidence intervals, and so on, must reflect the experimental frame of reference, and in this case, the biases of the investigator.

A second principle, which we shall note only briefly since it is closely related to the likelihood principle, is the *principle of sufficiency*. This principle states that any information about an experiment and its outcome over and above that provided by a sufficient statistic is irrelevant for inferences or decisions about the population parameter. Like the likelihood principle, Bayesians obviously also accept the principle of sufficiency.

Yet a third principle is the *principle of conditionality*, put forward by Birnbaum (1962). Suppose that there are several experiments, each of which provides some information about an unknown state parameter θ, and assume that we choose a particular experiment by a randomized device (that has nothing to do with θ) and observe the outcome. The principle of conditionality then states that the inference to be drawn from the outcome of this randomly chosen experiment shall *not* depend on the probabilities of the randomized device leading up to it.

Once again, it will be useful to illustrate with the example considered by Raiffa. Mrs. A visits her doctor wishing to know whether she is pregnant. Suppose that the physician can use either of two tests, neither of which is infallible, to ascertain whether this is the case: a frog test and a rabbit test. Assume that the physician decides to toss a coin to determine which test to use—heads, he will use the frog test, and tails, he will use the rabbit test. The coin comes up heads, and he proceeds to administer the frog test to Mrs. A and then analyzes the results. At this point, his rather energetic assistant enters the room and shouts, "Wait! Don't draw any inferences or conclusions yet! I have just examined the coin that was tossed and concluded that it is biased." Should knowledge that the coin that was used was biased influence the physician's evaluation of the evidence of the frog test that had just been applied? The principle of conditionality would say, "No."

We shall conclude this section by noting the relationships among the three principles that have just been discussed. To facilitate matters, let us denote the three principles by L, S, and C, respectively. To begin with, it is clear that in situations where a sufficient statistic exists, L implies S. In addition, L also implies C. Much less evident, however, is the fact that S and C together imply L.[5] This is to say that if we adopt the principle of sufficiency and the principle of conditionality, then we are implicitly adopting the likelihood principle.

22.3 STATISTICAL INFERENCE AND PLAUSIBLE REASONING

Our final effort in this concluding chapter will be to place statistical inference in as broad a perspective as possible, by viewing it as an aspect of the general problem of the interpretation of evidence. The discussion which follows is based primarily on Chapters XIII and XV of Polya (1968).

Let A be a conjecture and suppose that we wish to find out whether it is true. (For example, A may be a mathematical proposition.) Let B be a consequence of A. For the moment, we do not know whether A or B is true. However, suppose that we do know that A implies B. Accordingly, we undertake to check B. If B turns out to be false, we can conclude that A is false also. This is completely clear, for we have a classical elementary form of reasoning, the "modus tollens" of the classical syllogism:

$$\begin{array}{c} A \text{ implies } B \\ B \text{ false} \\ \therefore \\ A \text{ false.} \end{array} \tag{22.11}$$

Expression (22.11) is an example of *demonstrative inference* of a well-known type.

What happens, however, if B turns out to be true? There is no demonstrative conclusion, for the verification of B does not prove the truth of the conjecture A. Yet such verification clearly renders A more credible. In this case, we have an instance of *plausible inference*:

$$\begin{array}{c} A \text{ implies } B \\ B \text{ true} \\ \therefore \\ A \text{ more credible.} \end{array} \tag{22.12}$$

[5] For a proof, see Birnbaum (1962, pp. 284–285). However, see Durbin (1970) and Birnbaum (1970).

Expression (22.12) is sometimes referred to as the *fundamental inductive pattern.*

Let us now suppose that before B is examined, it is considered to be improbable. Assume, however, that it is found to be true. In this case, we would reason:

$$\begin{gathered} A \text{ implies } B \\ B \text{ improbable} \\ B \text{ true} \\ \therefore \\ A \text{ much more credible.} \end{gathered} \tag{22.13}$$

On the other hand, let us suppose that before it is examined, B is considered to be probable. Again, assume that it is found to be true. Now we would reason:

$$\begin{gathered} A \text{ implies } B \\ B \text{ probable} \\ B \text{ true} \\ \therefore \\ A \text{ somewhat more credible.} \end{gathered} \tag{22.14}$$

Expression (22.13) and (22.14) are examples of *shaded* plausible inference.

Next, let us suppose that instead of being a consequence of A, B is a ground for A. That is, we now have

$$B \text{ implies } A.$$

If B is true, then

$$\begin{gathered} B \text{ implies } A \\ B \text{ true} \\ \therefore \\ A \text{ true.} \end{gathered} \tag{22.15}$$

Like expression (22.9), this is an instance of demonstrative inference of a classic form. However, if B is false, then

$$\begin{gathered} B \text{ implies } A \\ B \text{ false} \\ \therefore \\ A \text{ less credible.} \end{gathered} \tag{22.16}$$

We have thus returned to the world of plausible inference. Finally, in the cases, where B is improbable and probable, respectively, we have analogous to (22.13) and (22.14):

$$\begin{array}{c} B \text{ implies } A \\ B \text{ improbable} \\ B \text{ false} \\ \therefore \\ A \text{ somewhat less credible.} \end{array} \tag{22.17}$$

$$\begin{array}{c} B \text{ implies } A \\ B \text{ probable} \\ B \text{ false} \\ \therefore \\ A \text{ much less credible.} \end{array} \tag{22.18}$$

Expressions (22.17) and (22.18) are clearly instances of shaded inference.

The reader will note that of the eight examples considered, only (22.11) and (22.15) lead to unequivocal conclusions; in all of the others, we can only conclude, after ascertaining the truth of B, that the conjecture A is either more or less credible. Deciding in these situations whether A is in fact true or false is, of course, a manifestation and the essence of the age-old problem of induction. In the study of physical, biological, and social phenomena, it is possible, as in (22.11), to prove demonstratively that a conjecture is false, but it is not possible, as in (22.15), to prove a conjecture to be correct. That is, we can demonstratively prove something to be wrong, but we cannot demonstratively prove it to be right. Proof in these endeavors can only proceed along inductive lines. The modern theory of probability has contributed materially to a clarification of the issues involved in inductive (or plausible) reasoning, and it is to this that we shall now turn.

We begin by letting the symbol $P(A)$ denote the strength of the evidence for (i.e., the credibility of) a conjecture A. We shall assume P to be a real-valued function which satisfies the axioms of probability listed in Chapter 3. For the moment, we shall not assume the actual numerical value of $P(A)$ to be known, unless A is known to be true, in which case we would have $P(A) = 1$; or known to be false, in which case we would have $P(A) = 0$. Similarly, let us assume that there are other conjectures B, C, and so on, to which we attach credibilities $P(B)$, $P(C)$, and so on.

As at the beginning of this section, let us now assume that B is known to be a consequence of A. Neither A nor B is known to be true, but we believe in them to a certain degree, and this is expressed by $P(A)$ and $P(B)$. Suppose that we tire of trying to establish the truth of A, and turn our attention to B. As we work with B, we sometimes see indications that it is true and at other times see indications that it is false. $P(B)$ varies accordingly. How does this affect our belief in A? The laws of probability provide the answer.

Since $P(AB) = P(BA)$, we have

$$P(A)P(B|A) = P(B)P(A|B). \tag{22.19}$$

However, since B is assumed to be a consequence of A,

$$P(B|A) = 1, \tag{22.20}$$

so that

$$P(A) = P(B)P(A|B). \tag{22.21}$$

This last equation answers our question, for it states that so long as we believe that B is a consequence of A, the credibility of A varies directly with the credibility of B.

Moreover, since $0 < P(B) < 1$, we can derive from Equation (22.21) the inequality

$$P(A) < P(A|B). \tag{21.22}$$

Now $P(A)$ and $P(A|B)$ represent the credibility of A before and after the proof of B, respectively; hence, this inequality is the formal statement of the principle expressed in (22.12): *The verification of a consequence renders a conjecture more credible.*

Finally, note from Equation (22.21) we also have

$$P(A|B) = \frac{P(A)}{P(B)}, \tag{21.23}$$

which is to say that the credibility of A after the verification of B is given by the ratio of the prior credibility of A to the prior credibility of B. If B is very credible to begin with, then the increase in credibility of A is less than what it would be if B were not very credible. This formalizes, it will be noted, the shaded inference arguments of expressions (22.13) and (22.14).

Now, let us assume that rather than being a consequence of A, B is a ground for A; that is, we now assume A to be a consequence of B. As before, suppose that in our investigations of B, its credibility varies. How does this affect the credibility of A? Let B' be the complement of B. Then

$$\begin{aligned} P(A) &= P(AB) + P(AB') \\ &= P(B)P(A|B) + P(B')P(A|B') \\ &= P(B)P(A|B) + [1 - P(B)]P(A|B'). \end{aligned} \tag{22.24}$$

However, since B implies A, $P(A \mid B) = 1$, this becomes

$$P(A) = P(B) + [1 - P(B)]P(A \mid B'), \tag{22.25}$$

so that

$$P(A \mid B') = \frac{P(A) - P(B)}{1 - P(B)}. \tag{22.26}$$

The left-hand side of this last expression represents the credibility of A after B (which is a possible ground for A) has been refuted, while the right-hand side refers to the situation before the refutation. The latter can be reformulated so as to give

$$P(A \mid B') = P(A) - P(B)\frac{1 - P(A)}{1 - P(B)}, \tag{22.27}$$

from which we can conclude that

$$P(A \mid B') < P(A). \tag{22.28}$$

Both sides of this inequality represent the credibility of A, the right-hand side before the refutation of B, and the left-hand side after the refutation. Equality (22.28) thus formalizes the rule implicit in expression (22.16): *Our confidence in a conjecture can only diminish when a possible ground for the conjecture has been exploded.*

However, this does not yet answer the question originally raised, which is the effect on our confidence in A when the credibility of B varies. To answer this, we hold $P(A)$ constant and let $P(B)$ vary. Now, $P(B)$ can be very small, but cannot be arbitrarily large, since it can never exceed $P(A)$. Hence, the range within which $P(B)$ can vary is $0 < P(B) \leq P(A)$. Consequently, we see from Equation (22.26) that as $P(B)$ increases from 0 to $P(A)$, $P(A \mid B')$ diminishes from $P(A)$ to 0. In other words, *the more confidence we place in a possible ground for a conjecture, the greater will be the loss of faith in that conjecture when the possible ground is refuted.* This rule formalizes the shaded inference arguments in expressions (22.17) and (22.18).

At this point, the reader should be reasonably convinced that the probability calculus provides a powerful tool for sharpening and systematizing our patterns of reasoning in situations where demonstrative proof cannot be attained. Moreover, it is clear that nothing has been said in this section that requires a full acceptance of the idea of subjective probability. The only requirement is that the latter be accepted in an ordinal sense, in that one be willing to say that evidence favorable to a conjecture increases his confidence that the conjecture is true, while unfavorable evidence reduces this confidence. The strong Bayesian is, of course, willing to go much further and actually assign numerical values to the credibilities involved.[6]

In the framework of the two models of plausible inference that we have analyzed in this section (that of examining a consequence of a conjecture and that of examining a possible ground), it is clear that statistical inference as we have studied it in the last several chapters will frequently be of use in two places: (1) deciding on the truth (or falsity) of the consequence or ground, and then (2) deciding when to accept the conjecture as proven or disproven. No matter what attitude toward probability and statistical inference is adopted, it is clear that these two issues must be confronted and come to grips with.

However, it is the author's considered opinion that in a student's first contact with probability and mathematical statistics these are issues that he should not lose any sleep over. His initial goal should be to learn the laws of probability, the basic theorems of mathematical statistics, *and* the Neyman-Pearson theory of hypothesis testing. For the laws of probability and the basic results of mathematical statistics are the same for the Bayesian as for the Classicist, and in the end if the student is going to reject the Neyman-Pearson framework, he must know and understand what it is that he is rejecting. Ideally, the hope is, of course, that the student can form his own opinion as to the *proper framework for him* free of the biases of his professors and the textbooks from which he has studied. The present chapter has hopefully contributed to this end.

REFERENCES

Birnbaum, A., "On the Foundations of Statistical Inference," *Journal of the American Statistical Association*, vol. 57, June 1962.

Birnbaum, A., "On Durbin's Modified Principle of Conditionality," *Journal of the American Statistical Association*, vol. 65, March 1970.

Blalock, H. M., Jr., *Causal Inferences in Nonexperimental Research*, The University of North Carolina Press, 1961, 1964.

[6] There are two schools of thought among Bayesians as to how the numerical values of credibilities arise. The smaller school, which is identified at the present time with Sir Harold Jeffreys [see Jeffreys (1961), especially the first and last chapters], views probability as measuring degree of reasonable belief, but maintains that it is nevertheless an objective quantity in that the logic of the situation dictates what the probability should be. On this view, two individuals with the same facts at their disposal about a conjecture will assign the conjecture identical credibilities. The second school takes the view that probability is strictly subjective, and that two individuals in identical circumstances regarding a conjecture will not necessarily assign identical credibilities. For a discussion, see Savage (1954, chap. 4).

Durbin, J., "On Birnbaum's Theorem on the Relation Between Sufficiency, Conditionality, and Likelihood," *Journal of the American Statistical Association*, vol. 45, March 1970.

Jeffreys, H., *Theory of Probability*, 3rd ed., Oxford University Press, 1961, chaps. I and VIII.

Polya, G., *Patterns of Plausible Inference*, Revised ed., Princeton University Press, 1968.

Raiffa, H., *Decision Analysis: Introductory Lectures on Choices Under Uncertainty*, Addison-Wesley, 1968, chap. 10.

Savage, L. J., *The Foundations of Statistics*, Wiley, 1954, chaps. 1–4.

Savage, L. J., *Statistical Inference*, Wiley, 1962.

Zellner, A., *An Introduction to Bayesian Inference in Econometrics*, Wiley, 1971, Chaps. I–III.

APPENDIX 1

THE ALGEBRA OF THE SUBSETS OF A SET

The notion of a set is one of the most fundamental concepts in mathematics. It is usually taken as part of the basic language of the discipline, and, therefore, is left undefined. For our purposes, we take a *set* to be a collection of arbitrary objects. Examples are:

1. the countries of Latin America
2. all females in the world
3. all consumers in the United States
4. all female consumers in the United States
5. the integers between 1 and 10
6. the even numbers between 1 and 10, and so forth.

We will denote a set by a capital letter and define it in the following way:

$$A = \{x: \quad x \text{ is a member of } A\}. \tag{A.1.1}$$

In words, what (A.1.1) says is that A is the set of all x such that x is a member of A. For example, in example 5 above, we would have

$$A = \{x: \quad x = 1, 2, \ldots, 10\}. \tag{A.1.2}$$

We can also write this set by enumerating all of its elements, namely,

$$A = \{1, 2, 3, \ldots, 10\}.$$

To denote that an element x belongs to a set A, we write

$$x \in A, \tag{A.1.3}$$

where $\in$ means "belongs to" or "is a member of."

We say that a set A is a *subset* of a set B if every element included in A is also included in B, and we denote this by

$$A \subset B \qquad \text{or} \qquad B \supset A, \tag{A.1.4}$$

where the symbol "$\subset$" means "is included in." In the above examples, it is evident that example 4 is a subset of example 3, for the set consisting of all consumers in the United States includes all female consumers in the United States, and similarly it is obvious that example 6 the even integers between 1 and 10, is included in example 5, the integers between 1 and 10. On the other hand, neither example 3 nor 2 are subsets of example 4.

A set can include any number of elements, from zero to infinity. The set with zero elements is called the *empty* (or *vacuous*) set, while the set of all real numbers betwen 0 and 1 is an example of a set with an infinity of elements. In particular, we need to distinguish between the set which consists of a single element and that element—that is, between $A = \{a\}$ and a.

We will now define the two fundamental operations of the algebra of sets. These operations are, respectively, intersection and union.

A.1.1 INTERSECTION

The *intersection* of two sets A and B is defined as follows:

$$A \cap B = \{x: \quad x \in A \text{ and } x \in B\}, \tag{A.1.5}$$

or in words, the intersection of A and B, read as A and B (or the product of A and B), is the set of all x that are included in *both* A and B. Loosely speaking, intersection in the algebra of sets corresponds to multiplication in the algebra of real numbers. Often $A \cap B$ will be written simply as AB.

A.1.2 UNION

The *union* of two sets A and B is defined as

$$A \cup B = \{x: \quad x \in A \text{ or } x \in B \text{ (or both)}\}, \tag{A.1.6}$$

or in words, the union of A and B, read as A or B, is the set of all x that are included in A or in B or in both. Note that the logical "or" is used in

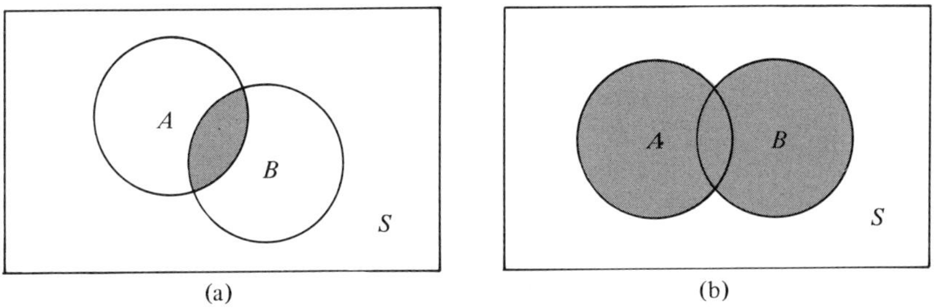

FIGURE A.1.1 (a) $A \cap B$; (b) $A \cup B$.

the definition. Again, speaking loosely, we can say that union in the algebra of sets corresponds to addition in simple algebra.

Fortunately, these two definitions can be illustrated very simply and conveniently through a device called the *Venn diagram*. The shaded areas designate the sets in question. (See Figure A.1.1.)

Thus, it is seen that the intersection of A and B corresponds to the overlap between the two sets, while their union corresponds to their projection on a screen.

As has been done in these two figures, it is convenient to designate the entire area within a diagram by the letter S. Similarly, we will denote the empty set by the symbol $\varnothing$. Several results follow immediately:

$$A \cap S = A, \tag{A.1.7}$$

$$A \cup S = S, \tag{A.1.8}$$

$$A \cap \varnothing = \varnothing, \tag{A.1.9}$$

$$A \cup \varnothing = A, \tag{A.1.10}$$

$$\varnothing \cap S = \varnothing, \tag{A.1.11}$$

$$\varnothing \cup S = S, \tag{A.1.12}$$

$$A \cap B = B \cap A, \tag{A.1.13}$$

$$A \cup B = B \cup A. \tag{A.1.14}$$

We also have the distribution laws:

Distributive law for $\cap$ over $\cup$

$$A \cap (B \cup C) = (A \cap B) \cup (A \cap C) \tag{A.1.15}$$

and

$$(B \cup C) \cap A = (B \cap A) \cup (C \cap A). \tag{A.1.16}$$

Distributive law for $\cup$ ***over*** $\cap$

$$A \cup (B \cap C) = (A \cup B) \cap (A \cup C) \tag{A.1.17}$$

and

$$(B \cap C) \cup A = (B \cup A) \cap (C \cup A). \tag{A.1.18}$$

We define the *complement* of a set A to be

$$A' = \{x: \quad x \text{ is } \textit{not} \text{ in } A\}, \tag{A.1.19}$$

or in words, the complement of A, read A prime, is the set of all x that are not in A. It follows, therefore, that

$$A \cap A' = \varnothing \tag{A.1.20}$$

and

$$A \cup A' = S. \tag{A.1.21}$$

Next, we have the laws of de Morgan:

$$(A \cup B)' = A' \cap B', \tag{A.1.22}$$

$$(A \cap B)' = A' \cup B'. \tag{A.1.23}$$

And finally there are the useful laws of absorption:

$$A \cap (A \cup B) = A, \tag{A.1.24}$$

$$A \cup (A \cap B) = A. \tag{A.1.25}$$

The proofs of these expressions are straightforward, and can be done either algebraically or through the use of a Venn diagram. We will illustrate the latter method in proving the distributive law of $\cap$ over $\cup$. (See Figure A.1.2.)

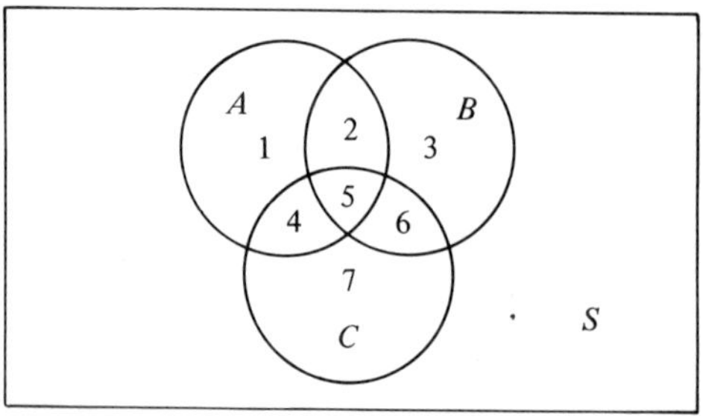

FIGURE A.1.2 Illustration of Distributive law of $\cap$ over $\cup$.

Note that we have numbered each of the subsets. We have

$$A = \{1, 2, 4, 5\}, \quad B = \{2, 3, 5, 6\}, \quad \text{and} \quad C = \{4, 5, 6, 7\}. \tag{A.1.26}$$

Therefore, since

$$B \cup C = \{2, 3, 4, 5, 6, 7\}, \tag{A.1.27}$$

it follows that

$$A \cap (B \cup C) = \{2, 4, 5\}. \tag{A.1.28}$$

On the other hand,

$$A \cap B = \{2, 5\} \tag{A.1.29}$$

and

$$A \cap C = \{4, 5\}, \tag{A.1.30}$$

so that

$$(A \cap B) \cup (A \cap C) = \{2, 4, 5\}. \tag{A.1.31}$$

Hence,

$$A \cap (B \cup C) = (A \cap B) \cup (A \cap C).$$

A.1.3 THE DIFFERENCE OF TWO SETS

An algebraic operation that is used frequently in probability theory is the *difference* of two sets, and it is defined as follows:

$$A - B = A \cap B', \tag{A.1.32}$$

or, in words, the difference between two sets A and B is equal to the intersection of A with the complement of B. Its Venn diagram is shown in Figure A.1.3.

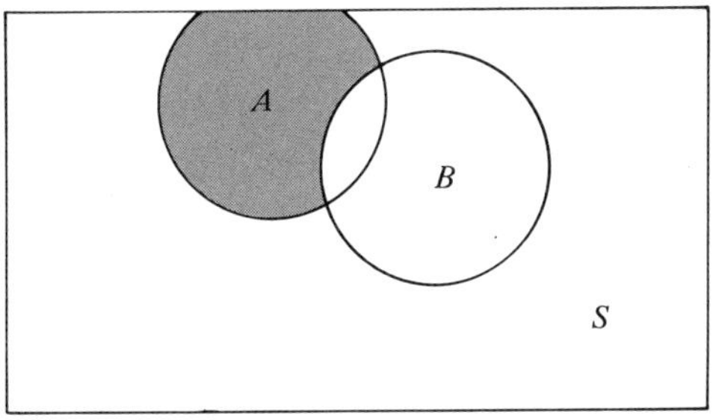

FIGURE A.1.3 $A - B$.

APPENDIX 2

BINOMIAL AND MULTINOMIAL THEOREMS

This appendix proves several theorems, beginning with the simple binomial theorem, that are needed for the study of the binomial and hypergeometric distributions in Chapter 7 and the multinomial distribution in Chapter 9. A proof of the Newton form of the binomial theorem is also provided.

THEOREM A.2.1 (Binomial Theorem)
If n is a positive integer and a and x are fixed constants, then

$$(a + x)^n = \binom{n}{0} a^n + \binom{n}{1} a^{n-1} x + \cdots + \binom{n}{r} a^{n-r} x^r + \cdots + \binom{n}{n} x^n. \qquad \text{(A.2.1)}$$

PROOF. For $n = 2$, we have

$$(a + x)^2 = a^2 + 2ax + x^2,$$

and for $n = 3$,

$$(a + x)^3 = a^3 + 3ax^2 + 3ax^2 + x^3.$$

If we were to continue for $n = 4, 5, 6, \ldots$, it would become apparent that each term in the expansion of $(a + x)^n$ is of degree n in the variables a and x. Ignoring its coefficient, the general term therefore has the form

$$a^{n-r} x^r, \qquad r = 0, 1, 2, \ldots, n.$$

This term is obtained by selecting x from r of the factors and a from the remaining $n - r$ factors. This selection can be made in $\binom{n}{r}$ ways. Hence the term $a^{n-r}x^r$ occurs $\binom{n}{r}$ times and its coefficient is thus $\binom{n}{r}$. Therefore, the complete general term is

$$\binom{n}{r}a^{n-r}x^r,$$

and the expansion of $(a + x)^n$ is that shown in (A.2.1).

In summation notation, the expansion of $(a + x)^n$ may be written

$$(a + x)^n = \sum_{r=0}^{n} \binom{n}{r} a^{n-r}x^r. \tag{A.2.2}$$

The coefficients $\binom{n}{r}$ are often called binomial coefficients. Moreover, since

$$\binom{n}{0} = 1, \qquad \binom{n}{1} = n, \qquad \binom{n}{2} = \frac{n(n-1)}{2!},$$

and so on, the binomial expansion may also be written as

$$(a + x)^n = a^n + \frac{n}{1!}a^{n-1}x + \frac{n(n-1)}{2!}a^{n-2}x^2 + \cdots + x^n. \tag{A.2.3}$$

THEOREM A.2.2 (Multinomial Theorem)
If n is a positive integer and $a_1, a_2, \ldots, a_k$ are fixed constants, then

$$(a_1 + a_2 + \cdots + a_k)^n = \sum \frac{n!}{x_1!x_2!\ldots x_x!} a_1^{x_1}a_2^{x_2}\cdots a_k^{x_k}, \tag{A.2.4}$$

where the sum on the right-hand side is to be understood to be over all possible combinations of $x_1, x_2, \ldots, x_k$ such that

$$x_1 + x_2 + \cdots + x_k = n.$$

PROOF. To begin with, let $k = 3$ and $n = 2$. Then

$$\begin{aligned}(a_1 + a_2 + a_3)^2 &\\ &= a_1^2 + a_2^2 + a_3^2 + 2a_1a_2 + 2a_1a_3 + 2a_2a_3\\ &= \sum_{x_1=0}^{2} \sum_{x_2=0}^{2-x_1} \frac{2!}{x_1!x_2!(2 - x_1 - x_2)!} a_1^{x_1}a_2^{x_2}a_3^{2-x_1-x_2}.\end{aligned} \tag{A.2.5}$$

It is thus clear that

$$(a_1 + a_2 + a_3)^n$$

$$= \sum_{x_1=0}^{n} \sum_{x_2=0}^{n-x_1} \frac{n!}{x_1!x_2!(n - x_1 - x_2)!} a_1^{x_1} a_2^{x_2} a_3^{n-x_1-x_2}. \qquad \text{(A.2.6)}$$

Now, let $k = 4$ and $n = 2$. Then

$$(a_1 + a_2 + a_3 + a_4)^2$$

$$= a_1^2 + a_2^2 + a_3^2 + a_4^2 + 2a_1a_2 + 2a_1a_3 + 2a_1a_4$$

$$+ 2a_2a_3 + 2a_2a_4 + 2a_3a_4$$

$$= \sum_{x_1=0}^{2} \sum_{x_2=0}^{2-x_1} \sum_{x_3=0}^{2-x_1-x_2} \frac{2!}{x_1!x_2!x_3!(2 - x_1 - x_2 - x_3)!}$$

$$\times a_1^{x_1} a_2^{x_2} a_3^{x_3} a_4^{(2-x_1-x_2-x_3)}.$$

Accordingly,

$$(a_1 + a_2 + a_3 + a_4)^n$$

$$= \sum_{x_1=0}^{n} \sum_{x_2=0}^{n-x_1} \sum_{x_3=0}^{n-x_1-x_2} \frac{n!}{x_1!x_2!x_3!(n - x_1 - x_2 - x_3)!}$$

$$\cdot a_x^{x_1} a_2^{x_2} a_3^{x_3} a_4^{(n-x_1-x_2-x_3)}. \qquad \text{(A.2.7)}$$

Continuing, we would eventually find for $(a_1 + a_2 + \cdots + a_k)^n$,

$$(a_1 + a_2 + \cdots + a_k)^n = \sum_{x_1=0}^{n} \sum_{x_2=0}^{n-x_1} \cdots \sum_{x_{k-1}=0}^{n-x_1-\cdots-x_{k-2}}$$

$$\frac{n!}{x_1!x_2! \cdots x_{k-1}!(n - x_1 - \cdots - x_{k-1})!} a_1^{x_1} a_2^{x_2} \cdots a_k^{n-x_1-\cdots-x_{k-1}},$$

$$\text{(A.2.8)}$$

which is (A.2.4) with $x_k = n - x_1 - \cdots - x_{k-1}$.

Let us now consider the Taylor series expansion in powers of x of the function $f(x) = (1 + x)^m$, where m is any real number. We have

$$\begin{aligned}
f(x) &= (1 + x)^m, \\
f'(x) &= m(1 + x)^{m-1}, \\
f''(x) &= m(m - 1)(1 + x)^{m-2}, \\
&\vdots \\
f^{(h)}(x) &= m(m - 1) \cdots (m - h + 1)(1 + x)^{m-h}, \\
&\vdots
\end{aligned} \qquad \text{(A.2.9)}$$

For $x = 0$, we have $f(0) = 1$, and for $h = 1, 2, \ldots,$

$$f^{(h)}(0) = m(m-1) + \cdots + (m-h+1). \tag{A.2.10}$$

Taylor's formula for $(1 + x)^m$ with a remainder is therefore

$$(1+x)^m = 1 + mx + \frac{m(m-1)}{2}x^2 + \cdots + \frac{m(m-1)\cdots(m-n+1)}{n!}x^n + R_{n+1}. \tag{A.2.11}$$

We now use Cauchy's formula for the remainder R_{n+1}, expressing it as[1]

$$R_{n+1} = \frac{f^{(n+1)}(\theta x)}{n!}x^{n+1}(1-\theta)^n, \tag{A.2.12}$$

for $0 < \theta < 1$.

If m is a positive integer or zero, then $f^{(h)}(x) \equiv 0$ when $h > m$. Thus, in this case we have $R_{m+1} \equiv 0$ and

$$(1+x)^m = 1 + mx + \frac{m(m-1)}{2}x^2 + \cdots + x^n, \tag{A.2.13}$$

which is just the ordinary binomial-expansion formula of Theorem A.2.1. However, when m is not a positive integer or zero, formula (A.2.12) for R_{n+1} becomes

$$\begin{aligned} R_{n+1} &= \frac{m(m-1)\cdots(m-n)(1+\theta x)^{m-n-1}}{n!}x^{n+1}(1-\theta)^n \\ &= \frac{m(m-1)\cdots(m-n)}{n!}\left(\frac{1-\theta}{1+\theta x}\right)^n (1+\theta x)^{m-1}x^{n+1}. \end{aligned} \tag{A.2.14}$$

Now, suppose that $-1 < x < 1$. Then

$$0 < \frac{1-\theta}{1+\theta x} < 1. \tag{A.2.15}$$

For $m > 1$, we have

$$\begin{aligned} 0 &< (1+\theta x)^{m-1} \\ &< (1+|x|)^{m-1}, \end{aligned} \tag{A.2.16}$$

while for $m < 1$, we have

$$\begin{aligned} (1+\theta x)^{m-1} &= \frac{1}{(1+\theta x)^{1-m}} \\ &< \frac{1}{(1-|x|)^{1-m}}, \end{aligned} \tag{A.2.17}$$

[1] See Taylor (1955, p. 115).

so that

$$0 < (1 + \theta x)^{m-1}$$
$$< (1 - |x|)^{m-1}. \tag{A.2.18}$$

Consequently, for $-1 < x < 1$,

$$|R_{n+1}| \leq \frac{m(m-1)\cdots(m-n+1)}{n!}(1 \pm |x|)^{m-1}|x|^{n+1}, \tag{A.2.19}$$

the choice of the double sign in $(1 \pm |x|)^{m-1}$ depending on the sign of $m - 1$.

From (A.2.19), we shall now show that $R_{n+1} \to 0$ as $n \to \infty$ if $|x| < 1$. Since $(1 \pm |x|)^{m-1}$ does not depend on n, it suffices to show that

$$\lim_{n\to\infty} \frac{m(m-1)\cdots(m-n+1)}{n!} x^{n+1} = 0. \tag{A.2.20}$$

Let

$$u_n = \frac{m(m-1)\cdots(m-n+1)}{n!} x^{n+1}, \tag{A.2.21}$$

and consider the series $\sum u_n$. Since, applying the ratio test,[2]

$$\lim_{n\to\infty}\left|\frac{u_{n+1}}{u_n}\right| = \lim_{n\to\infty}\left|\frac{m-n+1}{n+1}\right||x|$$
$$= |x|, \tag{A.2.22}$$

this series converges for $|x| < 1$. This being the case, it follows that $u_n \to 0$ as $n \to \infty$, and (A.2.10) is established. Consequently, we have proven Newton's form of the binomial theorem.

THEOREM A.2.3
For any m and $|x| < 1$,

$$(1 + x)^m = 1 + mx + \frac{m(m-1)}{2}x^2 + \cdots$$
$$+ \frac{m(m-1)\cdots(m-n+1)}{n!}x^n + \cdots. \tag{A.2.23}$$

[2] See Taylor (1955, p. 565) for a discussion of the ratio test for absolute convergence.

This series is referred to as the *binomial series*, and as already noted, reduces to the binomial-expansion formula of Theorem A.2.1 when m is a positive integer. If m is a positive integer and, in addition, x is equal to one, (A.2.23) can be shown to reduce to

$$2^m = \binom{m}{0} + \binom{m}{1} + \cdots + \binom{m}{m}. \tag{A.2.24}$$

REFERENCE

Taylor, A. E., *Advanced Calculus*, Ginn & Co., 1955.

APPENDIX 3

CHARACTERISTIC FUNCTIONS AND RELATED THEOREMS

It was noted in Chapter 6 that, since the moment-generating function of a distribution does not always exist, it is more useful for many purposes to work with the characteristic function. This appendix defines the characteristic function of a distribution, and shows how it can be used to obtain the distribution's moments. In addition, we shall prove several theorems which are of major importance to the development of the theory of statistics.

Although the discussion in this appendix assumes that the reader has had some contact with the theory of complex numbers, the actual demands in this regard are minimal. Familiarity with how a complex number is defined is the only essential. The purpose here is not to elucidate the theory of characteristic functions, for this is a task for advanced courses, but rather only to lay the groundwork for Theorem A.3.4, which establishes the uniqueness of the moment-generating function in situations where the function exists. Our discussion here will be restricted to distributions for which the density and distribution functions are related by $f(x) = F'(x)$.

Let X be a random variable with density function $f(x)$ and distribution function $F(x)$. The characteristic function $\phi(t)$ of the random variable X is then defined as

$$\begin{aligned}\phi(t) &\equiv E(e^{itX}) \\ &= \int_x e^{itx} f(x)\, dx. \qquad \text{(A.3.1)}\end{aligned}$$

where i is the imaginary unit, $i = \sqrt{-1}$, and t is real. Alternatively, $\phi(t)$ is frequently referred to as the characteristic function corresponding to the distribution $F(x)$.

Since, by definition,

$$e^{itx} = \cos tx + i \sin tx, \tag{A.3.2}$$

it follows that[1]

$$\begin{aligned} |\phi(t)| &\leq \int_x |e^{itx}| f(x)\, dx \\ &= 1. \end{aligned} \tag{A.3.3}$$

This implies that the integral in the definition of $\phi(t)$ will exist for any $F(x)$ for all t, and hence that, unlike the moment-generating function, the characteristic function of a random variable always exists.

Let us now assume that the rth moment of X exists. Differentiation of $\phi(t)$ with respect to t h times then yields

$$\phi^h(t) = i^h \int_x x^h e^{itx} f(x)\, dx, \tag{A.3.4}$$

which, upon valuation at $t = 0$, becomes

$$\begin{aligned} \phi^h(0) &= i^h \int_x x^h f(x)\, dx \\ &= i^h \mu_h', \qquad 0 \leq h \leq r, \end{aligned} \tag{A.3.5}$$

or

$$\mu_h' = \frac{\phi^h(0)}{i^h}. \tag{A.3.6}$$

This establishes that the characteristic function can be used as a moment-generating function.

Situations often arise in probability theory and its applications, especially to sampling theory, in which it is relatively easy to find the characteristic function of a function of the components of a multidimensional random variable, particularly a linear function. In these situations, the question then arises of how to find the distribution of the random variable from its characteristic function. In many cases, the answer to this question is answered by the following theorem due to Paul Levy.

THEOREM A.3.1 (Inversion Theorem)

Let X be a random variable with characteristic function $\phi(t)$ and distribution function $F(x)$. Assume that $F(x)$ is continuous at $x = x' \pm \delta$, $\delta > 0$. Then we have

$$F(x' + \delta) - F(x' - \delta) = \lim_{A \to \infty} \frac{1}{\pi} \int_{-A}^{A} \frac{\sin \delta t}{t} e^{-itx} \phi(t)\, dt. \tag{A.3.7}$$

[1] Let $z = a + ib$ be a complex number. Then the absolute value of z, $|z|$, is defined as the positive square root of $a^2 + b^2$. In the case, therefore, of e^{itx}, $|e^{itx}| = [\cos^2 tx + \sin^2 tx]^{1/2} = 1$.

Moreover, if

$$\int_{-\infty}^{\infty} \phi(t)\, dt < \infty, \qquad \text{(A.3.8)}$$

a density function $f(x)$ exists at $x = x'$ and

$$f(x') = \frac{1}{2\pi} \int_{-\infty}^{\infty} e^{-itx} \phi(t)\, dt. \qquad \text{(A.3.9)}$$

PROOF. See Wilks (1962, pp. 116–117).

The importance of the inversion theorem can hardly be overemphasized, for as an immediate consequence of it, we have the following.

THEOREM A.3.2
A necessary and sufficient condition for two random variables X_1 and X_2 to have identical distribution functions is for their characteristic functions to be identical.

PROOF. Note, first, that if X_1 and X_2 have identical distribution functions, it is evident from (A.3.1) that their characteristic functions will be identical.

Conversely, suppose that X_1 and X_2 both have the characteristic function $\phi(t)$. Then if $F_1(x)$ and $F_2(x)$ are the distribution functions of X_1 and X_2, it follows from (A.3.7) that if $x' \pm \delta$ is any interval such that $F_1(x)$ and $F_2(x)$ are continuous at the endpoints $x' - \delta$ and $x' + \delta$, then

$$F_1(x' + \delta) - F_1(x' - \delta) = F_2(x' + \delta) - F_2(x' - \delta), \qquad \text{(A.3.10)}$$

which, since F_1 and F_2 are distribution functions, implies that

$$F_1(x) \equiv F_2(x).$$

Next, let us note from (A.3.1) that if t in $\phi(t)$ is replaced by $t' = t/i$, then $\phi(t')$ is the moment-generating function for the random variable X. Consequently, in situations where the moment-generating function $M(t)$ $[=\phi(t/i)]$ exists, we have as a special case of Theorem A.3.2.

THEOREM A.3.3
Let X_1 and X_2 be two random variables with moment-generating functions $M_1(t)$ and $M_2(t)$. Then a necessary and sufficient condition for X_1 and X_2 to have identical distribution functions is for $M_1(t)$ to be identical with $M_2(t)$.

Our final result in this appendix deals with the uniqueness of the moments of a distribution.

THEOREM A.3.4
Let $F(x)$ be a distribution function with moments μ_r', $r = 0, 1, 2, \ldots,$ all of which are finite. If the series

$$\sum_{r=0}^{\infty} \frac{\mu_r'}{r!} c^r$$

is absolutely convergent for some $c > 0$, then $F(x)$ is the only distribution function having these moments.

PROOF. See Wilks (1962, pp. 125–126).

COROLLARY A.3.4.1
If X is a bounded random variable, then its distribution function $F(x)$ is uniquely determined by its moments μ_r', $r = 0, 1, 2, \ldots.$

PROOF. Since X is bounded, there therefore exist finite numbers a, b, $a > b$, such that $F(a) = 0$ and $F(b) = 1$. Let M denote the largest of $|a|$ and $|b|$. Then we have

$$\begin{aligned} |\mu_r'| &\leq \int_a^b |x|^r f(x)\, dx \\ &= \nu_r' \\ &\leq M^r \end{aligned} \tag{A.3.11}$$

and

$$\begin{aligned} \sum_{r=1}^{\infty} \left| \frac{\mu_r' c^r}{r!} \right| &\leq \sum_{r=0}^{\infty} \frac{|\mu_r'| c^r}{r!} \qquad \text{(A.3.12)} \\ &\leq \sum_{r=0}^{\infty} \frac{\nu_r' c^r}{r!} \\ &\leq \sum_{r=0}^{\infty} \frac{|Mc|^r}{r!} \\ &= \exp(|Mc|). \end{aligned}$$

However, $\exp(|Mc|)$ is finite for all c. The sufficient condition in Theorem A.3.4 is accordingly satisfied, thus establishing the result.

REFERENCES

Gnedenko, B. V., *Theory of Probability*, Chelsea Publishing Co., 1962, chap. VII.
Wilks, S. S., *Mathematical Statistics*, Wiley, 1962, chap. 5.

APPENDIX 4

GREEN'S THEOREM AND OTHER RESULTS FROM ADVANCED CALCULUS

In this appendix, we present several results that are needed in the discussion leading up to the theorem on a change of variables in a double integral in Chapter 10. The key theorem is Green's theorem in the plane. Also provided in this appendix is a derivation of the limit used in the proofs of the classical central limit theorem for the normal distribution in Chapter 14 and Appendix 6.

THEOREM A.4.1 (Green's Theorem in the Plane)
Let R be a closed and bounded region of the xy-plane. Let the boundary of R consist of a finite number of simple closed curves which are sectionally smooth.[1] Let $P(x, y)$ and $Q(x, y)$ be functions which are continuous and have continuous first partial derivatives in R. Let C denote the aggregate of curves forming the boundary of R, each curve being assumed to be oriented in such a way that the region is on the left as one advances along the curve in the positive direction. Then

$$\int_C P\,dx + Q\,dy = \iint_R \left(\frac{\partial Q}{\partial x} - \frac{\partial P}{\partial y}\right) dx\,dy. \qquad \text{(A.4.1)}$$

PROOF. We shall not prove this theorem in its full generality, but rather only for regions R which are of a particularly simple shape. The proof presented is the one of Taylor (1955, pp. 420–423).

[1] See footnote 3 of Chapter 10 for a definition of sectionally smooth.

Let us begin by observing that the functions P and Q are independent of one another, and in view of this, expression (A.4.1) can be broken down into

$$\int_C P\,dx = -\iint_R \frac{\partial P}{\partial y}\,dx\,dy, \tag{A.4.2}$$

$$\int_C Q\,dy = \iint_R \frac{\partial Q}{\partial x}\,dx\,dy. \tag{A.4.3}$$

Proof of the theorem is therefore equivalent to proofs of (A.4.2) and (A.4.3) separately.

It should also be mentioned that it is assumed that the x- and y-axes have their usual relation to one another—that is, x horizontal and y vertical. Moreover, positive direction is taken to mean that one advances along the boundary of R in a counterclockwise direction. These conventions account for the negative sign in (A.4.2) and its absence in (A.4.3).

We now assume that R has the simple form suggested by Figure A.4.1. That is, suppose that there is an interval $c \le x \le d$ such that for x' outside of the interval, the line $x = x'$ does not intersect R, while for $c < x' < d$, the line $x = x'$ intersects R in an interval. The lines $x = c$ and $x = d$ may intersect R in either a single point or an interval. The boundary of R can then be decomposed into a lower curve $y = Y_1(x)$, an upper curve $y = Y_2(x)$, and certain portions (either a point or a segment) of each of the lines $x = c$ and $x = d$. The positive orientation of the boundary of R is shown by the arrows in Figure A.4.1. Let C_1 and C_2 denote the oriented lower and upper curves, respectively.

If the region R has the simple form just described, let us call it *x-simple*. Similarly, we may define a *y-simple* region.

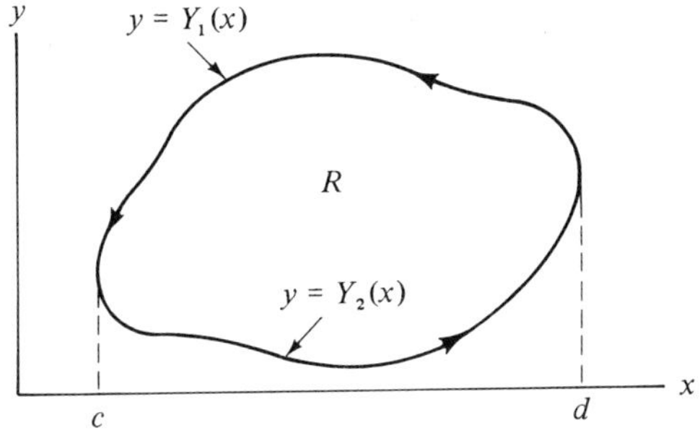

FIGURE A.4.1 x-simple R.

We now assume that R is x-simple, and let us consider expression (A.4.2). There are no contributions to the line integral from the portions of C on the lines $x = c$, $x = d$, since $dx = 0$ along any segment of either line. Consequently,

$$\int_C P\,dx = \int_{C_1} P\,dx + \int_{C_2} P\,dx. \tag{A.4.4}$$

However, on C_1 we can take x as a parameter and put $y = Y_1(x)$, so that

$$\int_{C_1} P(x, y)\,dx = \int_c^d P[x, Y_1(x)]\,dx. \tag{A.4.5}$$

Similarly,

$$\begin{aligned}\int_{C_2} P(x, y)\,dx &= \int_d^c P[x, Y_2(x)]\,dx \\ &= -\int_c^d P[x, Y_2(x)]\,dx.\end{aligned} \tag{A.4.6}$$

Hence,

$$\int_C P\,dx = -\int_c^d [P(x, Y_2) - P(x, Y_1)]\,dx. \tag{A.4.7}$$

Consider, now, the double integral in (A.4.2), which, using the iterated integral formula,[2] we can write as

$$\iint_R \frac{\partial P}{\partial y}\,dx\,dy = \int_c^d dx \int_{Y_1}^{Y_2} \frac{\partial P}{\partial y}\,dy. \tag{A.4.8}$$

The y integration may now be performed with x held constant, and the result is

$$\int_{Y_1}^{Y_2} \frac{\partial P}{\partial y}\,dy = P(x, Y_2) - P(x, Y_1). \tag{A.4.9}$$

On combining (A.4.9) with (A.4.8) and comparing the result with (A.4.7), we see the truth of (A.4.2) for an x-simple region R.

An entirely parallel proof can be given for formula (A.4.3) on the assumption that R is a y-simple R. Finally, if R is both x-simple and y-simple, (A.4.2) and (A.4.3) combine to yield (A.4.1).

An immediate consequence of Green's theorem follows.

[2] See Taylor (1955, p. 326).

THEOREM A.4.2
Let R be a region in the xy-plane with the properties described in Theorem A.4.1, and let A be the area of R. Then

$$A = -\int_C y\,dx = \int_C x\,dy = \frac{1}{2}\int_C -y\,dx + x\,dy. \qquad \text{(A.4.10)}$$

PROOF. Putting $P = -y$, $Q = x$ in Green's theorem, we get for the third formula,

$$\begin{aligned} \frac{1}{2}\int_C -y\,dx + x\,dy &= \frac{1}{2}\iint_R (1+1)\,dx\,dy \qquad \text{(A.4.11)} \\ &= A, \end{aligned}$$

and similarly for the other two formulas.

The next theorem, also needed in the theorem in Chapter 10 on a change of variables in a double integral, is stated without proof. Its proof is not difficult, but is tedious, and the reader is referred to Taylor (1955, p. 323).

THEOREM A.4.3
Let R be a closed and bounded region in the xy-plane and let $f(x, y)$ be a function continuous in R. Let R be divided in any manner into a finite number of subregions, of areas $\Delta A_1, \ldots, \Delta A_n$, the shapes being arbitrary (except that they be sufficiently regular that they can unambiguously be assigned an area). Let a point $P_k(x_k, y_k)$ be chosen arbitrarily in ΔA_k. Then

$$\iint_R f(x, y)\,dx\,dy = \lim_{n\to\infty} \sum_{k=1}^{n} f(x_k, y_k)\Delta A_k. \qquad \text{(A.4.12)}$$

Moreover, it is not essential that the areas $\Delta A_1, \ldots, \Delta A_k$ completely exhaust the area of R, provided that the amount of area omitted approaches zero in the limit.

The final theorem of this appendix relates to a limit that is needed in the proof of the classical central limit theorem for the normal distribution.

THEOREM A.4.4
If $\lim_{n\to\infty} f(n) = 0$, then

$$\lim_{n\to\infty}\left[1 + \frac{b}{n} + \frac{f(n)}{n}\right]^n = e^b. \qquad \text{(A.4.13)}$$

PROOF.[3] Taking the natural logarithm, we have

$$n\{\ln[n+b+f(n)]-\ln n\}=n\int_n^{n+b+f(n)}\frac{1}{x}\,dx. \qquad \text{(A.4.14)}$$

Applying the mean value theorem for integrals, the right-hand side can be written

$$n\int_n^{n+b+f(n)}\frac{1}{x}\,dx=n\left\{[n+b+f(n)-n]\frac{1}{n+\theta b+\theta f(n)}\right\}$$

$$=\frac{n}{n+\theta b+\theta f(n)}[b+f(n)], \qquad \text{(A.4.15)}$$

for some θ, $0<\theta<1$. We now take the limit of this last expression as $n\to\infty$. Since by assumption $\lim_{n\to\infty} f(n)=0$,

$$\lim_{n\to\infty}[b+f(n)]=b. \qquad \text{(A.4.16)}$$

Also,

$$\lim_{n\to\infty}\left[\frac{n}{n+\theta b+\theta f(n)}\right]=\lim_{n\to\infty}\left[1+\frac{\theta b+\theta f(n)}{n}\right]^{-1}$$

$$=1, \qquad \text{(A.4.17)}$$

since whatever the precise value of θ (depending on n), $\theta b+\theta f(n)$ remains bounded, so that

$$\lim_{n\to\infty}\frac{\theta b+\theta f(n)}{n}=0. \qquad \text{(A.4.18)}$$

Consequently,

$$\lim_{n\to\infty} n\{\ln[n+b+f(n)-\ln n]\}=b \qquad \text{(A.4.19)}$$

and

$$\lim_{n\to\infty}\left[1+\frac{b}{n}+\frac{f(n)}{n}\right]^n=e^b, \qquad \text{(A.4.20)}$$

as was to be shown.

REFERENCE

Taylor, A. E., *Advanced Calculus*, Ginn & Co., 1955.

[3] I am grateful to Peter S. Landweber of Rutgers University for providing the proof which follows.

APPENDIX 5

REAL SYMMETRIC, POSITIVE DEFINITE, AND IDEMPOTENT MATRICES

The purpose of this appendix is to draw together a number of results from linear algebra relating to positive definite matrices which are fundamental to normal distribution theory.

THEOREM A.5.1
The latent roots of a real symmetric matrix $\mathbf{A}$ are real.

PROOF. Assume the contrary. Since $\mathbf{A}$ is a real matrix, it follows from the characteristic equation $|\mathbf{A} - \lambda\mathbf{I}| = 0$ that the conjugate of any complex characteristic root λ is also a characteristic root, that is, if $\mathbf{Ax} = \lambda\mathbf{x}$ for λ complex, then also $\mathbf{A\bar{x}} = \bar{\lambda}\mathbf{\bar{x}}$, where $\mathbf{\bar{x}}$ and $\bar{\lambda}$ are the complex conjugates of $\mathbf{x}$ and λ. Then

$$\mathbf{\bar{x}'Ax} = \lambda\mathbf{\bar{x}'x}$$

and

$$\mathbf{x'A\bar{x}} = \bar{\lambda}\mathbf{x'\bar{x}}.$$

However, since $\mathbf{A}$ is symmetric,

$$\mathbf{\bar{x}'Ax} = \mathbf{x'A\bar{x}}$$

and

$$\mathbf{\bar{x}'x} = \mathbf{x'\bar{x}}.$$

Hence,

$$\lambda = \bar{\lambda},$$

which is true only if λ is real.

THEOREM A.5.2
Latent vectors associated with distinct latent roots of a real symmetric matrix **A** are orthogonal.

PROOF. Let δ and μ be distinct latent roots. From

$$\mathbf{Ax} = \lambda \mathbf{x},$$
$$\mathbf{Ay} = \mu \mathbf{y},$$

$\delta \neq \mu$, we obtain

$$\mathbf{y'Ax} = \lambda \mathbf{y'x},$$
$$\mathbf{x'Ay} = \mu \mathbf{x'y}.$$

However, since **A** is symmetric,

$$\mathbf{y'Ax} = \mathbf{x'Ay}$$

and

$$\mathbf{y'x} = \mathbf{x'y}.$$

Therefore,

$$(\lambda - \mu)\mathbf{x'y} = 0.$$

Whence, since $\lambda \neq \mu$,

$$\mathbf{x'y} = 0.$$

THEOREM A.5.3
Let $\mathbf{T}_1, \ldots, \mathbf{T}_n$ be the latent vectors associated with the latent roots $\lambda_1, \ldots, \lambda_n$, assumed to be distinct, of the real symmetric matrix **A**. Assume that the **T**'s are normalized, $\mathbf{T}_i'\mathbf{T}_i = 1$ for all i. Then

$$\mathbf{T'AT} = \hat{\boldsymbol{\lambda}}, \tag{A.5.1}$$

where $\hat{\boldsymbol{\lambda}}$ denotes the diagonal matrix

$$\hat{\boldsymbol{\lambda}} = \begin{bmatrix} \lambda_1 & 0 & \cdots & 0 \\ 0 & \lambda_2 & \cdots & 0 \\ \vdots & & \ddots & \vdots \\ 0 & 0 & \cdots & \lambda_n \end{bmatrix}. \tag{A.5.2}$$

PROOF. From the preceding theorem and the assumption that the **T**'s are normalized, we have

$$\begin{aligned} \mathbf{T'T} &= \mathbf{I} \\ &= \mathbf{TT'}. \end{aligned}$$

Also, since

$$\mathbf{AT}_i = \lambda_i \mathbf{T}_i,$$

we have

$$\mathbf{AT} = (\lambda_1 \mathbf{T}_1, \lambda_2 \mathbf{T}_2, \ldots, \lambda_n \mathbf{T}_n).$$

Then

$$\begin{aligned} \mathbf{T}'\mathbf{AT} &= (\lambda_i \mathbf{T}_i{}'\mathbf{T}_j) \\ &= (\lambda_i \delta_{ij}), \end{aligned}$$

where

$$\delta_{ij} = \begin{cases} 1, & i = j, \\ 0, & i \neq j. \end{cases}$$

Consequently,

$$\begin{aligned} \mathbf{T}'\mathbf{AT} &= \hat{\boldsymbol{\lambda}}\mathbf{I} \\ &= \hat{\boldsymbol{\lambda}}, \end{aligned}$$

as was to be shown.

DEFINITION 1

A square matrix $\mathbf{A}$ is *positive definite* if, for all vectors $\mathbf{x} \neq \mathbf{0}$, $\mathbf{x}'\mathbf{Ax} > \mathbf{0}$.

THEOREM A.5.4

A symmetric matrix $\mathbf{A}$ is positive definite if and only if all latent roots of $\mathbf{A}$ are positive.

PROOF. Let $\mathbf{T}$ be an orthogonal matrix which diagonalizes $\mathbf{A}$,

$$\mathbf{T}'\mathbf{AT} = \hat{\boldsymbol{\lambda}}.$$

Let $\mathbf{y} = \mathbf{T}'\mathbf{x}$, so that $\mathbf{x} = \mathbf{Ty}$; then

$$\begin{aligned} \mathbf{x}'\mathbf{Ax} &= \mathbf{y}'\mathbf{T}'\mathbf{ATy} \\ &= \mathbf{y}'\hat{\boldsymbol{\lambda}}\mathbf{y} \\ &= \sum \lambda_i y_i{}^2. \end{aligned}$$

If $\lambda_i > 0$, then clearly,

$$\begin{aligned} \mathbf{x}'\mathbf{Ax} &= \mathbf{y}'\hat{\boldsymbol{\lambda}}\mathbf{y} \\ &> 0. \end{aligned}$$

Conversely, assume that **A** is positive definite, and suppose that $\lambda_i < 0$. Let $\mathbf{y}^*$ be the vector with first element 1 and remaining elements 0, and let $\mathbf{x}^* = \mathbf{T}\mathbf{y}^*$. Since **T** is nonsingular, $\mathbf{x}^* \neq \mathbf{0}$. Then

$$\begin{aligned}\mathbf{x}^{*\prime}\mathbf{A}\mathbf{x}^* &= \mathbf{y}^{*\prime}\mathbf{T}'\mathbf{A}\mathbf{T}\mathbf{y}^* \\ &= \mathbf{y}^{*\prime}\hat{\boldsymbol{\lambda}}\mathbf{y}^* \\ &= \lambda_1 \\ &< 0,\end{aligned}$$

which contradicts the assumpton that **A** is positive definite. Hence, $\lambda_1 < 0$. Since λ_1 can be any λ_i, the theorem is proven.

THEOREM A.5.5
Let **A** be an $n \times n$ positive definite matrix. Then $|\mathbf{A}| > 0$.

PROOF. Let $\lambda_1, \ldots, \lambda_n$ be the latent roots of **A**, and let **T** be an orthogonal matrix that diagonalizes **A**. Then

$$\begin{aligned}|\hat{\boldsymbol{\lambda}}| &= |\mathbf{T}'\mathbf{A}\mathbf{T}| \\ &= |\mathbf{T}'|\,|\mathbf{A}|\,|\mathbf{T}| \\ &= |\mathbf{A}|,\end{aligned}$$

since **T** is orthogonal and therefore has determinant equal to 1. Hence,

$$\begin{aligned}|\mathbf{A}| &= \prod_{i=1}^{n} \lambda_i \\ &> 0.\end{aligned}$$

THEOREM A.5.6
If **A** is an $n \times n$ positive definite matrix and **P** is an $n \times m$ matrix of rank m, then $\mathbf{P}'\mathbf{A}\mathbf{P}$ is also positive definite.

PROOF. Let **y** be any nonzero $m \times 1$ vector. Then

$$\mathbf{y}'(\mathbf{P}'\mathbf{A}\mathbf{P})\mathbf{y} = \mathbf{x}'\mathbf{A}\mathbf{x},$$

where $\mathbf{x} = \mathbf{P}\mathbf{y}$. Since **A** is positive definite, $\mathbf{x}'\mathbf{A}\mathbf{x} > 0$. Hence, $\mathbf{P}'\mathbf{A}\mathbf{P}$ is also positive definite.

THEOREM A.5.7
If **A** is positive definite, then so is $\mathbf{A}^{-1}$.

PROOF. Let $\mathbf{P} = \mathbf{A}^{-1}$. Then

$$\begin{aligned} \mathbf{P}'\mathbf{A}\mathbf{P} &= (\mathbf{A}^{-1})'\mathbf{A}\mathbf{A}^{-1} \\ &= (\mathbf{A}^{-1})'. \end{aligned}$$

But $\mathbf{P}'\mathbf{A}\mathbf{P}$ is positive definite, hence so is $\mathbf{A}^{-1}$.

THEOREM A.5.8

If $\mathbf{P}$ is an $n \times m$ matrix of rank m, $\mathbf{P}'\mathbf{P}$ is positive definite.

PROOF. We know that if $\mathbf{A}$ is positive definite, $\mathbf{P}'\mathbf{A}\mathbf{P}$ is also. Let $\mathbf{A} = \mathbf{I}$. Then

$$\begin{aligned} \mathbf{P}'\mathbf{A}\mathbf{P} &= \mathbf{P}'\mathbf{I}\mathbf{P} \\ &= \mathbf{P}'\mathbf{P}. \end{aligned}$$

DEFINITION 2

Let $\mathbf{A}$ be an $n \times n$ symmetric matrix of rank $m \leq n$ such that

$$\mathbf{A} \cdot \mathbf{A} = \mathbf{A}. \tag{A.5.3}$$

Such a matrix is said to be *idempotent*.

THEOREM A.5.9

Let $\mathbf{A}$ be an idempotent matrix. Then the latent roots of $\mathbf{A}$ are either 1 or 0.

PROOF. Let λ be a latent root of $\mathbf{A}$. Then

$$\mathbf{A}\mathbf{x} = \lambda\mathbf{x}$$

and

$$\mathbf{A} \cdot \mathbf{A}\mathbf{x} = \lambda\mathbf{A}\mathbf{x}.$$

However, $\mathbf{A} \cdot \mathbf{A} = \mathbf{A}$ and $\mathbf{A}\mathbf{x} = \lambda\mathbf{x}$, so that

$$\mathbf{A}\mathbf{x} = \lambda^2\mathbf{x}.$$

This being the case, λ can only be 1 or 0.

DEFINITION 3

The *trace* of a matrix is the sum of the diagonal elements.

It is to be noted that the trace is defined for square matrices only. Usually, the trace of a matrix $\mathbf{A}$ is denoted by $\mathrm{Tr}(\mathbf{A})$.

THEOREM A.5.10

Let $\mathbf{A}$ be an idempotent matrix. Then the trace of $\mathbf{A}$ is equal to its rank.

PROOF. This theorem is an immediate consequence of Theorem A.5.9.

THEOREM A.5.11
Let **A**, **B**, and **C** be matrices such that

$$\mathbf{A} = \mathbf{B} + \mathbf{C}. \tag{A.5.4}$$

Then

$$\operatorname{Tr}(\mathbf{A}) = \operatorname{Tr}(\mathbf{B}) + \operatorname{Tr}(\mathbf{C}). \tag{A.5.5}$$

In addition, for matrices **D** and **E** such that the products **DE** and **ED** are both defined,

$$\operatorname{Tr}(\mathbf{DE}) = \operatorname{Tr}(\mathbf{ED}). \tag{A.5.6}$$

PROOF. Proof of (A.5.5) is immediate from the fact that the trace is defined as a linear operation. For (A.5.6),

$$\begin{aligned}\operatorname{Tr}(\mathbf{DE}) &= \sum_i \sum_j d_{ij} e_{ji} \\ &= \sum_j \sum_i e_{ji} d_{ij} \\ &= \operatorname{Tr}(\mathbf{ED}).\end{aligned}$$

It is sometimes convenient to obtain the inverse of a matrix in partitioned form. The following theorem is then applicable.

THEOREM A.5.12
Let **A** be an $n \times n$ nonsingular matrix, which is partitioned as

$$\mathbf{A} = \begin{bmatrix} \mathbf{E} & \mathbf{F} \\ \mathbf{G} & \mathbf{H} \end{bmatrix}, \tag{A.5.7}$$

where **E** is $n_1 \times n_1$, **F** is $n_1 \times n_2$, **G** is $n_2 \times n_1$, and **H** is $n_2 \times n_2$, with $n_1 + n_2 = n$. Let $\mathbf{D} = \mathbf{H} - \mathbf{G}\mathbf{E}^{-1}\mathbf{F}$, and suppose that **D** and **E** are nonsingular. Then

$$\mathbf{A}^{-1} = \begin{bmatrix} \mathbf{E}^{-1}(\mathbf{I} + \mathbf{F}\mathbf{D}^{-1}\mathbf{G}\mathbf{E}^{-1}) & -\mathbf{E}^{-1}\mathbf{F}\mathbf{D}^{-1} \\ -\mathbf{D}^{-1}\mathbf{G}\mathbf{E}^{-1} & \mathbf{D}^{-1} \end{bmatrix}. \tag{A.5.8}$$

PROOF. The validity of (A.5.8) is established by showing that the product of the right-hand sides of (A.5.8) and (A.5.7) results in the identity matrix. The details are left to the reader.

APPENDIX 6

THE LINDEBERG-LEVY CENTRAL LIMIT THEOREM

Theorem 14.2 presents a proof of the asymptotic normality of a sum of independently distributed standardized random variables under the assumption that the parent distribution possesses a moment-generating function. However, as was noted, Theorem 14.2 is unnecessarily restrictive in that its conclusion holds irrespective of whether the parent distribution possesses a moment-generating function. The purpose of this appendix is to supply a proof for this more general case. As a starter, we require the characteristic function for the normal distribution with mean 0 and variance 1.

By definition, this is given by

$$\phi(t) = \frac{1}{\sqrt{2\pi}} \int_{-\infty}^{\infty} e^{ity} e^{-y^2/2}\, dy. \tag{A.6.1}$$

Expressing e^{ity} in its power series,

$$e^{ity} = \sum_{k=0}^{\infty} \frac{(ity)^k}{k!}, \tag{A.6.2}$$

$\phi(t)$ becomes

$$\phi(t) = \frac{1}{\sqrt{2\pi}} \int_{-\infty}^{\infty} \sum_{k=0}^{\infty} \frac{(it)^k}{k!} y^k e^{-y^2/2}\, dy. \tag{A.6.3}$$

Since the function $e^{ity - y^2/2}$ is bounded for all values of y in the range $-\infty < y < \infty$, it is integrable over this range, which means that we can integrate the sum in (A.6.3) term-by-term. Accordingly, we can write

$$\phi(t) = \sum_{k=0}^{\infty} \frac{(it)^k}{k!} \frac{1}{\sqrt{2\pi}} \int_{-\infty}^{\infty} y^k e^{-y^2/2}\, dy$$

$$= \sum_{k=0}^{\infty} \frac{(it)^k}{k!} \mu_k, \tag{A.6.4}$$

μ_k being the kth moment of the $N(0, 1)$ variable y. Since $\mu_k = 0$ for $k = 1, 3, 5, \ldots$, $\phi(t)$ can be further reduced to

$$\phi(t) = \sum_{k=0}^{\infty} \frac{(it)^{2k}}{(2k)!} \mu_{2k}. \tag{A.6.5}$$

By repeated differentiation of the moment-generating function for the $N(0, 1)$ variable,

$$M(t) = e^{t^2/2}, \tag{A.6.6}$$

it can be established that

$$\mu_{2k} = \frac{(2k)!}{2^k k!} \tag{A.6.7}$$

Inserting this result into (A.6.5), we now find for $\phi(t)$,

$$\phi(t) = \sum_{k=0}^{\infty} \frac{(it)^{2k}}{(2k)!} \frac{(2k)!}{2^k k!}$$

$$= \sum_{k=0}^{\infty} \frac{(it)^{2k}}{2^k k!}$$

$$= \sum_{k=0}^{\infty} \frac{(-1)^k (t^2/2)^k}{k!}, \tag{A.6.8}$$

since $i^2 = -1$. The last sum, however, will be recognized as the power series expansion of $e^{-t^2/2}$; hence, we finally find for the characteristic function of the $N(0, 1)$ random variable,

$$\phi(t) = e^{-t^2/2}. \tag{A.6.9}$$

THEOREM A.6.1 (Lindeberg-Levy Theorem)
Let $X_1, \ldots, X_n$ be a sequence of independently and identically distributed random variables that have mean μ and variance σ^2. Let

$$Y_n = \frac{\overline{X}_n - \mu}{\sigma/\sqrt{n}}$$

$$= \sum_{i=1}^{n} \left(\frac{X_i - \mu}{\sigma\sqrt{n}} \right). \tag{A.6.10}$$

Then, as $n \to \infty$, $y_n \to N(0, 1)$.

PROOF. The proof of this theorem parallels the proof of Theorem 14.2, except that the characteristic function is employed instead of the moment-generating function.

Let $\phi(t)$ be the characteristic function for $Y = (X - \mu)/\sigma$. By definition, this is given by

$$\phi(t) = \int_y e^{ity} f(y)\, dy. \tag{A.6.11}$$

As above, we can use the power-series expansion for e^{ity} to write $\phi(t)$ as

$$\begin{aligned} \phi(t) &= \int_y \sum_{k=0}^{\infty} \frac{(it)^k}{k!} y^k f(y)\, dy \\ &= \sum_{k=0}^{\infty} \frac{(it)^k}{k!} \int_y y^k f(y)\, dy \\ &= \sum_{k=0}^{\infty} \frac{(it)^k}{k!} \mu_k . \end{aligned} \tag{A.6.12}$$

Since Y is standardized, $\mu_1 = 0$ and $\mu_2 = 1$, hence we can write

$$\phi(t) = 1 - \frac{1}{2} t^2 + \sum_{k=3}^{\infty} \frac{(it)^k}{k!} \mu^k . \tag{A.6.13}$$

Next, since the X_i are independently and identically distributed, the characteristic function $\phi(t, n)$ for $\bar{Y}_n$ will be equal to

$$\begin{aligned} \phi(t, n) &= \left[\phi\left(\frac{t}{\sqrt{n}}\right)\right]^n \\ &= \left[1 - \frac{t^2}{2n} + \sum_{k=3}^{\infty} \frac{(it)^k}{n^{k/2} k!} \mu_k\right]^n \\ &= \left[1 - \frac{t^2}{2n} + \psi(n)\right]^n , \end{aligned} \tag{A.6.14}$$

where

$$\psi(n) = \sum_{k=3}^{\infty} \frac{(it)^k}{n^{k/2} k!} \mu_k . \tag{A.6.15}$$

Note that, since k is always greater than 2, we have

$$\lim_{n \to \infty} \psi(n) = 0. \tag{A.6.16}$$

This being the case, it follows from Theorem A.4.4 that

$$\lim_{n\to\infty} \phi(t, n) = \lim_{n\to\infty} \left[1 - \frac{t^2/2}{n} + \psi(n)\right]^n$$
$$= e^{-t^2/2}. \tag{A.6.17}$$

Since, from (A.6.9), $e^{-t^2/2}$ is the characteristic function of a $N(0, 1)$ random variable, the theorem is proven.

INDEX

(*Italics* denote pages of primary discussion.)

74 75 76 77 78 9 8 7 6 5 4 3 2 1